THE MOTHS OF AMERICA NORTH OF MEXICO

The Moths of America North of Mexico

INCLUDING GREENLAND

FASCICLE 27.2

NOCTUOIDEA

NOCTUIDAE (PART)

NOCTUINAE (Part—*Euxoa*)

J. DONALD LAFONTAINE
BIOSYSTEMATICS RESEARCH CENTRE
RESEARCH BRANCH, AGRICULTURE CANADA

COLOR PHOTOGRAPHS BY
KJELL B. SANDVED
SMITHSONIAN INSTITUTION

MONOCHROME PHOTOGRAPHS BY
T. STOVELL, C. E. BEDDOE, AND W. LUKEY

DRAWINGS BY
J. DONALD LAFONTAINE

1987
WASHINGTON

THE WEDGE ENTOMOLOGICAL RESEARCH FOUNDATION

This work is to be cited as
Lafontaine, J. D., 1987. *Noctuoidea: Noctuidae* (part)
in Dominick, R. B., et al.,
The Moths of America North of Mexico, fasc. 27.2

DISTRIBUTORS

THE WEDGE ENTOMOLOGICAL RESEARCH FOUNDATION
% National Museum of Natural History,
MRC-127, Washington, D.C. 20560, U.S.A.

E. W. CLASSEY LTD
P.O. Box 93, Faringdon, Oxfordshire SN7 7DR, England

BIOQUIP PRODUCTS
P.O. Box 61, Santa Monica, California 90406, U.S.A.

ENTOMOLOGICAL REPRINT SPECIALISTS
P.O. Box 77224, Dockweiler Station
Los Angeles, California 90007, U.S.A.

ISBN 0-933003-03-X

LIBRARY OF CONGRESS CATALOG NUMBER 87-50657

Printed in the United States of America by Allen Press, Inc., Lawrence, Kansas

Color plates by Lanman Progressive Co., Washington, D.C.

To *DAVID F. HARDWICK*
Lepidopterist, Teacher, Friend

CONTENTS

PREFACE

This is the first of the many parts required to treat the fauna of the large family Noctuidae of America north of Mexico. It is the first of three projected parts on the subfamily Noctuinae, treating the large genus *Euxoa* whose species are frequently difficult to identify. Several factors combine to create taxonomic and identification problems: large number of species; subtle differences between species; and variability within species combined with parallel geographical variation among different species. These problems were largely overcome by the large amount of material available for study; about 200,000 specimens were examined and about 12,000 genital slides studied, almost all of which had the vesica everted, or the bursa inflated. Some species complexes, however, can be resolved only by taxonomic experiments on living material. The former "Experimental Taxonomy Unit" in Ottawa (see acknowledgements) conducted experiments on life history and behavior, hybridization, enzyme electrophoresis, blood cell morphology, and egg morphology. Data from these studies helped resolve the specific identities of the *declarata, comosa, ridingsiana, mimallonis, obeliscoides,* and *aequalis* species complexes. The *infausta* and *costata* complexes remain incompletely known.

This revision of *Euxoa* is the culmination of many years of research on the genus by workers in Ottawa, first by J. H. McDunnough, later by D. F. Hardwick, and finally by me and members of the "Experimental Taxonomy Unit."

The main goal of this volume is identification of species of *Euxoa.* I have not been able to make their identification easy—the species are too variable and the differences among them too subtle—but I hope I have made it possible. With practice, most species can be recognized without dissection; although, it frequently is necessary to remove scales from the end of the abdomen with a small artist's brush so that the shape of the ovipositor lobes (papillae anales) in females, or length and shape of the protruding valves and saccular extensions in males, can be examined under a dissecting microscope. The appearance of *Euxoa tessellata* should be learned with this technique because it is easy to recognize; however, it is by far the most frequently misidentified *Euxoa* species.

I did not find it practical to devise a key to species based on external appearance; the color plates will serve this function. Identification of species must be based on genital characters in conjunction with the keys or on external appearance through comparison of specimens with those in the color plates. The text is designed to assist the identification process by discussing species most frequently confused by giving diagnostic genital characters for the species. For example, although *Euxoa perpolita* is almost identical in genitalia to other species in the *ridingsiana* group, the account of *perpolita* concentrates on differences between it and *velleripennis,* the species most frequently confused with it.

For simplicity, I tried to design the keys based on male genitalia so that characters of the vesica need not be used. Instead, I provide an alternate, although somewhat awkward, key to subgenera and include characters of the vesica only near the end so that the alternative couplets may be followed, if necessary. I cannot overemphasize the importance of the vesica as a character system in *Euxoa,* and in the Noctuidae in general, as it makes possible the recognition of species, species-groups and genera. Many species may most easily, and some only, be distinguished by characters of the vesica. Such characters are remarkably constant, and our laboratory hybridization experiments showed that differences in the vesica between species are more indicative of inability to hybridize than are differences in the shape of the valves. Differences in the shape of the vesica usually correspond to differences in the shape of the bursa copulatrix in the female.

Much remains to be done in North America before our knowledge of classification, distribution, or natural history of noctuids approaches that of the butterflies. A major factor in this discrepancy has been the significant role of the amateur Lepidopterist in butterfly research. This situation is slowly changing as identification aids encourage more amateurs to expand their interests to moths. Several dozen noctuid species have been added to the fauna of America north of Mexico since the publication of the *Check List of the Lepidoptera of America North of Mexico,* through descriptions of new species and by discovery of Asian species in the northwest and Mexican species in the southwest. I hope that this volume on *Euxoa* will serve as a catalyst to encourage more collectors and researchers to expand their horizons to include the Noctuidae.

AUTHOR'S ACKNOWLEDGEMENTS

I wish to thank the following persons for loan of specimens, for loan of type specimens, and for assistance during field trips and museum work: P. H. Arnaud, Jr., E. U. Balsbaugh, Jr., A. and M. E. Blanchard, B. Bowen, J. R. Byers, C. V. Covell, Jr., J. P. Donahue, D. C. Ferguson, C. D. Ferris, R. L. Fischer, L. Handfield, J. H. Hannemann, J. R. Heitzman, R. Holland, M. R. Honey, C. P. Kimball (deceased), E. C. Knudson, V. S. Kononenko, R. Leuschner, G. E. Lewis, B. Mather, T. L. McCabe, E. H. Metzler, K. Mikkola, S. E. Miller, A. Moberg, M. C. Nielson, K. W. Philip, R. W. Poole, J. A. Powell, E. L. Quinter, J. E. Rawlins, C. L. Remington, F. H. Rindge, D. F. Schweitzer, J. H. Shepard, I. L. Sukhareva, E. L. Todd (deceased), J. P. Walas, and B. Wright.

I thank my associates in Ottawa for technical assistance: E. W. Rockburne made most of the genital preparations; W. H. Forrest reared the larvae and prepared the genital photographs for the monochrome plates; T. Stovell, C. E. Beddoe, and W. Lukey photographed the genitalia. K. B. Sandved took the color photographs of the moths.

I thank D. C. Ferguson, J. G. Franclemont, R. W. Hodges, T. L. McCabe, E. G. Munroe, and R. W. Poole for reviewing the manuscript and providing many helpful comments and suggestions. R. W. Hodges spent a great deal of time editing the manuscript; his efforts are most appreciated.

I am indebted to D. F. Hardwick, formerly of the Biosystematics Research Centre, for his guidance and encouragement in the preparation of this revision of *Euxoa*. He initiated revisionary work on the genus in North America, collected the bulk of the specimens used in this study, and provided me with all of his field and museum notes and photographs of type specimens.

I thank my wife, Herma, who accompanied me on many field trips and has been a constant source of encouragement.

THE MOTHS OF AMERICA NORTH OF MEXICO

SUPERFAMILY
NOCTUOIDEA (continued)

FAMILY
Noctuidae (continued)

SUBFAMILY
Noctuinae (continued)

TRIBE
Agrotini (continued)

GENUS
Euxoa Hübner

Euxoa Hübner, [1821], *Verzeichniss Bekannter Schmettlinge* [sic], 209.
Type species: *Noctua decora* [Denis and Schiffermüller], 1775. Designated by Hampson, 1903, *Catalogue of the Lepidoptera Phalaenae in the British Museum,* **4**: 153.

Mimetes Hübner, [1821], *Verzeichniss Bekannter Schmettlinge* [sic], 210.
Type species: *Noctua nebulosa* Hübner, [1808], considered to be a synonym of *Noctua decora* [Denis and Schiffermüller], 1775. Designated by Hampson, 1903, *Catalogue of the Lepidoptera Phalaenae in the British Museum,* **4**: 153.
NOTE—*Mimetes* Hübner, [1821] is a homonym of *Mimetes* Leach, 1820 in the Coleoptera.
NOTE—*Mimetes* was misspelled as *Mimetis* by Hampson, 1903.

Metaxyja Hübner, [1821], *Verzeichniss Bekannter Schmettlinge* [sic], 223.
Type species: *Phalaena vitta* Esper, 1789. Designated by Hampson, 1903, *Catalogue of the Lepidoptera Phalaenae in the British Museum,* **4**: 153.
NOTE—*Metaxyja* was misspelled as *Metaxya* by Walker, 1857, and as *Metaxyia* by Hampson, 1903.

Exarnis Hübner, [1821], *Verzeichniss Bekannter Schmettlinge* [sic], 225.
Type species: *Noctua ruris* Hübner, [1809]. Designated by Hampson, 1903, *Catalogue of the Lepidoptera Phalaenae in the British Museum,* **4**: 153.

Brotis Hübner, [1821] *Verzeichniss Bekannter Schmettlinge* [sic], 226.
Type species: *Noctua fumosa* [Denis and Schiffermüller], 1775, considered to be a synonym of *Noctua nigricans* Linnaeus, 1761. Designated by Hampson, 1903, *Catalogue of the Lepidoptera Phalaenae in the British Museum,* **4**: 153.

Telmia Hübner, [1821], *Verzeichniss Bekannter Schmettlinge* [sic], 228.
Type species: *Noctua sagitta* Hübner, [1813]. Designated by Hampson, 1903, *Catalogue of the Lepidoptera Phalaenae in the British Museum,* **4**: 153.

Pleonectopoda Grote, 1873, *Bull. Buffalo Soc. Nat. Sci.,* **1**: 136.
Type species: *Pleonectopoda lewisi* Grote, 1873. Monotypy.
NOTE—*Pleonectopoda* is treated as a subgenus of *Euxoa* following Hardwick, 1970.

Carneades Grote, 1883, *Can. Ent.,* **15**: 4.
Type species: *Carneades moerens* Grote, 1883. Monotypy.
NOTE—*Carneades* Grote, 1883 is a homonym of *Carneades* Bates, 1869 in the Mammalia.

Chorizagrotis Smith, 1890, *Bull. U. S. Natl. Mus.,* **38**: 98.

Type species: *Agrotis auxiliaris* Grote, 1873. Original designation.

NOTE—*Chorizagrotis* is treated as a subgenus of *Euxoa* following Hardwick, 1970.

Paragrotis Dyar, [1903], *Bull. U. S. Natl. Mus.*, **52**: 140.
Type species: *Carneades moerens* Grote, 1883. Original designation.

NOTE—*Paragrotis* is a replacement name for *Carneades* Grote, 1883.

Orosagrotis Hampson, 1903, *Catalogue of the Lepidoptera Phalaenae in the British Museum*, **4**: 135.
Type species: *Agrotis montana* Morrison, 1875. Original designation.

NOTE—*Orosagrotis* is treated as a subgenus of *Euxoa* following Hardwick, 1970.

NOTE—*Orosagrotis* was misspelled as *Prosagrotis* by Warren (1911: 251).

Mesoeuxoa Corti, 1932, *in* Seitz, *Die Gross-Schmetterlinge der Erde*, **3** Suppl.: 38.
Type species: *Phalaena lidia* Stoll, 1782. Designated by Kozhanchikov, 1937, *Fauna of the USSR (Insecta) Lepidoptera*, **13**: 521.

NOTE—*Mesoeuxoa* is a junior subjective synonym of the subgeneric name *Chorizagrotis*.

Menada Kozhanchikov, 1937, *Fauna of the USSR (Insecta) Lepidoptera*, **13**: 593.
Type species: *Agrotis nomas* Erschov, 1874. Original designation.

NOTE—*Menada* is a junior subjective synonym of the subgeneric name *Orosagrotis*.

Longivesica Hardwick, 1970, *Mem. Ent. Soc. Can.*, **67**: 44.
Type species: *Agrotis divergens* Walker, 1857. Original designation.

NOTE—*Longivesica* is treated as a subgenus of *Euxoa* following Hardwick, 1970.

Palaeoeuxoa Lafontaine, NEW SUBGENUS (p. 31).
Type species: *Agrotis mimallonis* Grote, 1873.

Heteroeuxoa Lafontaine, NEW SUBGENUS (p. 34).
Type species: *Mamestra septentrionalis* Walker, 1865.

DESCRIPTION OF STRUCTURAL CHARACTERS. Adults of the genus *Euxoa* are medium-sized, stout-bodied noctuids. They can be recognized as members of the subfamily Noctuinae, tribe Agrotini, by the following combination of characters: eye without surface hair; eye without tuft of lashlike scales in front of eye below antenna; frons bulging, with a raised ringlike tubercle in most species (text figure 1 *c*); hindwing venation trifid (i.e., vein M_3 reduced and arising at middle of cell) (text figure 1 *b*); all tibiae with spinelike setae; middle and hind tarsi with fourth row of setae laterally on outside of first segment (text figure 1 *e, f*).

Members of the genus *Euxoa* can be distinguished from those of other noctuine genera by male and female genital characters. Males can most easily be recognized by the sclerotized extension of the sacculus that extends posteriorly just below the lower margin of the cucullus. This extension varies greatly in length and shape, but its presence is diagnostic. It usually can be observed by removing some of the scales from the lower margin of the valve with a small brush. The female genitalia of species of *Euxoa* differ from those of other noctuine genera in having elongate sclerotized plates in the dorsal and ventral wall of the ductus bursae that extend anteriorly from the sclerotized ostium bursae.

Other characters of the male genitalia are: valves bilaterally symmetrical, or slightly asymmetrical with respect to length and shape of right and left saccular extensions; saccular extensions cylindrical in most species (flattened and bladelike, or apically spatulate, in some), about as stout as, or slightly stouter than harpes and 0.5–1.5 times harpe length (occasionally very short to more than twice harpe length); sacculus slightly shorter than cucullus to slightly longer, crescentic in outline with dorsal margin concave in most species (oval in some with dorsal margin convex or sinuate); a small lightly sclerotized clavus usually present at base of sacculus; cucullus expanded and foot shaped apically with apical corona of stout setae; harpe usually C-shaped, projecting beyond dorsal margin of valve (in some species less strongly curved, or slightly S-shaped, lying along inner margin of valve parallel to dorsal margin); surface of harpe densely pubescent with very small setae in some species; juxta usually flat and shield shaped (rarely with transverse sclerotized band; in subgenus *Chorizagrotis* frequently with raised sclerotized longitudinal ridge); uncus in most species thin and cylindrical, slightly wider near middle, with long hairlike setae mesially and shorter stouter setae apically (in some species uncus broad and spatulate apically, or rarely broad and dorsoventrally flattened; in these species apical setae on ventral surface of uncus usually short and stout);

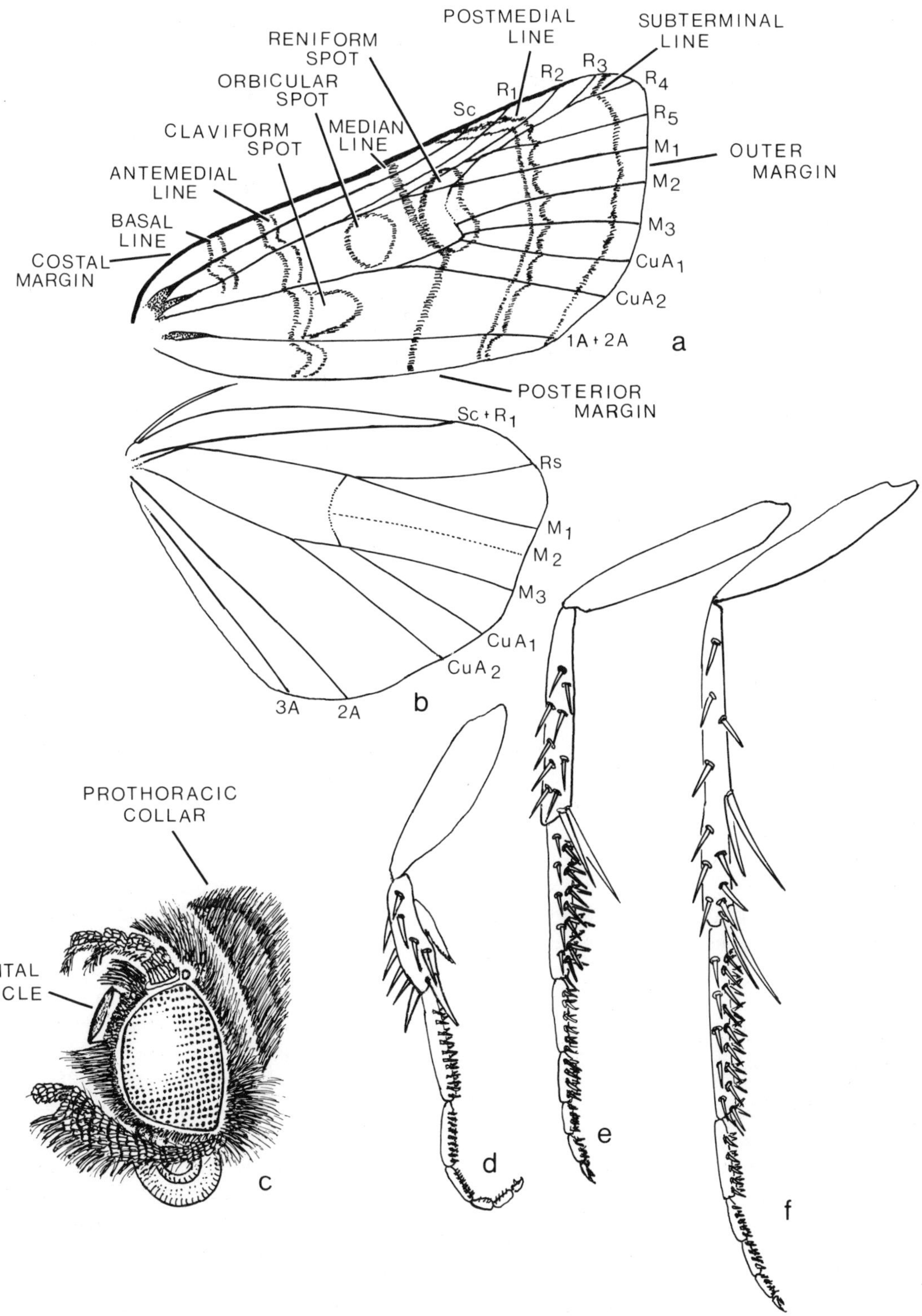

FIGURE 1: STRUCTURAL CHARACTERS OF TYPICAL SPECIES OF *EUXOA*
a. Forewing venation and generalized pattern; *b.* Hindwing venation; *c.* Lateral view of head; *d.* Foreleg; *e.* Midleg; *f.* Hindleg.

vesica projecting dorsally beyond apex of aedoeagus in subgenus *Chorizagrotis* and most species in subgenus *Euxoa,* to right in subgenera *Palaeoeuxoa, Heteroeuxoa, Pleonectopoda, Longivesica,* and some *Euxoa,* and to left in subgenus *Orosagrotis*; vesica one to two times length of aedoeagus in all subgenera except *Longivesica* in which it is 2.5–3.0 times as long; vesica with large subbasal diverticulum, curved median diverticulum and small bubblelike apical diverticulum (some species with additional subbasal diverticula and two apical diverticula); subbasal diverticulum in vesica with 1–2 small cornuti in most species; in some subgenera vesica with a subbasal loop or coil.

The bursa copulatrix is bisaccate in one-third of our species, unisaccate in two-thirds; in those with a bisaccate bursa the appendix bursae is dorsal, or to the left, of the corpus bursae. The shapes and relative positions of the ductus, corpus, and appendix bursae vary greatly from group to group within *Euxoa* and provide the best characters for distinguishing subgenera and species-groups. The length and shape of sclerotized plates in the ductus bursae and shape and form of vestiture of the ovipositor lobes (papillae anales) also vary greatly within *Euxoa,* but these characters are most useful for distinguishing species within species-groups. In about one-half of the North American species of *Euxoa* the ovipositor lobes are about twice as long as wide, somewhat bullet shaped, tapered and rounded apically; they are densely clothed with short fine-tipped setae and have a row of about 12 long setae subbasally. The ovipositor lobes in the other half of the *Euxoa* fauna have been modified in the following ways: 1) shape may be shortened and truncated apically to elongate and pointed apically; 2) apical setae may be short and conical or long, stout, and spikelike; 3) subbasal row of long setae may be reduced or absent, increased to a dense border of setae, or stout and spikelike; 4) the lobes may have a narrow, sclerotized apical process, a dorsal ridge, or a rim along the outer margin; and 5) in some groups the individual dorsal margins of the two are fused together. The various modifications of the ovipositor lobes are illustrated in plates EE and FF. Other characters of the female genitalia of taxonomic value in certain groups are length and vestiture of the sclerotized ring formed by abdominal segment eight and relative lengths of the posterior and anterior apophyses.

The genital characters used in the keys to species of *Euxoa* are labeled in text figure 2.

LARVA. The larvae of *Euxoa* are true cutworms as defined by the following combination of characters: body tapered anteriorly, with head relatively small; hypopharynx usually with transverse dorsal groove, surface spined at least to base of labial palpi (to base of spinneret in most Noctuinae); spinneret dorsally grooved or flattened (shorter than wide in Agrotini); prolegs present on abdominal segments 3, 4, 5, 6, and 10; crochets uniordinal; body with primary setae only; 2 SV setae on abdominal segment 1 (3 on segment 2); setae SD-3 on meso- and metathorax and SD-1 on abdominal segment 9 thin and hairlike (in a black saucer-shaped depression in *Euxoa* and *Agrotis* group of genera).

The larvae of the *Euxoa* and *Agrotis* group of genera can be distinguished from those of other cutworm genera by a combination of three characters: larval skin granulose; spinneret much shorter than its width; epicranial suture essentially absent (i.e., adfrontals extending to occipital foramen). Other characters of this group are: seta L-2 on abdominal segment 6 closer to seta L-1 than to seta L-3; seta SD-1 on meso- and metathorax not connected to muscle attachment by a sclerotized bar.

Crumb (1956: 85) distinguished the larvae of species of *Euxoa* from those of species of the *Agrotis* group of genera by "... reticulation of head replaced entirely by black or brown flecks like freckles; submedian arcs usually vestigial; skin finely pavement granulose ..." (*Euxoa*) versus "... head more or less reticulate with black or brownish and without freckles except in early instars of *Agrotis ipsilon*; submedian arcs usually fully developed; skin often bearing coarse, isolated granules ..." (the *Agrotis* group). In this system *Agrotis vetusta* Walker and *A. orthogonia* Morrison key to *Euxoa.* I have found that the most reliable character for distinguishing members of the two groups is the presence of two pairs of pigmented spots on the frons in species of *Euxoa.* One pair is located about one quarter of the distance from the base of the frons between the frontal setae and the margin and the second pair about three quarters of the distance up the frons on the margin adjacent to the adfrontal punctures. These spots have been observed in all 40 species of *Euxoa* examined (the upper pair was faint in two species) and in none of the 15 species of *Agrotis* Ochsenheimer, *Feltia* Walker and *Trichosilia* Hampson examined (lower pair of spots were faintly present in *A. orthogonia* Morrison).

Within the genus *Euxoa* body color and pattern vary greatly interspecifically but no consistent struc-

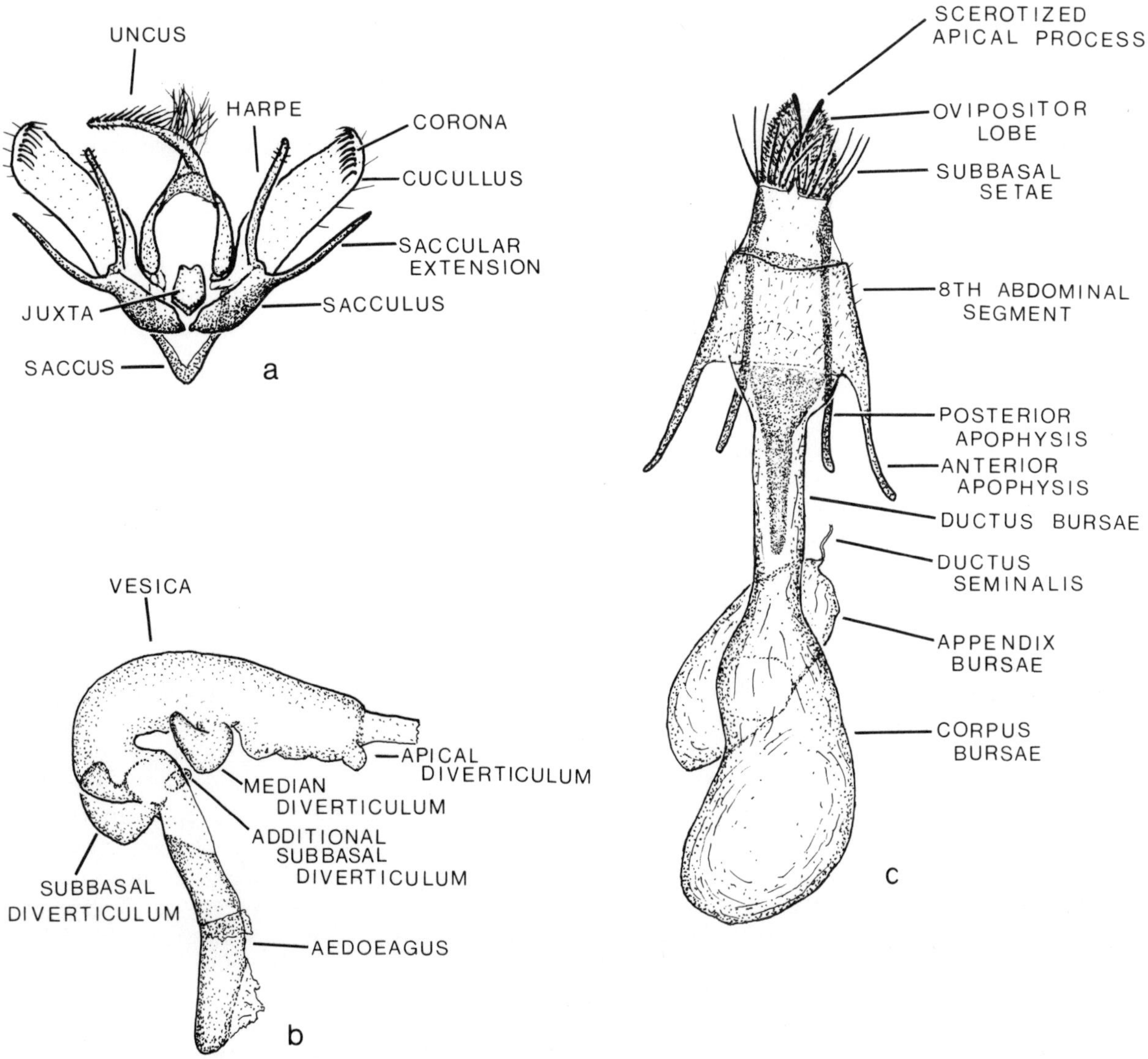

FIGURE 2: GENITALIA OF TYPICAL SPECIES OF *EUXOA*

a. Male genital capsule; *b.* Aedoeagus and vesica; *c.* Female genitalia.

tural characters were observed that would enable me to develop a key to species based on larvae.

No characters are known by which the eggs or pupae of species of *Euxoa* can be distinguished from those of other cutworm genera. Variation in size and surface sculpturing of eggs among species of *Euxoa* is illustrated in a series of papers by Salkeld (1975, 1976, 1977, 1984).

DISTRIBUTION AND DIVERSITY. *Euxoa* is primarily a genus of arid and semiarid habitats of the northern hemisphere. About 130 species occur in the Old World with the greatest diversity, in terms of numbers of species, in the Soviet Union, Mongolia, and northern China. Relatively few species occur in Europe, northern Africa, the Middle East, and the Far East. Only a handful of species occur in sub-Saharan Africa (as far south as Tanzania) and as far south in Asia as northern India and central China; none occurs in southeast Asia or Australia. The palearctic fauna includes a few species in the subgenera *Chorizagrotis, Pleonectopoda,* and *Orosagrotis,* but the vast majority of species are in the subgenus *Euxoa.*

In the New World, 175 species have been recorded, primarily in the western United States and Canada; 171 of these are treated in this fascicle. The ranges of only 10 species are known to extend into Mexico; although 40 more occur in southern Arizona and southern California and may occur in northern Mexico. The four species that do not occur in America north of Mexico are *Euxoa sorella* (Schaus) (subgenus *Chorizagrotis*), which occurs from central Mexico southward to Costa Rica, *Euxoa scortea* (Schaus) (subgenus *Chorizagrotis*), known only from central Mexico, *Euxoa cartagensis* Schaus (subgenus *Euxoa*; *bostoniensis* group), endemic to Costa Rica, and *Euxoa fauna* (Morrison) (subgenus *Orosagrotis*), endemic to Guadaloupe Island off the west coast of Mexico.

One *Euxoa* species, *E. cineraria* (Blanchard), has been reported from South America but the single known specimen, nominally from Chile, is probably mislabeled because it appears to belong to the *E. tritici* group of the Old World.

The 175 species of *Euxoa* in the New World are arranged in seven subgenera and 52 species-groups; all subgenera and species-groups are treated in this fascicle with the exception of the *fauna* group (subgenus *Orosagrotis*), which includes only *Euxoa fauna* of Guadaloupe Island, Mexico.

HABITATS AND HABITS. *Euxoa* is primarily a genus of dry temperate habitats. The North American fauna divides almost evenly between species associated with open aridland habitats, such as grassland and temperate desert areas, and dry forested habitats, particularly open pine parkland, dry fir, spruce and aspen forests, and piñon-juniper woodland. Less than 10 percent of the fauna is widely distributed in both open and forested habitats.

Aridland species are most abundant, in terms of both species and individuals, in areas where the ground vegetation is low and somewhat sparse, covering about one-half to three-quarters of the soil surface. This habitat is most widespread in shortgrass prairie in the Great Plains and in sagebrush (*Artemisia* spp.) areas in the northern Great Basin and Colorado Plateau. Individuals of *Euxoa* are often just as abundant, although species diversity decreases, in even drier habitats where the ground vegetation is sparse, such as badland areas in the Great Plains, and open shadescale (*Atriplex confertifolia* (Torrey and Frémont) Watson) flats of the central and southern Great Basin. Less than a dozen species occur in the hot deserts of the American southwest and those that occur there fly in the early spring or late fall. Individual species, however, may dominate the noctuid fauna; in the Mojave Desert in southern California, 3–4,000 specimens of *Euxoa recula* may be collected at a single light in one night. About a dozen species of *Euxoa* occur in the more mesic tallgrass prairie of the eastern Great Plains; only three aridland species occur on the Atlantic Seaboard.

The *Euxoa* fauna associated with forested habitats is most abundant and diverse in open pine parkland and piñon-juniper woodland at lower elevations in the mountain ranges of the west. About 30 species occur in more mesic forests of fir, spruce, and aspen in the boreal forest zone and at higher elevations in montane areas of the West. Only five species of *Euxoa* occur in the eastern deciduous forest region.

A detailed biogeographic analysis of the North American *Euxoa* fauna has been published (Lafontaine, 1982a).

Species of *Euxoa* have one generation per year. In most of western North America north of 40° latitude the adult flight season is most abundant from late June until early September. To the south of this the spring and early summer species fly progressively earlier southward while late summer and fall species fly progressively later. Hardwick (1971) recognized July 31 as the end of the spring flight and start of the fall flight and devised a formula for calculating the flight periods of species according to the increasing length of the summer season southward. The late summer and fall flight of most species of *Euxoa* is closely correlated with the blooming season of rabbit-brush (*Chrysothamnus* spp.), which is an important food source for the adults. Adults frequently can be collected on the flowers of rabbit-brush and gumweed (*Grindelia* spp.) in the late evening just before dark. Adults of most species of *Euxoa* come readily to lights and bait.

Adults of some species in the subgenus *Chorizagrotis* are migratory. They emerge in the spring; migrate from desert and grassland areas to higher elevations in the mountains to pass the summer, thereby avoiding the hottest, driest part of the summer; then return to the lowlands in the fall to mate and lay their eggs.

Most species of *Euxoa* overwinter as fully formed larvae within the egg; in some species a portion of the eggs hatch in the fall and the species overwinters both in the egg stage and as first instar larvae. The larvae of some species that occur as adults in the spring, hatch from the egg in late summer and pass the winter as partly grown larvae. All known *Euxoa* adults lay their eggs in the soil, either just below the surface, or slightly deeper. Some species, such as the darksided cutworm (*E. messoria*) lay their eggs singly, whereas others, such as the striped cutworm (*E. tessellata*), lay them in clusters.

The larvae of most species of *Euxoa* are surface feeding cutworms; although some species, such as *E. messoria, E. scandens,* and occasionally *E. tessellata,* can be destructive climbing cutworms. The larvae spend the day buried in the soil just under the surface. When the larvae are fully grown, normally after six or seven instars, there is a period of arrested development. This prepupal diapause usually is short in spring flying species and in species that occur in forested habitats and usually is long in aridland species. Most aridland species complete larval feeding in the spring when the vegetation is still relatively lush, then aestivate as mature larvae through the hot dry summer, to pupate and emerge as adults in late summer and fall when the desert Compositae are coming into bloom. Pupation takes place in an earthen cell in the ground with adult emergence occurring three to five weeks later. As far as is known, larvae of all species of *Euxoa* are general feeders; most species prefer to feed on broad leaved plants with a smaller number of species, such as the army cutworm (*E. auxiliaris*), that prefer to feed on grasses. On the basis of limited field data, it appears that soil conditions such as moisture, soil texture, and exposure, are more important factors in determining the range of a species than is the presence of specific larval foodplants.

CLASSIFICATION AND PHYLOGENY OF THE NOCTUINAE. An overview of the classification of the subfamily Noctuinae is included here so that I can discuss the position of *Euxoa* within the subfamily and polarize the characters used in the classification of subgenera within *Euxoa.*

The genera associated with the subfamily Noctuinae can be arranged in two tribes, Noctuini and Agrotini, each with different trends in modification of structural characters. The more primitive lineages of the two tribes lack some of the more conspicuous modifications of the tribes, and as a result, were included in a separate group (group II) by Forbes (1934) in a reclassification of the Agrotini (*recte* Noctuinae). Forbes' arrangement of genera was followed in the *Check List of the Lepidoptera of America North of Mexico* with "group II" genera referred to the tribe Aniclini. The Aniclini is not a natural group because it combines the more primitive elements of the Agrotini and Noctuini.

The Noctuini show a general trend toward the reduction and loss of rows of setae on the legs, first the fourth row of setae on the first segment of the middle and hind tarsi, then the outer and inner rows of setae on the foretibia. The male genitalia show a trend toward modification of the sacculus and harpe (ampulla of the clasper), coupled with the loss of the corona of the valve and a narrowing of the apex of the valve. The most primitive lineage in this tribe, including the genera *Diarsia* Hübner, *Ochropleura* Hübner, and most African genera of Noctuinae, retain the corona at the apex of the valve. In the female genitalia the ductus bursae and ostium bursae are sclerotized and have a narrow membranous band, or suture, between these two areas. Signa usually are present in the corpus bursae. In the larva, the epicranial suture is two or three times the length of the adfrontal suture; on the head capsule the genal seta (G_1), as in other noctuid subfamilies, is as close to the posterior margin of the head as is the genal puncture (G_a) so that the distance between G_1 and G_a is about ¼ or ⅓ that between G_1 and seta SO_3; the frontal punctures are located below the level of the frontal setae; the mandible has a basal tooth on the inner surface (secondarily lost in *Spaelotis* Boisduval, *Choephora* Grote and Robinson, *Cerastis* Ochsenheimer, and some species of *Xestia* Hübner); the larval skin is smooth; and the spinneret is longer than wide in most species.

The Agrotini exhibit quite different trends in structural modifications from those of the Noctuini. The genera included in this tribe can be arranged in four major lineages. The most primitive lineage of the Agrotini includes 24 genera, all primarily neotropical, of which four genera occur northward into our area, namely *Peridroma* Hübner, *Euagrotis* McDunnough, *Anicla* Grote, and *Hemieuxoa* McDunnough. The members of this lineage combine a number of primitive character states (shared with the Noctuini), with derived character states characteristic of the Agrotini. In some genera, including the four that reach our area, the fourth row of setae on the midtarsus is lost, and the outer row of setae on the foretibia is reduced to a few apical setae. The foretibia, as in the Noctuini, is longer than the first tarsal segment. In the female genitalia the ostium bursae is sclerotized, and this sclerotization extends along the sides of the posterior portion of the ductus bursae, but there is no separate sclerite in the ductus bursae. Signa are present as elongate bands in the corpus bursae of many species. Larval characters associate this lineage with the tribe Agrotini. On the head capsule the epicranial suture is equal in length, or less than, that of the adfrontal suture; seta G_1 is positioned farther anteriorly so that the distance between G_a and G_1 is more than ⅔ of that between G_1 and SO_3; the frontal punctures are located between and at the same level as the frontal setae. In the Agrotini the mandible lacks a basal tooth on the inner surface; the larval skin may be either smooth or granulose; and the spinneret is usually shorter than its width.

The remaining three lineages of the Agrotini are associated by the male vesica, in which there is a basal diverticulum with a cornutus, a subbasal coil in the vesica, and an apical diverticulum, and by the female genitalia, which have an entirely membranous ductus bursae (except for the condition in *Euxoa* described below). There is a complete row of setae on the inner and outer sides of the foretibia, and all genera except *Protogygia* McDunnough and some *Dichagyris* Lederer have the fourth row of tarsal setae. The corpus bursae lacks signa. This group of genera exhibits several trends that are apparently related to occurrence in aridlands, particularly subterranean pupation, and to assist the adult in crawling out of the soil upon emergence. The frons in the adult is thickened and often bulging or roughened and frequently has a sclerotized tuberclelike process; the foretibia tends to be shortened and usually is similar in length, or shorter than, the first tarsal segment; and the apical setae of the foretibia are frequently stout and clawlike. The first of the three lineages in this group includes the genera *Protexarnis* McDunnough, *Actebia* Stephens, *Parexarnis* Boursin, *Hemiexarnis* Boursin, and *Perissandria* Warren; these genera have sometimes been included in the Noctuini because the corona has been lost on the valve of the male. All other characters of both larvae and adults, however, indicate a placement within the Agrotini.

The two remaining lineages in the Agrotini are *Euxoa* and its sister group: a large number of genera associated with *Agrotis,* including *Feltia* Walker, *Trichosilia* Hampson, *Copablepharon* Harvey, *Protogygia* McDunnough, *Dichagyris* Lederer, and *Richia* Grote. In both lineages the larval skin is roughened, usually granulose. In *Euxoa* females the ostium bursae is sclerotized and this extends onto the ductus bursae as narrow sclerotized plates in the dorsal and ventral walls. In the *Agrotis* lineage the ostium bursae and ductus bursae are entirely membranous, or there may be a small sclerotized plate in the ventral wall of the ostium bursae.

All four lineages in the tribe Agrotini exhibit a trend in the male genitalia in which the harpe and uncus are modified. In the basal group in each of the four lineages, namely, *Peridroma, Protexarnis, Euxoa* subgenus *Chorizagrotis,* and *Richia,* the harpe projects dorsally at a right angle to the axis of the valve, and the uncus is relatively narrow. In each lineage there is a trend for the harpe to shift in position and lie in a groove along the inner surface of the valve parallel to the axis of the valve, and for the uncus apparently to become more important in the copulation process; it tends to become flattened and enlarged, more densely setose apically, or have short spinelike setae or a hook apically. This trend occurs several times independently within *Euxoa* (e.g., *edictalis, cinereopallida,* subgenus *Orosagrotis*) where the shift in harpe orientation is correlated with a shortening of the saccular extension.

CLASSIFICATION AND PHYLOGENY OF *EUXOA*. The sequence of subgenera and species-groups used in this fascicle follows the phylogenetic sequence discussed in detail previously (Lafontaine, 1981). Several trends in the development of genital characters can be traced along this sequence; the direction of these trends was determined by comparison among character states in *Euxoa* with those in related genera in the Agrotini. Two primitive character states, a short harpe and a central ridge or spine on the juxta, are retained in *Euxoa* only in the subgenus *Chorizagrotis.* The subbasal coil, or loop, in the vesica is retained in the subgenera *Palaeoeuxoa, Heteroeuxoa, Longivesica,* and *Pleonectopoda* but has been lost in the subgenera *Chorizagrotis, Euxoa,* and *Orosagrotis.* A bisaccate corpus bursae is found in all related genera in the Agrotini and in most primitive lineages within *Euxoa.* A unisaccate corpus bursae has been derived independently at least eight times within *Euxoa.* Most commonly, a unisaccate corpus bursae is oval with the ductus bursae entering the corpus bursae about one-third of the way along the right wall from the posterior end. In some species-groups the corpus bursae is T-shaped, or dumbbell shaped, with the ductus bursae at the posterior end of the corpus bursae.

Very few external structural characters vary enough to be used to characterize subgenera and species-groups. But, the length of male antennal serrations frequently is valuable to distinguish closely related species and sometimes to characterize groups. For example, the male antennae of species in the subgenus *Chorizagrotis* are only slightly serrate, almost filiform, and those of the species in the *edictalis, vetusta, camalpa,* and *serotina* groups are deeply biserrate, almost plumose. The eye is round and globular in most *Euxoa* species, slightly reduced in some subarctic and subalpine species, and markedly reduced and ellipsoid in three diurnal species of arctic and alpine areas. The tuberclelike process on the frons, characteristic of most species of *Euxoa* and *Agrotis,* has been secondarily lost in six species of *Euxoa* that occur in dry sandy habitats.

The paucity of external structural differences among species of *Euxoa* results in the necessity to distinguish species, and determine relationships, on the basis of genital characters. Structural characters of taxonomic and phylogenetic importance in the male genitalia include: size and shape of sacculus, saccular extension, cucullus, harpe, juxta, and uncus; presence or absence of pubescence on the harpe; shape of the vesica; and shape and position of accessory pouches of the vesica. In the female genitalia the shape of the corpus bursae is the most important character for distinguishing species-groups, and determining relationships among groups. The shape of the ovipositor lobes, distribution, length and thickness of their setae, and presence or absence of sclerotized processes on them, provide important characters for distinguishing species.

HISTORICAL REVIEW. The first species of *Euxoa* to be named in North America were *tessellata* and *messoria,* described in 1841 by T. W. Harris in connection with a study of the pest insects of Massachusetts. In the latter half of the nineteenth century many additional species were described, primarily by A. Guenée, F. Walker, H. K. Morrison, A. R. Grote, L. F. Harvey, J. B. Smith, and H. Strecker; most were described in the genus *Agrotis* because the species presently associated with *Euxoa* were not recognized as forming a distinctive group. This changed in 1890 when Smith showed that one group of species within *Agrotis* formed a natural group that could readily be distinguished from true *Agrotis* by the form of the clasper of the male genitalia; Smith used the generic name *Carneades* for this group of species (although the species that lack a frontal tubercle were not included). In a revision of the Agrotinae (=Noctuinae) of the world, G. F. Hampson (1903) transferred these species to *Euxoa* but took a backward step when he defined the genus on the basis of the foretibial spining and presence of a frontal tubercle. J. McDunnough (1923) was the first North American worker to use male genital characters to distinguish *Euxoa* species, and in 1950 he based a revision of eastern North American *Euxoa* species largely on characters of the female genitalia. This followed the pioneering work of I. Kozhanchikov (1929, 1937) who made extensive use of both male and female genital characters in revisions of the species of *Euxoa* of the Soviet Union. D. F. Hardwick (1970) demonstrated the value of the characters of the male vesica, not only for distinguishing closely related species, but also for arranging species into species-groups and subgenera. This subdivision of *Euxoa* into more manageable units, which could be studied intensively both in the field and in the laboratory, initiated a series of studies on selected subgenera and species-groups by Hardwick and later by me. The information accumulated in these studies, both published and unpublished, forms the basis for this revision of the genus *Euxoa* in America north of Mexico.

PREPARATION OF GENITALIA. The technique of macerating the abdomen overnight in a 5–10% solution of potassium hydroxide, then cleaning and slide mounting the genitalia, is well known and will not be repeated here. Preparation of Lepidoptera genitalia for slide mounting was described in detail by Hardwick (1950). The technique of everting and inflating the vesica of the male and inflating the bursa of the female was also described by Hardwick (1950) but is not widely used. I redescribe this technique here because these structures contain such important characters for taxonomic work in *Euxoa* and many other groups of Lepidoptera.

In the male the aedoeagus should be removed from the remainder of the genitalia and the vesica carefully prodded to the apex of the aedoeagus with a blunt pin or probe (a short piece of hair can be used in species with a very narrow aedoeagus). In species with numerous spines or stout cornuti in the vesica (absent in *Euxoa*) it usually is advisable to push the vesica beyond the apex of the aedoeagus to free the spines. A hypodermic syringe with a small needle (27–30 gauge) is inserted into the basal half of the aedoeagus through the membranous opening at the base of the aedoeagus and held onto the needle with a pair of fine-tipped forceps (e.g., watchmakers' forceps). The vesica is then everted by squirting isopropyl alcohol from the syringe through the aedoeagus and vesica while immersed in a small dish of isopropyl alcohol. If the vesica is allowed to come into contact with water after being everted, either during cleaning or staining, the membrane will absorb water and collapse. The concept of inflating the vesica carries the connotation of variation in size and shape depending on inflation pressure. The size and shape of the vesica, however, is fixed and remarkably constant within a species; the effect of the inflation pressure simply determines whether the wall of the vesica is fully distended or wrinkled. The corpus bursae in the female is inflated by inserting the needle of the hypodermic syringe into the corpus bursae via the ductus bursae and inflating it with isopropyl alcohol while the bursa is immersed in alcohol. As with the vesica, the female genitalia should not come in contact with any solution that contains water after the bursa has been inflated. When the genitalia are mounted on a microscope slide, the cover slip should be raised with small pieces of polyethylene tubing so that the shapes of the vesica and bursa are not distorted in mounting.

CRITERIA FOR SPECIES AND SUBSPECIES. Species usually are defined as groups of actually or potentially interbreeding populations. Within *Euxoa,* and other noctuid genera, structural differences in male and/or female genitalia, or structure of male antennae, usually have been taken as indicative of species status. Hybridization experiments have shown the general validity of this assumption. Some species, however, do not interbreed in the field and yet do not differ in structural characters. Indication that more than one species is involved usually results from several kinds of evidence: two color or size forms occurring together without evidence of interbreeding; differences in behavior, habitat, season, or larval foodplants; lack of cross attractancy in pheromones. These kinds of discrepancies are most often noticed in areas where the species occur together, and the status of the forms usually can be resolved with careful field studies. The status of geographically isolated populations that differ in these kinds of ways usually requires laboratory experiments for resolution.

I use the rank of species for distinctive populations that overlap in range, yet show no evidence of hybridization in nature, and for isolated distinctive populations that differ in characters that are not highly variable within populations and are indicative of species status in related, yet sympatric species.

I use the rank of subspecies for isolated yet distinctive populations when the characters by which the subspecies is recognized tend to be variable within populations. I also use the rank of subspecies for populations that appear to be as distinct as species and yet are completely interfertile in laboratory experiments, or have a zone of intergradation where their ranges meet or approach each other. I do not use subspecific names to try to describe geographical variation in species that differ in size or color from habitat to habitat.

Undoubtedly, future experiments will reveal that some highly variable species (such as *idahoensis*) are complexes of sibling species, whereas other species complexes (such as some of the members of the *infausta* group) represent a balanced polymorphism, or habitat forms within a species.

KEY TO THE SUBGENERA OF *EUXOA* OF AMERICA NORTH OF MEXICO BASED ON MALE GENITALIA

1. Right and left saccular extensions flattened and cupped or spatulate apically (plate A, figures 1–8) 2
— Right and left saccular extensions cylindrical, or bladelike, tapering to apex (plate B, figures 1–8) 3

2. Harpe much shorter than saccular extension (plate A, figure 1) .... subgenus *Chorizagrotis* p. 28
— Harpe as long as, or longer than, saccular extension (plate A, figure 5) .. subgenus *Palaeoeuxoa* p. 31

3. Right saccular extension with a finlike structure projecting from inner surface at ⅓ length; saccular extension with a 90° twist mesially, flattened and bladelike apical to twist (plate A, figures 9, 10) .... *Heteroeuxoa* p. 34

– Right saccular extension tapering evenly from base to apex with at most a ridge extended onto base of saccular extension (plate B, figures 2–8) 4

4. Vesica with a submedian twist or coil (plate C, figures 1–10) . . . . subgenus *Pleonectopoda* p. 40

– Vesica variably curved or bent but without submedian twist or coil (plate E, figures 1–10) 5

5. Vesica 2.5–3.0 times as long as aedoeagus (plate B, figures 3, 4) . . . . subgenus *Longivesica* p. 37

– Vesica, at most, twice as long as aedoeagus 6

6. Vesica projected to left beyond apex of aedoeagus (plate O, figures 3–6) . . . . subgenus *Orosagrotis* p. 147

– Vesica projected dorsally, or to right, beyond apex of aedoeagus (plate E, figures 1–10) . . . . subgenus *Euxoa* p. 58

KEY TO SUBGENERA *PLEONECTOPODA*, *LONGIVESICA*, *OROSAGROTIS* AND *EUXOA* OF AMERICA NORTH OF MEXICO (FOR USE WHEN CHARACTERS OF VESICA UNAVAILABLE)

1. Antenna bipectinate, almost plumose (plate 2, figure 14); occurring in Californian deserts and on Pacific coast (*vetusta* and *fuscigera*) . . . . subgenus *Pleonectopoda* (part) p. 40

– Antenna biserrate or bipectinate; if bipectinate, then occurring east of Sierra Nevada/Cascades axis, or at high elevations in Sierra Nevada 2

2. Right harpe lying along inner side of cucullus parallel to dorsal margin; uncus not markedly flattened or spatulate apically 3

– Right harpe projected dorsally beyond dorsal margin of cucullus, at least at apex, but if lying along inner side of cucullus, then with uncus markedly flattened or spatulate apically 5

3. Antenna slightly to moderately biserrate, about 2× width of central region of shaft . . . . subgenus *Orosagrotis* p. 147

– Antenna bipectinate, almost plumose, about 5–6× width of central region of shaft 4

4. Apex of right saccular extension curved inward (plate B, figure 4) (*edictalis*) . . . . subgenus *Longivesica* (part) p. 37

– Apex of right saccular extension straight or curved dorsally toward cucullus (plate D, figure 10; plate E, figure 1) . . . . subgenus *Euxoa* (part) p. 58

5. Harpes markedly asymmetrical (plate D, figures 5–8) (*atomaris* and *pleuritica* groups) . . . . subgenus *Pleonectopoda* (part) p. 40

– Harpes not markedly asymmetrical 6

6. Right saccular extension either less than ¾ length of right harpe, or almost as long as right harpe but much stouter 7

– Right saccular extension longer than right harpe or if slightly shorter, then about as stout as harpe 9

7. Frontal tubercle absent (*westermanni* group, in part) . . . . subgenus *Pleonectopoda* (part) p. 40

– Frontal tubercle present 8

8. Occurring in subarctic or at high elevations (*westermanni* group, in part) . . . . subgenus *Pleonectopoda* (part) p. 40

– Occurring at lower elevations in pine parkland or grassland areas in southern Canada and western United States . . . . . subgenus *Euxoa* (part) p. 58

9. Juxta prominently constricted mesially, about ⅔ as wide mesially as anteriorly and posteriorly (plate B, figure 2) (*messoria*) . . . . subgenus *Longivesica* (part) p. 37

– Juxta not prominently constricted mesially, often much narrower anteriorly than posteriorly 10

10. Forewing ground color whitish gray (plate 2, figures 11, 12); hindwing dirty white with pale smoky-gray band on outer third; male genitalia with saccular extensions very stout, slightly longer than harpes and bending ventrally near apex (plate D, figure 2) (*tristicula*) . . . . subgenus *Pleonectopoda* (part) p. 40

– Not as described above 11

11. Forewing dark brown, or blackish brown with

large, prominent claviform spot and pale, contrasting shading on cubital vein from wing base to reniform spot (plate 1, figure 31) (*divergens*) .... subgenus *Longivesica* (part) p. 37

— Not as described above 12

12. Forewing dark blackish gray with paler gray reniform and orbicular spots and costa (plate 1, figures 32, 33); small species with forewing length of 12–14 mm; distributed in boreal forest zone of eastern Canada and northern Michigan (*sinelinea*) .... subgenus *Longivesica* (part) p. 37

— Not as described above 13

13. Sacculus oval, dorsal margin convex; saccular extensions slightly longer than harpes and essentially straight, tapered abruptly from stout base to apex; harpes short and incurved, about ⅔ length of sacculus (plate D, figure 1); occurring at higher elevations in Cascades and Sierra Nevada Mountains (*extranea*) .... subgenus *Pleonectopoda* (part) p. 40

— Not as described above .... subgenus *Euxoa* (part) p. 58

KEY TO SUBGENERA AND SPECIES-GROUPS OF GENUS *EUXOA* OF AMERICA NORTH OF MEXICO (FEMALES)

1. Ductus seminalis arising from right side of corpus bursae (subgenus *Orosagrotis*) 2

— Ductus seminalis arising from left side of corpus bursae, or from left sac if bursa bisaccate 4

2. Ovipositor lobe without sclerotized process at apex .... *wilsoni* group p. 154

— Ovipositor lobe with sclerotized process at apex 3

3. Sclerotized processes at apex of ovipositor fused .... *ridingsiana* group p. 149

— Sclerotized processes at apex of ovipositor not fused .... *nomas* group p. 149

4. Corpus bursae bisaccate 5

— Corpus bursae unisaccate 32

5. Left sac about twice as long and large as right sac .... subgenus *Longivesica* p. 37

— Left sac, at most, slightly longer or larger than right sac 6

6. Dorsal margins of ovipositor lobes fused by a narrow sclerotized band (plate EE, figures 11, 12) .... *punctigera* group p. 83

— Dorsal margins of ovipositor lobes not fused, with membrane between lobes 7

7. Apex of ovipositor lobes with sclerotized processes 8

— Apex of ovipositor lobes without sclerotized processes 9

8. Apex of ovipositor lobe with sclerotized rim; base of lobe with dense border of several hundred fine setae (plate EE, figure 5) .... *serricornis* group p. 73

— Apex of ovipositor lobe with a sclerotized fingerlike projection; base of lobe with single row of about 20 stout setae (plate FF, figure 3) .. .. *obeliscoides* group p. 115

9. Sclerotized plate in ventral wall of ductus bursae asymmetrical, more heavily sclerotized and bulging on left side 10

— Sclerotized plate in ventral wall of ductus bursae symmetrical, long and lanceolate, extending down middle of ductus bursae 13

10. Sclerotized plate in ventral wall of ductus bursae tongue shaped, rounded anteriorly (plate P, figure 7) .... subgenus *Heteroeuxoa* (part) p. 34

— Sclerotized plate in ventral wall of ductus bursae more or less triangular, pointed anteriorly 11

11. Appendix bursae oriented dorsoventrally, ductus seminalis arising at dorsal end (plate P, figure 5) .... subgenus *Palaeoeuxoa* p. 31

— Appendix bursae oriented anteroposteriorly, ductus seminalis arising at posterior end 12

12. Appendix bursae oblong, extending as far posteriorly from its junction with corpus bursae as anteriorly (plate V, figure 8) .... *declarata* group (part) p. 78

— Appendix bursae somewhat triangular, extending much farther anteriorly from its junction with corpus bursae than posteriorly (plate P, figure 6) .... subgenus *Heteroeuxoa* (part) p. 34

13. Apex of ovipositor lobe clothed with short, blunt-tipped, almost conical setae, or with stout setae at apex 14
– Apex of ovipositor lobe clothed with fine or fine-tipped setae 21

14. Base of ovipositor lobe with one to several rows of long curving setae, these about ¾ length of lobe 15
– Base of ovipositor lobe with only a few scattered long setae, these usually about ½ length of lobe 16

15. Appendix bursae about as large as corpus bursae and extending almost as far anteriorly; genitalia 7.5–8.5 mm long (plate T, figure 1) ... . *hardwicki* group p. 63
– Appendix bursae usually smaller than corpus bursae and not extending as far anteriorly as corpus bursae; genitalia 9 to 12 mm long (plate V, figure 1) .... *pluralis* group p. 75

16. Bisaccate condition of corpus bursae apparently resulting from an invagination of the left wall of bursa; appendix rounded or triangular; if oblong, then extending anteriorly from its junction with corpus bursae 18
– Bisaccate condition of corpus bursae well defined, appendix bursae oblong, about twice as long as wide and extending as far posteriorly from its junction with corpus bursae as anteriorly 17

17. Apex of ovipositor lobe truncate, lobe somewhat rectangular in outline (plate EE, figure 9); abdominal segment VIII with setae thin and inconspicuous (plate W, figure 2) .... *silens* group (part) p. 80
– Apex of ovipositor lobe acute, lobe triangular in outline (plate EE, figure 10); abdominal segment VIII bristly looking, setae stout and conspicuous (plate W, figure 3) .... *simulata* group p. 82

18. Ovipositor lobe much longer than wide and acute at apex .... *westermanni* group (part) p. 44
– Ovipositor lobe about as long as wide, apex blunt and rounded 19

19. Ductus seminalis arising on anterolateral surface of appendix bursae; sclerotized plate in ventral wall of ductus bursae extending about ⅚ length of ductus bursae .... *pleuritica* group p. 56
– Ductus seminalis arising at posterior end of appendix bursae; sclerotized plate in ventral wall of ductus bursae extending about ⅔ length of ductus bursae 20

20. Ovipositor lobe semicircular in outline with at most a slight angle at apex (plate EE, figure 7); orbicular spot a pale spot on forewing (plate 3, figures 26–32) .... *setonia* group p. 76
– Ovipositor lobe truncate, somewhat rectangular in outline with two angled points toward apex (plate EE, figure 9); orbicular spot at least partially outlined in black (plate 3, figures 43–49) .... *silens* group p. 80

21. Posterior apophyses 2.75–3.00 times as long as anterior apophyses .... subgenus *Pleonectopoda* (part) p. 40
– Posterior apophyses less than 2.75 times as long as anterior apophyses 22

22. Anterior apophyses longer than abdominal segment VIII; sclerotized plate in ventral wall of ductus bursae less than half length of ductus bursae (plate T, figure 2) .... *camalpa* group p. 64
– Anterior apophyses as long as, or shorter than, abdominal segment VIII; sclerotized plate in ventral wall of ductus bursae more than half length of ductus bursae 23

23. Corpus bursae very large, its length two or three times length of ductus bursae .... *vetusta* group p. 54
– Corpus bursae smaller, its length less than twice length of ductus bursae 24

24. Appendix bursae rounded or triangular, extending mainly laterally or anteriorly from its junction with corpus bursae 25
– Appendix bursae elongate, two or three times as long as wide, extending as far posteriorly from its junction with corpus bursae as anteriorly 28

25. Corpus bursae bending to left anteriorly around anterior end of appendix bursae; forewing reddish brown with contrasting gray basal area (plate 6, figures 28–30) .... *basalis* group p. 117

— Corpus bursae not bending to left anteriorly; forewing without a contrasting gray basal area 26

26. Sclerotized plate in dorsal wall of ductus bursae about 1/8 longer than plate in ventral wall; ground color of forewing pale gray . . *tristicula* group p. 53

— Sclerotized plates in dorsal and ventral walls of ductus bursae similar in length; ground color of forewing not pale gray 27

27. Ductus seminalis arising on anterolateral surface of appendix bursae; sclerotized plate in ventral wall of ductus bursae extending about 5/6 length of ductus bursae .... *pleuritica* group p. 56

— Ductus seminalis arising at posterior end of appendix bursae; sclerotized plate in ventral wall of ductus bursae extending about 2/3 length of ductus bursae .... *westermanni* group (part) p. 44

28. Ovipositor lobe with subbasal row of 10–15 setae, these about 2/3 length of lobe; forewing with reniform and orbicular spots inconspicuous, not set off by darker surrounding area; space between reniform and orbicular spots not darker than remainder of forewing except possibly by presence of a dark median line ... 29

— Ovipositor lobe with subbasal row of 10–15 setae but these short, about 1/3–1/2 length of lobe; forewing with reniform and orbicular spots prominent, usually surrounded by dark-brown or black shading, including space between these spots 30

29. Appendix bursae at least 4.0 mm long; if shorter, then with forewing maculation obscure except for conspicuous black antemedial and postmedial lines .... *bostoniensis* group (part) p. 65

— Appendix bursae usually shorter than 3.5 mm long; forewing with maculation obscure or prominent but antemedial and postmedial lines not particularly conspicuous ......*fuscigera* group p. 55

30. Forewing with antemedial line consisting of a black line preceded by a pale-yellow line, reniform and orbicular spots ringed interiorly and partially filled with pale-yellow or cream-colored scales, usually a black basal dash present (plate 2, figures 43–46) ....*bostoniensis* group (part) p. 65

— Forewing with antemedial line consisting of two black lines, space between lines, and space inside reniform and orbicular spots, similar in color to remainder of forewing 31

31. Corpus bursae elongate, constricted mesially; appendix bursae dorsal to corpus bursae, oval, its longitudinal axis oblique to that of corpus (plate Z, figure 8); forewing with basal space and reniform and orbicular spots paler than remainder of forewing and with a violet suffusion (plate 6, figures 26, 27) ...... *lillooet* group p. 116

— Corpus bursae oval, appendix bursae elongate and to left of corpus bursae, its longitudinal axis roughly parallel to that of corpus bursae (plate V, figure 7); forewing with basal area and reniform and orbicular spots frequently paler than remainder of forewing but without a pale-violet suffusion (plate 3, figures 33–42) .... *declarata* group (part) p. 78

32. Posterior margin of ovipositor lobe with a sclerotized rim 33

— Ovipositor lobe with at most a sclerotized flangelike projection at apex, or on dorsal margin of lobe, but without a sclerotized rim around posterior margin 35

33. Base of ovipositor lobe with a single row of 6–10 setae (plate EE, figure 3) ..... *scholastica* group p. 70

— Base of ovipositor lobe with a dense border of about a hundred long setae 34

34. Anterior apophyses about as long as abdominal segment VIII (plate U, figure 5) .... *terrena* group p. 71

— Anterior apophyses about twice as long as abdominal segment VIII (plate U, figure 3) ... . *annulipes* group p. 70

35. Dorsal margins of ovipositor lobes fused by a narrow sclerotized band, or flangelike projections at apex of lobes fused 36

— Dorsal margins of ovipositor lobes, or flangelike projections at apex of lobes, not fused but with a membrane between them 39

36. Ductus seminalis arising at middle of corpus bursae on a small pouch directed laterally on left side (plate DD, figure 8) .... *luctuosa* group p. 147

— Ductus seminalis arising at posterior end of corpus bursae, or on a pouch at posterior end 37

37. Dorsal margin of ovipositor lobes with a sclerotized, finlike structure projecting dorsally from lobes (plate EE, figure 19) (*satiens*) .... *infausta* group (part) p. 103

– Dorsal margin of ovipositor lobes without a raised ridge, at most, with a flangelike projection at apex projecting posteriorly 38

38. Forewing longitudinally streaked; reniform and orbicular spots usually fused (plate 8, figures 19, 20) .... *atristrigata* group (part) p. 144

– Forewing not longitudinally streaked; reniform and orbicular spots not fused (plate 7, figures 58–64) .... *aequalis* group (part) p. 138

39. Apex of ovipositor lobe without a sclerotized flangelike projection 40

– Apex of ovipositor lobe with a sclerotized flangelike projection 60

40. Setae along dorsal margin, or near apex, of ovipositor lobe longer and much stouter than remainder of setae on lobe 41

– Setae on dorsal margin, or near apex, of ovipositor lobe sometimes short and conical, but not long and stout 42

41. Frontal tubercle absent .. *westermanni* group (part) p. 44

– Frontal tubercle present (reduced and hidden by overhanging vestiture in *detersa*) .... *detersa* group (part) p. 118

42. Frontal tubercle absent .... *violaris* group p. 109

– Frontal tubercle present 43

43. Apical third of ovipositor lobe clothed with short, blunt-tipped, conical or spikelike setae 44

– Ovipositor lobe clothed with hairlike, fine-tipped setae, occasionally a few short setae near apex 46

44. Ovipositor lobe clothed with stout, spikelike setae, these 4 or 5 times longer than wide, more numerous toward apex of lobe but not densely crowded near apex (plate EE, figure 15) .... *cincta* group p. 91

– Ovipositor lobe with setae near apex much shorter and more conical than remainder of setae on lobe, these slightly longer than wide and densely crowded toward apex of lobe 45

45. Setae on basal half of ovipositor lobe mostly long and fine tipped; ovipositor lobe usually rounded or pointed toward apex (plate FF, figures 4, 5) .... *detersa* group (part) p. 118

– Setae on basal half of ovipositor lobe sparse, these consisting of numerous short, almost conical setae intermixed with moderately long setae, latter fine tipped; ovipositor lobe conspicuously truncated toward apex (plate EE, figure 13) (*tessellata*) .... *tessellata* group (part) p. 87

46. Length of corpus bursae from anterior end to junction with ductus bursae 2–3 × length of ductus bursae 47

– Length of corpus bursae from anterior end to junction with ductus bursae 1.25–1.75 × length of ductus bursae 54

47. Sclerotized plate in ventral wall of ductus bursae triangular or semicircular, about as wide as long and without a tongue-shaped extension at anterior end (plate P, figures 1, 4) .... subgenus *Chorizagrotis* (part) p. 28

– Sclerotized plate in ventral wall of ductus bursae with an elongate, tongue-shaped extension at anterior end extending along ductus bursae 48

48. Ovipositor lobe with long setae, similar in length to those in subbasal row of setae, scattered over entire surface of lobe; northern species .... *westermanni* group (part) p. 44

– Ovipositor lobe with setae as long as those in subbasal row of setae essentially confined to basal third of lobe 49

49. Corpus bursae markedly constricted mesially or submesially by ⅓–½ of its width; bursa appearing swollen both posteriorly and anteriorly 50

– Corpus bursae enlarged posteriorly by presence of one or two diverticula but anterior half of corpus bursae evenly tapered 54

50. Anterior end of corpus bursae wider than posterior end, or corpus bursae with a diverticulum on ventral surface at junction with ductus bursae (plate P, figures 2, 3) .... subgenus *Chorizagrotis* (part) p. 28

— Anterior end of corpus bursae no wider than posterior end, posterior end usually wider; corpus bursae without a diverticulum on ventral surface, at most, a slight bulge in wall of bursa at posterior end 51

51. Ductus bursae swollen at anterior end of sclerotized plates in dorsal and ventral walls of ductus bursae (plate P, figure 8) (*olivia*) .... subgenus *Heteroeuxoa* (part) p. 34

— Ductus bursae not swollen mesially 52

52. Anterior apophyses similar in length, or shorter than, abdominal segment VIII (plate Z, figure 4) .... *cursoria* group (part) p. 109

— Anterior apophyses about 1.25× length of abdominal segment VIII 53

53. Forewing with a prominent basal dash, reniform and orbicular spots fused (plate 4, figures 44–46) .... *hollemani* group (part) p. 92

— Forewing without a basal dash, reniform and orbicular spots not fused (plate 2, figure 31) .... *serotina* group p. 65

54. Anterior apophyses about 0.6 mm long .... *westermanni* group (part) p. 44

— Anterior apophyses about 0.9–1.5 mm long ... 55

55. Corpus bursae, from anterior end to junction with ductus bursae at least 2× length of ductus bursae (plate Z, figure 3); ovipositor lobe with an inconspicuous row of 6–9 long setae subbasally, these about half length of lobe (plate EE, figure 20) .... *cursoria* group (part) p. 109

— Corpus bursae less than 2× length of ductus bursae; if corpus bursae, from anterior end to junction with ductus bursae, 2× length of ductus bursae, then subbasal row of setae on ovipositor lobe conspicuous, consisting of 15–20 long setae, these about ⅔ length of lobe 56

56. Forewing with a long, black basal dash extending about ¼ length of wing; antemedial and postmedial lines absent; cubital vein not pale and contrasting; reniform and orbicular spots fused in most specimens (plate 4, figures 44–46) .... *hollemani* group (part) p. 92

— If forewing with a long dash, then either transverse lines present, or cubital vein pale and contrasting; reniform and orbicular spots not fused 57

57. Sclerotized plate in dorsal wall of ductus bursae ending before posterior opening of ductus bursae; ductus seminalis arising on small diverticulum to left of, or dorsal to, posterior end of corpus bursae (plate Y, figure 6) .... *infausta* group (part) p. 103

— Sclerotized plate in dorsal wall of ductus bursae continuing posteriorly beyond posterior end of ductus bursae to sclerotized wall of abdominal segment VIII; ductus seminalis usually arising at posterior end of corpus bursae; ductus bursae appearing to enter corpus bursae part way down right side 58

58. Forewing with a prominent basal dash, or inner edge of terminal area with a W-mark where pale-lined veins CuA_1 and M_3 project into it (plate 7, figures 34–38) .... *detersa* group (part) p. 118

— Forewing without a basal dash, occasionally with an indistinct dark smudge near wing base in species with a very dark forewing; inner edge of terminal area without a W-mark at veins CuA_1 and M_3 59

59. Hindwing white at base except for light-brown shading on veins (plate 4, figures 14–16) .... *pallipennis* group p. 85

— Hindwing buff or brown at base .... *rufula* group; *intrita* group; *albipennis* group (part: *henrietta*); *detersa* group (part: *dodi, difformis*) pp. 68, 69, 89, 118

60. Sclerotized flangelike structure on ovipositor lobe with two finlike projections dorsally, one at anterior end and one at posterior end (plate FF, figure 16) .... *cinereopallida* group p. 145

— Sclerotized flangelike structure on ovipositor lobe straight or rounded along dorsal margin .. 61

61. Ovipositor lobe with a sclerotized rim along dorsal margin, this tapering out at apex of lobe (plate FF, figure 1) .... *siccata* group p. 114

— Ovipositor lobe with a sclerotized ridge on dorsal margin, this ending in a pointed, or fingerlike, projection near apex of lobe 62

62. Base of ovipositor lobe with a dense border of

about a hundred fine setae (plate EE, figure 16) .... *catenula* group p. 95

– Base of ovipositor lobe with a row of 15–20 long setae 63

63. Corpus bursae somewhat pear shaped, larger posteriorly than anteriorly, with bulge to right as well as diverticulum to left; junction of ductus bursae and corpus bursae in middle of posterior end of corpus bursae 64

– Corpus bursae oval, not enlarged posteriorly except by presence of diverticulum to left, without bulge to right at posterior end; junction of ductus bursae and corpus bursae on right or ventral side of corpus bursae 67

64. Ovipositor lobe with long setae in subbasal row of setae much stouter on ventral surface than on dorsal surface (plate FF, figure 2) .... *choris* group p. 114

– Ovipositor lobe with long setae in subbasal row of setae, thin and hairlike on both dorsal and ventral surfaces 65

65. Forewing finely streaked longitudinally; a black basal dash present; orbicular spot elongate 66

– Forewing evenly colored, or mottled, but not longitudinally streaked; basal dash usually absent, occasionally a trace of a dash present in species with a very dark forewing; orbicular spot usually round or slightly oval .... *comosa* group (part) p. 96

66. Sclerotized plates in dorsal and ventral walls of ductus bursae extending much farther anteriorly than anterior apophyses (plate X, figure 1); forewing with antemedial, and usually postmedial lines present (plate 4, figures 29–31) (*plagigera*) .... *tessellata* group (part) p. 87

– Sclerotized plates in dorsal and ventral walls of ductus bursae extending no farther anteriorly than anterior apophyses (plate X, figure 7); forewing with transverse lines absent or barely traceable (plate 4, figures 47, 48) (*xasta*) .... *hollemani* group (part) p. 92

67. Sclerotized plate in ventral wall of ductus bursae extending much farther anteriorly than anterior apophyses, usually by at least half length of apophyses 68

– Sclerotized plate in ventral wall of ductus bursae usually extending about as far anteriorly as anterior apophyses, occasionally extending farther anteriorly than anterior apophyses by about ⅓ length anterior apophyses 70

68. Forewing with black basal dash; or forewing streaked longitudinally (plate 7, figures 39–44) .... *detersa* group (part) p. 118

– Forewing without basal dash; forewing not streaked longitudinally 69

69. Sclerotized flangelike structure on ovipositor lobe extending along at least posterior third of lobe as a raised, finlike ridge (plate FF, figure 11) .... *detersa* group (part) p. 118

– Sclerotized flangelike structure on ovipositor lobe a rounded projection at apex of lobe, this tapering out along dorsal surface .... *comosa* group (part); *albipennis* group (part); *detersa* group (part: *murdocki, moerens, latro*) p. 89, 96, 118

70. Forewing longitudinally streaked, with a black, basal dash in most specimens 71

– Forewing not longitudinally streaked and without a basal dash 73

71. Forewing dark gray with pale shading along costal and cubital veins distal to claviform spot and in subterminal area (plate 7, figures 36–38) (*quadridentata*) .... *detersa* group (part) p. 118

– Forewing longitudinally streaked along or between veins but not particularly pale on cubital and costal veins or in subterminal area 72

72. Sclerotized flangelike structure on ovipositor lobe rounded at apex of lobe, then tapering out along dorsal margin; sclerotized plate in ventral wall of ductus bursae triangular toward anterior end and bending slightly to left (plate DD, figure 3) .... *misturata* group (part) p. 143

– Sclerotized flangelike structure on ovipositor lobe heavily sclerotized and finlike on dorsal margin of lobe; sclerotized plate in ventral wall of ductus bursae tongue shaped toward anterior end and not bending to left (plate DD, figure 6) .... *atristrigata* group (part) p. 144

73. Sclerotized flangelike structure on ovipositor lobe forming a raised finlike plate along posterior half of dorsal margin of lobe 74

– Sclerotized flangelike structure on ovipositor lobe forming a rounded projection at apex of lobe, this tapering out evenly along dorsal margin 75

74. Forewing with median line indistinct, less conspicuous than antemedial and postmedial lines; ovipositor lobe with a large sclerotized plate extending laterally from dorsal margin (plate FF, figure 19) .... *mitis* group p. 146

— Forewing with median line distinct, much more prominent than antemedial and postmedial lines; ovipositor lobe with a sclerotized ridge extending above, not lateral to, dorsal margin (plate FF, figure 10) (*teleboa*) .. *detersa* group (part) p. 118

75. Sclerotized plate in ventral wall of ductus bursae short and triangular toward anterior end, extending less than half length of ductus bursae (plate DD, figure 3) .... *misturata* group (part) p. 143

— Sclerotized plate in ventral wall of ductus bursae long and tongue shaped toward anterior end, about ⅔ as long as ductus bursae 76

76. Forewing with dark shading in lower end of reniform spot and often between reniform and orbicular spots; orbicular spot large, wider than the space between it and reniform spot; corpus bursae oval (*munis*) (plate DD, figure 2) .... *aequalis* group (part) p. 138

— Forewing with, at most, median line between reniform and orbicular spots, reniform spot pale filled or dark filled but without a dark spot at lower end; orbicular spot small, usually not as wide as space between reniform and orbicular spots; corpus bursae elongate, much longer than wide in most specimens (plate Y, figure 1) .... *comosa* group (part) p. 96

SUBGENUS *Chorizagrotis* Smith

Chorizagrotis Smith, 1890.

Mesoeuxoa Corti, 1932.

Chorizagrotis is a small holarctic subgenus containing six species: one holarctic and five nearctic. Of the six North American species, four occur north of Mexico; two occur in southern Mexico and Central America (Hardwick, 1970).

Males of this subgenus are most easily recognized by presence of long, apically spatulate, saccular extensions and short harpes. Other character states restricted to *Chorizagrotis* males are a posteriorly constricted juxta; juxta with a central process (absent in *auxiliaris*); vesica with a large, multilobed, saclike subbasal diverticulum; aedoeagus with a narrow, sclerotized band extended onto subbasal diverticulum of the vesica. The relatively narrow forewings of most species, and very weakly biserrate antennae are useful charcters for sorting males of *Chorizagrotis* from those of the other subgenera except *Orosagrotis.*

Females of *Chorizagrotis* can usually be distinguished from those of other subgenera by the dorsoventrally flattened abdomen. They can also be recognized by genital characters given in the key.

Adults of *auxiliaris,* and probably those of the other southern species in the subgenus, emerge in the spring, migrate to higher elevations to aestivate through the summer and return in the fall. Few *Euxoa* (sensu lato) species occur in the desert regions of southwestern United States and northern Mexico where members of *Euxoa* (*Chorizagrotis*) are abundant. It is probable that extreme summer temperatures are a limiting factor in preventing species that pass the summer months as larvae in the soil from surviving in these areas. The habit of spring emergence and movement out of the desert regions during the hot summer months probably allows species of the subgenus *Chorizagrotis* to survive in these areas.

The subgenus *Chorizagrotis* in North America, including Mexico, was revised by Hardwick (1970); my treatment of the subgenus relies heavily on his revision.

KEY TO NORTH AMERICAN SPECIES OF SUBGENUS *CHORIZAGROTIS*

1. Males 2

— Females 5

2. Juxta with broad, posterior portion flat; harpe short, about 3 times as long as wide (plate A, figure 2) .... *auxiliaris* p. 30

— Juxta with broad, posterior portion with a central rounded, triangular, or finlike process; harpe longer, 5–8 times as long as wide in most specimens 3

3. Juxta with median carina (plate A, figure 4) .... *terrealis* p. 31

– Juxta with rounded or conical median process 4

4. Juxta with shallow, rounded, median process (plate A, figure 1) .... *lidia* this page

– Juxta with prominent, conical, median process (plate A, figure 3) .... *inconcinna* p. 31

5. Sclerotized plate in ventral wall of ductus bursae linear, about half as wide as ductus, expanded abruptly at abdominal segment VIII (plate P, figure 3) .... *inconcinna* p. 31

– Sclerotized plate in ventral wall of ductus bursae triangular or U-shaped, expanded gradually toward abdominal segment VIII 6

6. Sclerotized plate in ventral wall of ductus bursae broad and rounded anteriorly (plate P, figure 4) .... *terrealis* p. 31

– Sclerotized plate in ventral wall of ductus bursae V-shaped, narrowed anteriorly 7

7. Sclerotized plate in ventral wall of ductus bursae lightly sclerotized and inconspicuous anteriorly; corpus bursae with pouch ventrally at junction with ductus bursae (plate P, figure 2) .... *auxiliaris* p. 30

– Sclerotized plate in ventral wall of ductus bursae heavily sclerotized and ending abruptly anteriorly; corpus bursae without ventral pouch at junction with ductus bursae (plate P, figure 1) .... *lidia* this page

Euxoa (Chorizagrotis) lidia (Stoll)

PL. 1, FIGS. 1–5; PL. A, FIG. 1; PL. P, FIG. 1 (RWH 10730).

Phalaena lidia Stoll, 1782, *in* Cramer, *Uitlandsche Kapellen,* **4**: 222.
Type locality: [Europe].

Agrotis adumbrata Eversmann, 1842, *Bull. Soc. Nat. Moscow,* **3**: 543. SUBSPECIES.
Type locality: U.S.S.R. [Zoological Institute, Leningrad]

Agrotis drewseni Staudinger, 1857, *Stettiner Ent. Zeit.,* **18**: 302. NEW SYNONYMY, SUBSPECIES.
Type locality: Greenland. [Zoological Museum, Copenhagen]

Agrotis norwegica Staudinger, 1861, *Stettiner Ent. Zeit.,* **22**: 383.
Type locality: Norway.

Agrotis polygonides Staudinger, 1874, *Stettiner Ent. Zeit.,* **35**: 94.
Type locality: Caucasus, U.S.S.R. [HUMB]

Agrotis polygonides ab. *obscura* Staudinger, 1874, *Stettiner Ent. Zeit.,* **35**: 94. Infrasubspecific name.
Type locality: Caucasus, U.S.S.R. [HUMB]

Porosagrotis thanatologia Dyar, 1904, *Proc. U. S. Natl. Mus.,* **27**: 883. NEW SYNONYMY, SUBSPECIES.
Type locality: British Columbia. [USNM]

Chorizagrotis sordida Smith, 1908, *Jour. New York Ent. Soc.,* **16**: 86. NEW SYNONYMY.
Type locality: Kaslo, British Columbia. [AMNH]

Chorizagrotis boretha Smith, 1908, *Jour. New York Ent. Soc.,* **16**: 86. NEW SYNONYMY.
Type locality: Kaslo, British Columbia. [USNM]

Euxoa thanatologia var. *perfida* Dod, 1916, *Can. Ent.,* **48**: 64. NEW SYNONYMY.
Type locality: Calgary, Alberta. [CNC]

Euxoa norwegica ab. *obscura* Nordström, 1937, *Svenska Fjarilar,* **2**: 90. Infrasubspecific name.
Type locality: Norway.

Euxoa (Chorizagrotis) drewseni form *pseudovitta* Boursin, 1959, *Zeits. Wiener Ent. Ges.,* **44**: 169. NEW SYNONYMY.
Type locality: Greenland. [HUMB]

NOTE—Type material was discussed by Hardwick (1970: 72).

Populations of this species can be arranged into four subspecies. The nominate subspecies, with dark-brown forewings and contrasting maculation, occurs in central and eastern Europe. *Euxoa lidia adumbrata,* with dark-filled reniform and orbicular spots and dark costal area of the forewing, occurs in Scandinavia, Siberia, and Mongolia. The two North American subspecies are discussed below.

Euxoa (Chorizagrotis) lidia drewseni (Staudinger), NEW STATUS

PL. 1, FIG. 1.

Agrotis drewseni Staudinger, 1857.

Euxoa drewseni form *pseudovitta* Boursin, 1959.

Specimens of *drewseni* can be recognized by the pale, mouse-gray color of the forewings and by their small size; wing length varies from 10 to 14 mm. The immature stages are unknown.

Subspecies *drewseni* is known only from the southern half of Greenland; it has been collected on both the east and west coasts during July and August (Wolff, 1964: 55).

Euxoa (Chorizagrotis) lidia thanatologia (Dyar), NEW STATUS
PL. 1, FIGS. 2–5; PL. A, FIG. 1; PL. P, FIG. 1.

Porosagrotis thanatologia Dyar, 1904.

Chorizagrotis sordida Smith, 1908.

Chorizagrotis boretha Smith, 1908.

Euxoa thanatologia var. *perfida* Dod, 1916.

Specimens of *thanatologia* are variable in color and maculation of the forewing although the relatively evenly colored forms are by far the most common. This subspecies is sexually dimorphic with reddish-brown males and blackish-brown females, but a few females are reddish brown and a few males blackish brown. Specimens of *thanatologia* usually are much larger than are those of *drewseni.* Forewing length varies from 16 to 19 mm in most of the range of *thanatologia*; northern specimens are smaller, usually 13 to 16 mm. The reddish-brown color form frequently is mistaken for the corresponding form of *ochrogaster.* Specimens of the two species can easily be distinguished by the genitalia; differences in shape of the saccular extensions can be observed by brushing away the scales at the end of the abdomen.

The immature stages have been reared and are preserved in the Canadian National Collection, but the habits of the larvae are unknown.

Subspecies *thanatologia* occurs from northern Quebec westward to western Alaska and southward to the northern tier of states; in the Rocky Mountain region it occurs as far south as Colorado. Specimens have been collected from mid-June until early September.

The two North American subspecies have been treated as a species distinct from *lidia* of the Old World. I consider both Old and New World taxa as subspecies of a single polytypic species; they are allopatric and are indistinguishable by structural characters. Structural differences discussed for Old World subspecies by Nordström (1946) are not valid when more material is considered; the range of variation in *thanatologia* specimens encompasses that reported to exist between the Old World subspecies. Many specimens of *thanatologia* from western Canada are indistinguishable from Siberian specimens of *lidia adumbrata,* but the later subspecies is not sexually dimorphic.

Euxoa (Chorizagrotis) auxiliaris (Grote) (Army Cutworm*; Légionnaire Grise, f., Fr.)
PL. 1, FIGS. 6–8; PL. A, FIG. 2; PL. P, FIG. 2 (RWH 10731).

Agrotis auxiliaris Grote, 1873, *Bull. Buffalo Nat. Hist. Soc.,* **1**: 96.
Type locality: Colorado. [BMNH]

Agrotis introferens Grote, 1875, *Proc. Acad. Nat. Sci. Philadelphia,* **27**: 423.
Type locality: Texas. [BMNH]

Agrotis auxiliaris var. *agrestis* Grote, 1877, *Bull. U. S. Geol. Surv.,* **3**: 118.
Type locality: Colorado. [BMNH]

Agrotis auxiliaris ab. *tegularis* Strand, [1916] *Archiv für Naturgeschichte,* **81A12**: 144.
Type locality: United States. [BMNH]

Chorizagrotis auxiliaris form *montanus* Cook, 1930, *Can. Ent.,* **62**: 149.
Type locality: Malta, Montana. [Montana Experiment Station, Bozeman]

NOTE—Type material of *auxiliaris* and its synonyms was discussed by Hardwick (1970: 80).

Euxoa auxiliaris is the most common species of the subgenus *Chorizagrotis* and one of the most common species of *Euxoa* in western North America. It is the only member of the subgenus that occurs in much of western United States; however, in the southwestern United States its range overlaps that of *inconcinna,* and specimens of the two species frequently are difficult to distinguish by external appearance. Specimens of *auxiliaris* can readily be identified by genital characters; they are variable in maculation, but most can be put in one of three basic forms. The forewing of the most common form (plate 1, figure 6) has a black subbasal dash, black between the reniform and orbicular spots, a contrasted pale-gray or yellow costa, and a pale streak distal to the claviform spot. A less common form (plate 1, figure 7) has a more evenly colored pale-gray or pale-brown forewing. Specimens of this form frequently are confused with those of *inconcinna.* A third, uncommon form, has a dark-brown forewing with white-outlined reniform and orbicular spots (plate 1, figure 8). This form resembles the dark form of *thanatologia*. *Euxoa auxiliaris* is the largest species of *Chorizagrotis*; forewing length varies from 17 to 22 mm, but in most specimens it is 19 to 20 mm long.

The larva, commonly known as the army cutworm, occasionally reaches outbreak proportions and causes extensive damage to crops, particularly cereal crops, in southwestern Canada and western United States. The larva is a surface-feeding cutworm.

Euxoa auxiliaris occurs throughout western North America from the southern Northwest Territories, Canada, southward to northern Mexico and as far east as Michigan, Missouri and Texas. Adults are long lived; they emerge in the spring, March in the south, later farther north, and most migrate to high elevations in the mountains to aestivate during the summer months. Adults return to the lowlands in late summer and fall to mate and lay their eggs. The migration of *auxiliaris* was first documented by Pruess (1967).

Euxoa (*Chorizagrotis*) *inconcinna* (Harvey)
PL. 1, FIGS. 9, 10; PL. A, FIG. 3; PL. P, FIG. 3 (RWH 10732).

Agrotis inconcinna Harvey, 1875, *Bull. Buffalo Soc. Nat. Sci.*, **3**: 5.
Type locality: Clifton, Bosque Co., Texas. [BMNH]

Agrotis mercenaria Grote, 1878, *Bull. U. S. Geol. Surv.*, **4**: 171.
Type locality: Texas. [BMNH]

Agrotis sorror Smith, [1888], *Proc. U. S. Natl. Mus.*, **10**: 453.
Type locality: "Montana." [USNM]
NOTE—Type material was discussed by Hardwick (1970: 86).

Most specimens of *inconcinna* (plate 1, figure 10) resemble the pale form of *auxiliaris*. Specimens of *inconcinna* usually are paler and the forewings are disproportionately narrower than in *auxiliaris* specimens; however, only the genital characters given in the key are completely reliable. A rarer form of *inconcinna* (plate 1, figure 9) resembles a washed-out version of the streaked form of *auxiliaris*. Where the two species occur together, *inconcinna* specimens can frequently be identified by their smaller size; forewing length varies from 14 to 20 mm, but in most specimens it is about 16 to 18 mm long.

The immature stages of *inconcinna* are unknown.

The range of *inconcinna* extends from central Mexico (Durango) northward to Oklahoma, Colorado, southern Nevada, and southern California. A few specimens have been collected in Florida; it is not known whether these represent an established, disjunct population, or are individual specimens that have straggled to Florida during migration.

Long adult life and occurrence of adults at high elevations during the summer months, suggest that this species, like *auxiliaris*, is migratory. Adults have been collected from early March until late October.

Euxoa (*Chorizagrotis*) *terrealis* (Grote)
PL. 1, FIGS. 11–13; PL. A, FIG. 4; PL. P, FIG. 4 (RWH 10733).

Agrotis terrealis Grote, 1883, *Trans. Kansas Acad. Sci.*, **8**: 47.
Type locality: Las Vegas, New Mexico. [USNM]
NOTE—The lectotype of *terrealis* was designated by Hardwick (1970: 89).

This species can easily be identified by the genital characters given in the key. Most specimens, can be distinguished from those of other *Chorizagrotis* species by smaller size and reddish-brown coloration of the forewing. Forewing length varies from 15 to 18 mm. In most specimens the area between the reniform and orbicular spots is shaded in black; the costa may be pale (plate 1, figure 12) or concolorous with the ground color (plate 1, figures 11, 13). Occasional specimens have a more unicolorous reddish-brown forewing.

The immature stages have been reared and are preserved in the Canadian National Collection, but the habits of the larvae are unknown.

Euxoa terrealis has been collected throughout New Mexico and Arizona; it probably occurs in western Texas and northern Mexico as well. Specimens have been collected from early June until mid-October.

SUBGENUS
Palaeoeuxoa Lafontaine, NEW SUBGENUS

Gender: feminine.
Type species: *Agrotis mimallonis* Grote, 1873.

Palaeoeuxoa is a small nearctic subgenus containing four species. The phylogenetic position of the group relative to other subgenera of *Euxoa* was discussed previously (Lafontaine, 1981: 54).

Characters of the subgenus are as follows: external structural characters similar to those of other subgenera; male genitalia with sacculus large; saccular extension short, spatulate apically; harpe longer than saccular extension, without pubescence; juxta with transverse, sclerotized thickening; vesica forming loop above apex of aedoeagus; female genitalia with ovipositor lobes clothed with fine setae, lobes not fused and without sclerotized apical process; sclerotized plate in ventral wall of ductus bursae enlarged at middle, slightly asymmetrical, more heavily sclerotized and bulging on left side; corpus bursae bisaccate with appendix bursae oriented dorsoventrally and with ductus seminalis at dorsal end rather than anteroposteriorly with ductus seminalis at posterior end.

Males of the subgenus *Palaeoeuxoa* can be recognized by the short, spatulate saccular extension that is shorter than the harpe. In all other subgenera except *Chorizagrotis,* the saccular extension is tapered and cylindrical or bladelike apically. In *Chorizagrotis* males the saccular extension is spatulate apically but is several times the length of the harpe. Females of *Palaeoeuxoa* can be recognized by the peculiar configuration of the corpus bursae (plate P, figure 5).

Palaeoeuxoa larvae can be distinguished from those of the other subgenera by the texture of the skin of the last instar larva: each polygon of the skin has a central, large, raised bump; this gives the skin a dull, coarse appearance. In larvae of other subgenera the polygons of the last instar larvae are flat or slightly domed; the skin has a smooth, shiny appearance. Retention of a bumpy skin in the last instar larvae may be a neotenic condition. It occurs in all species of *Palaeoeuxoa* (J. R. Byers, personal communication).

The name *Palaeoeuxoa* refers to the relatively primitive phylogenetic position of this lineage within *Euxoa.*

KEY TO SPECIES OF SUBGENUS *PALAEOEUXOA*

1. Hindwing of male buff near base, brown on marginal third of wing; hindwing of female brown 2
— Hindwing of male white; hindwing of female white near base, brown on marginal third of wing 3

2. Male genitalia with harpe less than 1.5 mm long; apex of saccular extension spoon shaped (plate A, figure 5); forewing of female dark red .... *shasta* this page
— Male genitalia with harpe more than 1.5 mm long; apex of saccular extension flattened but not broad and spoon shaped (plate A, figure 6); forewing of female dark reddish brown .. .. *biformata* p. 33

3. Forewing red or gray .... *mimallonis* p. 33
— Forewing black or dark brown with maroon suffusion .... *intermontana* p. 33

Euxoa (*Palaeoeuxoa*) *shasta* Lafontaine

PL. 1, FIG. 14; PL. A, FIG. 5 (RWH 10734, 10735).

Euxoa shasta Lafontaine, 1975, *Can. Ent.,* **107**: 163.
Type locality: Mt. Shasta, McBride Springs, Siskiyou County, California. [CNC]

Euxoa condita Lafontaine, 1975, *Can. Ent.,* **107**: 163. NEW SYNONYMY, NEW STATUS, SUBSPECIES.
Type locality: Forestville, Quebec. [CNC]

Most specimens of *shasta* can be recognized by the dark-red color of the forewing with prominent antemedial and postmedial lines in combination with the brown hindwings. The two sexes differ markedly in size: forewing length in males varies from 15 to 20 mm; that of females varies from 17 to 22 mm. Males can be recognized by the spoon-shaped saccular extension, short harpe (length: 1.37 ± 0.06 mm, N = 12) and large sacculus (length: 2.50 ± 0.10 mm). The female genitalia are indistinguishable from those of *biformata* (plate P, figure 5).

The immature stages are known only from laboratory reared material. The egg was illustrated by Salkeld (1975: 1143, figures 1–3).

Euxoa shasta has been found in three, apparently disjunct, regions: northern California, southwestern British Columbia, and the Gulf of St. Lawrence region in eastern Quebec. Adults have been collected from late July until mid-September.

The populations of *shasta* may be arranged in two subspecies.

Euxoa (*Palaeoeuxoa*) *shasta shasta* Lafontaine

PL. 1, FIG. 14; PL. A, FIG. 5.

Euxoa shasta Lafontaine, 1975.

Specimens of the nominate subspecies are larger than are those of *condita* (forewing length: males 17–20 mm; females 19–22 mm) and in the male genitalia the sacculus is disproportionately larger (length: 2.70 ± 0.10 mm, N = 6) in comparison to harpe length (1.40 ± 0.02 mm).

The subspecies occurs at middle and high elevations in montane areas of southwestern British Columbia and northern California.

Euxoa (*Palaeoeuxoa*) *shasta condita* Lafontaine, NEW STATUS

Euxoa condita Lafontaine, 1975.

Specimens of *condita* are smaller than are those of *shasta* (forewing length: males 15–17 mm; females 17–19 mm) and in the male genitalia the sacculus is disproportionately smaller (length: 2.30 ± 0.10 mm, N = 6) in comparison to harpe length (1.33 ± 0.06 mm).

Subspecies *condita* is known only from the Gulf of St. Lawrence region in eastern Quebec.

Euxoa (*Palaeoeuxoa*) *biformata* Smith
PL. 1, FIGS. 15, 16; PL. A, FIG. 6; PL. P, FIG. 5 (RWH 10736).

Euxoa biformata Smith, 1910, *Trans. Amer. Ent. Soc.*, **36**: 261.
Type locality: Sierra Nevada, California. [AMNH]

Specimens of *biformata* can be recognized by the dark reddish-brown color of the forewings and the brown hindwings. Forewing length varies from 17 to 19 mm; males are not markedly smaller than females as in *shasta*. This species is intermediate both in wing color and in genital characters, between *shasta* and *mimallonis*. The harpe is long, as in *mimallonis* (length: 1.65 ± 0.03 mm, N = 6) but is stouter; the sacculus is more massive (length: 2.59 ± 0.07 mm) than is that of *mimallonis* males. The female genitalia (plate P, figure 5) are indistinguishable from those of *shasta* and *mimallonis*.

The immature stages are known only from laboratory reared material. The egg was illustrated by Salkeld (1975: 1145, figures 13–15).

Euxoa biformata occurs from south-central British Columbia southward in the Cascades and Sierra Nevada Mountains to the Transverse Ranges of southern California. Adults have been collected from late July until mid-September. Adults occur at middle to high elevations in pine and fir forests.

This species appears to be intermediate between *shasta* and *mimallonis* and may have arisen as a hybrid population. Its range, however, extends farther south than does that of either *shasta* or *mimallonis*. Both *biformata-shasta* and *biformata-mimallonis* hybrids have been obtained in the laboratory. The *biformata-shasta* cross is probably sterile because multiple chromosome assemblages are produced at meiosis (Fontana, 1976: 540, figures 5–6). The *biformata-mimallonis* cross is at least partially fertile; about 5 percent of the eggs of the progeny of the first generation of hybrids were fertile, and a partial second generation of hybrids was reared (J. R. Byers, personal communication).

Euxoa (*Palaeoeuxoa*) *intermontana* Lafontaine
PL. 1, FIGS. 17, 18; PL. A, FIG. 7 (RWH 10737).

Euxoa intermontana Lafontaine, 1975, *Can. Ent.*, **107**: 157.
Type locality: Lee Vining, California, 7 mi WSW, 9,600 ft. [CNC]

This Great Basin species can be distinguished from other members of the subgenus both by wing color and male genitalia. Color differences are given in the key. The male genitalia of *intermontana* are most similar to those of *mimallonis*; however, the sacculus is disproportionately more massive (length: 2.52 ± 0.10 mm, N = 20), and the harpe is shorter (length: 1.65 ± 0.08 mm) and markedly bent near the middle rather than evenly curved; in the vesica the subbasal arc above the aedoeagus has a conical bulge, and the median diverticulum is T-shaped with pouches directed posteriorly as well as anteriorly. The female genitalia are indistinguishable from those of *biformata*.

The immature stages are known only from laboratory reared material. The egg was illustrated by Salkeld (1975: 1143, figures 4–6).

Euxoa intermontana occurs in the intermontane region from central Washington eastward to western Montana and Colorado and southward to eastern California, northern Arizona and northern New Mexico (Lafontaine, 1975: 158). Adults occur from late July until late September.

Euxoa (*Palaeoeuxoa*) *mimallonis* (Grote)
PL. 1, FIGS. 19–21; PL. A, FIG. 8 (RWH 10738).

Agrotis mimallonis Grote, 1873, *Bull. Buffalo Soc. Nat. Sci.*, **1**: 98.
Type locality: New York. [BMNH]

Agrotis gagates Grote, 1875, *Can. Ent.*, **7**: 69. SUBSPECIES.
Type locality: Colorado. [BMNH]

Agrotis rufipennis Grote, 1875, *Can. Ent.*, **7**: 83.
Type locality: New York. [BMNH]

Agrotis caenis Grote, 1879, *North Amer. Ent.*, **1**: 44.
Type locality: Colorado. [BMNH]

Agrotis muscosa Grote, 1883, *Can. Ent.*, **15**: 26.
Type locality: Oak Creek Cañon, Colorado. [USNM]

Euxoa lenola Smith, 1910, *Trans. Amer. Ent. Soc.*, **36**: 260.
Type locality: Glenwood Springs, Colorado. [AMNH]

NOTE—The male holotypes of *mimallonis* and *rufipennis* and the female holotypes of *gagates, caenis,* and *muscosa* have been dissected. The female lectotype of *lenola,* designated by Todd (1982: 120), has been dissected.

This is the most common and widespread species in the subgenus. The red, or gray, coloration of the forewing can be used to distinguish *mimallonis* from the other species of the subgenus. Forewing length varies from 16 to 20 mm.

Males can be recognized by a combination of small sacculus (length: 2.23 ± 0.10 mm, N = 50) and long, evenly incurved harpe (length: 1.70 ± 0.07 mm). Female genitalia are indistinguishable from those of other species in the subgenus.

The immature stages have been reared and are preserved in the Canadian National Collection, but the habits of the larvae are unknown.

Euxoa mimallonis occurs from Nova Scotia westward across southern Canada to British Columbia. In the East it occurs as far south as the northern tier of states but in the West it occurs southward to western Nebraska, central New Mexico, central Arizona and east-central California (Lafontaine, 1975: 158). Adults are in flight during August and September.

Populations of *mimallonis* can be arranged in the following two subspecies.

Euxoa (*Palaeoeuxoa*) *mimallonis mimallonis* (Grote)
PL. 1, FIG. 19; PL. A, FIG. 8.

Agrotis mimallonis GROTE, 1873.

Agrotis rufipennis GROTE, 1875.

Specimens of this subspecies can be recognized by the even red color of the forewing with dark-red contrasting antemedial and postmedial lines and orbicular and reniform spots.

This subspecies occurs from Nova Scotia westward across Canada and northern United States to the Rocky Mountains. In the western Great Plains it occurs from central Alberta southward to northeastern New Mexico.

Euxoa (*Palaeoeuxoa*) *mimallonis gagates* (Grote)
PL. 1, FIGS. 20, 21.

Agrotis gagates Grote, 1875.

Agrotis muscosa Grote, 1883.

Euxoa lenola Smith, 1910.

The western subspecies of *mimallonis* occurs in two color forms. A gray color form (plate 1, figure 20) occurs in open aridlands; a form with a red forewing dusted with gray scales (plate 1, figure 21) occurs in forested habitats. In both forms the transverse lines and orbicular and reniform spots are not dark and contrasted as in the eastern subspecies.

Subspecies *gagates* occurs in the intermontane region from southern British Columbia southward to central California and central New Mexico. This subspecies occurs east of the Rocky Mountains in two areas where its range overlaps that of the eastern subspecies. Specimens typical of both subspecies occur in Wyoming and western South Dakota; in this region, however, the two subspecies appear to intergrade since most specimens are intermediate. The range of *gagates* extends eastward south of the Rocky Mountains into the southwestern Great Plains region in northeastern New Mexico. In this region specimens typical of both subspecies occur together, but there is no evidence of intergradation.

SUBGENUS
Heteroeuxoa Lafontaine, NEW SUBGENUS

Gender: feminine.
Type species: *Mamestra septentrionalis* Walker, 1865.

Heteroeuxoa is a small nearctic subgenus that contains three species. The phylogenetic position of the lineage with respect to other subgenera was discussed by Lafontaine (1981: 54).

Characters of the subgenus are as follows: external structural characters similar to those of other subgenera; male genitalia with sacculus large; saccular extension asymmetrical; right saccular extension stout, a finlike structure projecting from inner surface about ⅓ from base, extension with 90° twist mesially, the apical portion flattened and bladelike; left saccular extension thinner, with only trace of finlike structure and mesial twist; harpe shorter than saccular extension, without pubescence, left harpe slightly shorter than right harpe; juxta shield shaped, without central protuberance or sclerotized thickening; aedoeagus heavily sclerotized apically; vesica differing markedly in each taxon, primitive condition (*vernalis*) with loop above apex of aedoeagus and anvil-shaped subbasal diverticulum as in subgenus *Palaeoeuxoa*; loop of vesica reduced in *septentrionalis,* lost in *olivia*; subbasal diverticulum lost in *olivia* males, replaced by pouch in vesica, its surface spinulose; female genitalia with ovipositor lobes clothed with fine setae, lobes not fused and without sclerotized apical process; sclerotized plate in ventral wall of ductus bursae asymmetrical, more heavily sclerotized and bulging on left side; corpus bursae bisaccate or unisaccate, ductus seminalis ventrolateral to opening of ductus bursae.

Males of the subgenus *Heteroeuxoa* can be recognized by the asymmetrical saccular extensions and by the configuration of the vesica. Females can be recognized by the shape of the bursae from which the ductus seminalis arises. The corpus bursae of an *olivia* female is similar to that of members of the *ochrogaster* group (subgenus *Euxoa*), but the ductus seminalis arises ventrolateral to the ductus bursae rather than dorsal or dorsolateral to it.

The immature stages of only one species are known (*olivia*); these are similar to those of species in other subgenera.

The subgeneric name *Heteroeuxoa* refers to the atypical and diverse forms of the genital characters in members of the subgenus.

KEY TO SPECIES OF THE SUBGENUS *HETEROEUXOA*

1. Males 2
– Females 4

2. Right saccular extension massive with finlike ridge on basal ⅔; apical portion broad and triangular (plate A, figure 10); adults in late summer and fall .... *septentrionalis* p. 36
– Right saccular extension bladelike with finlike structure at ⅓ from base, apical portion elongate, dagger shaped (plate A, figure 9); adults in spring or fall 3

3. Harpe C-shaped, markedly curved in apical half through angle of about 90°; vesica with subbasal diverticulum absent, replaced by spiny pouch of vesica (plate B, figure 1); adults in fall .... *olivia* p. 36
– Harpe elbowed subbasally, bent through an angle of about 45°; apical half straight; vesica with subbasal diverticulum present, its surface smooth (plate A, figure 9); adults in spring .... *vernalis* this page

4. Corpus bursae unisaccate (plate P, figure 8) .... *olivia* p. 36
– Corpus bursae bisaccate (plate P, figures 6, 7) 5

5. Plate in ventral wall of ductus bursae rounded anteriorly (plate P, figure 7); adults in late summer and fall .... *septentrionalis* p. 36
– Plate in ventral wall of ductus bursae pointed anteriorly (plate P, figure 6); adults in spring .... *vernalis* this page

Euxoa (*Heteroeuxoa*) *vernalis* Lafontaine

PL. 1, FIG. 22; PL. A, FIG. 9; PL. P, FIG. 6 (RWH 10740).

Euxoa vernalis Lafontaine, 1976, *Can. Ent.*, 108: 1276.

Type locality: Fort Valley, Arizona, 7½ mi NW Flagstaff, Coconino County, 7,350 ft. [JGF]

In general appearance, specimens of *vernalis* resemble dark specimens of *messoria* (subgenus *Longivesica*) and *septentrionalis*; the forewing is gray, heavily dusted with black. Transverse lines and orbiclar and reniform spots are outlined in black but are obscure because of dark shading on the forewing. Forewing length varies from 16 to 18 mm. Specimens of *vernalis* are most easily distinguished from those of similar species by the early flight period; *vernalis* flies in the spring while similar species fly in the late summer and fall.

The male genitalia are similar to those of *olivia* but can be distinguished by the shape of the harpe and by characters of the vesica. The female genitalia of *vernalis* can be distinguished from those of *septentrionalis* by the shape of the plate in the ductus bursae, and from those of *messoria* by the much shorter appendix bursae.

The immature stages of *vernalis* are unknown.

This species is rare in collections; it has been collected in Coconino County, in central Arizona and in Durango and Hidalgo states in northern and central Mexico respectively (Lafontaine, 1976c: 1279). Adults have been collected from mid-April until early July.

Euxoa (*Heteroeuxoa*) *septentrionalis* (Walker)
PL. 1, FIGS. 23, 24; PL. A, FIG. 10; PL. P, FIG. 7 (RWH 10739).

Mamestra septentrionalis Walker, 1865, *List of the Specimens of Lepidopterous Insects in the Collection of the British Museum,* **32**: 660.
Type locality: Vancouver Island, British Columbia. [BMNH]

Carneades incubita Smith, 1900, *Proc. U. S. Natl. Mus.,* **22**: 435.
Type locality: Pullman, Washington. [USNM]
NOTE—The male lectotype was designated by Todd (1982: 104).

Specimens of *septentrionalis* frequently are confused with those of *messoria* (subgenus *Longivesica*). In both species the forewing is brown or brownish gray with the maculation outlined in black. Specimens of *septentrionalis* tend to be larger (forewing length 15 to 19 mm) and the forewing more heavily dusted with black than that of *messoria* so that the maculation appears somewhat smudged. In most *septentrionalis* the orbicular spot is round; it is oval in most *messoria.* The species can be distinguished by genital characters. The apex of the right saccular extension of *septentrionalis* males is flattened and appears shovel shaped; the saccular extensions of *messoria* males are tapered apically. This difference can be observed by removing scales from the genitalia with a small brush. In the female genitalia the appendix bursae of *septentrionalis* is about as long as the remainder of the corpus bursae; in *messoria* females it is very long and curved around the anterior end of the corpus bursae. Specimens of *septentrionalis* can be distinguished from those of *vernalis* by characters given in the key.

The immature stages of *septentrionalis* are unknown.

Euxoa septentrionalis occurs in western North America from southern British Columbia southward to central Colorado, southwestern Arizona and southern California (Lafontaine, 1976c: 1278). It has been collected east of the Continental Divide only in east-central Colorado. Specimens from arid areas in the Great Basin are paler than are those from the surrounding areas. This form was described as a subspecies by Barnes and Benjamin (1926); the holotype chosen, however, is a specimen of *messoria.* Adults of *septentrionalis* have been collected from late July until mid-October.

Euxoa (*Heteroeuxoa*) *olivia* (Morrison)
PL. 1, FIGS. 25–28; PL. B, FIG. 1; PL. P, FIG. 8 (RWH 10741).

Agrotis olivia Morrison, 1876, *Proc. Boston Soc. Nat. Hist.,* **18**: 238.
Type locality: Utah. [MSU]

Agrotis lacunosa Grote, 1878, *Bull. U. S. Geol. Surv.,* **4**: 172.
Type locality: California. [BMNH]

Carneades segregata Smith, 1894, *Trans. Amer. Ent. Soc.,* **21**: 47.
Type locality: Colorado. [AMNH]
NOTE—The male lectotype of *segregata* was designated by Todd (1982: 192).

Carneades enteridis Smith, 1900, *Proc. U. S. Natl. Mus.,* **22**: 427.
Type locality: Pullman, Washington. [USNM]

Carneades vanidicus Smith, 1900, *Proc. U. S. Natl. Mus.,* **22**: 453.
Type locality: Pullman, Washington. [USNM]
NOTE—The male lectotype of *vanidicus* was designated by Todd (1982: 221).

Euxoa anacosta Smith, 1905, *Jour. New York Ent. Soc.,* **13**: 196.
Type locality: Stockton, Utah. [AMNH]
NOTE—The female lectotype of *anacosta* was designated by Todd (1982: 16).

Euxoa zembla Smith, 1905, *Jour. New York Ent. Soc.,* **13**: 198.
Type locality: Stockton, Utah. [AMNH]
NOTE—The female lectotype of *zembla* was designated by Todd (1982: 228).

Euxoa fieldii Dyar, 1908, *Proc. Ent. Soc. Washington,* **10**: 54.
Type locality: San Diego, California. [USNM]
NOTE—A specimen labeled "San Diego, Cal. 11.12, Geo. H. Field," "Type No. 11697 USNM," "*Euxoa fieldii* Type Dyar," "♂ genitalia slide Mar. 1966 E.L.T. 2162" is here designated lectotype. The specimen is a male in good condition.

Specimens of *olivia* can be arranged in two basic forms on the basis of forewing maculation. In one (plate 1, figure 27), the forewing is evenly colored with transverse lines prominent and defined in black. In the second form (plate 1, figures 25, 26, 28), the forewing is streaked longitudinally with the costal, subterminal, and posterior portions of the forewing pale and the transverse lines obscure; there is also a pale streak distal to the claviform spot. Forewing length varies from 13 to 17 mm. In most males the forewing ground color is yellow brown or orange brown; the hindwing is white with brown shading near the terminal line. In most females the forewing ground color is gray or silver gray; the hindwing is pale pearl gray with the veins darker gray.

The male genitalia of *olivia* are similar to those of *vernalis* but can be distinguished by characters given in the key. *Euxoa olivia* is the only species in *Euxoa* that has lost the subbasal diverticulum of the vesica; its position is indicated by a small, round area near the base of the vesica that lacks spinules and has the cornutus that is in the diverticulum of most other species. The subbasal diverticulum is replaced functionally by a subbasal pouch of the vesica. Unlike typical subbasal diverticula, its surface is covered with spinules like those on the remainder of the vesica; in other *Euxoa* species the spinules are present on the other parts of the vesica but not on the subbasal diverticulum.

The female genitalia are similar to those of *nostra* and *ochrogaster* (subgenus *Euxoa*), but the ductus seminalis arises ventrolateral to the ductus bursae rather than dorsolateral to it.

Larvae have been found on strawberry plants and corn in Nebraska (Walkden, 1950: 17). The species overwinters in the egg; the larvae are reported to have a long summer diapause (Hinks and Byers, 1976: 1349). The egg was illustrated by Salkeld (1975: 1148, figures 34–36).

Euxoa olivia occurs in arid regions of western North America from southern Manitoba westward to southern British Columbia and southward to northern Texas, central New Mexico, southern Arizona, and southern California (Lafontaine, 1976c: 1278). Adults have been collected from early September until late October.

SUBGENUS *Longivesica* Hardwick

Longivesica Hardwick, 1970

The subgenus *Longivesica* is a small, exclusively nearctic subgenus that contains four species.

Males can be recognized by the very long vesica; this is about 2.5–3.0× the length of the aedoeagus. If the vesica has not been everted, males can be recognized only by characters of the individual species.

Females of *Longivesica* can be recognized by the very long appendix bursae; this is about twice as long as the corpus bursae. In other subgenera the appendix bursae is shorter than the corpus bursae.

The subgenus *Longivesica* was revised by Hardwick (1970); my treatment of the subgenus is based largely on his revision.

KEY TO SPECIES OF SUBGENUS *LONGIVESICA*

1. Male with right saccular extension about ½ length of right harpe (plate B, figure 4); female with corpus bursae about ¼ size of appendix bursae and arising near middle of appendix bursae (plate Q, figure 3) .... *edictalis* p. 40

– Male with right saccular extension longer than right harpe; female with corpus bursae about ½ size of appendix bursae and arising at posterior end of appendix bursae 2

2. Male with juxta markedly constricted mesially; vesica with three subbasal diverticula (plate B, figure 2); female with appendix bursae curving to right around anterior end of corpus bursae (plate Q, figure 1) .... *messoria* this page

– Male with juxta not constricted mesially; vesica with one subbasal diverticulum (plate B, figure 3); female with appendix bursae not curving to right anteriorly (plate Q, figure 2).... 3

3. Cubital vein of forewing pale and contrasting with ground color; forewing length 14 to 16 mm; widespread .... *divergens* p. 39

– Cubital vein not pale and contrasting; forewing length 12 to 14 mm; occurring in eastern boreal forest zone from Newfoundland westward to western Ontario and northern Michigan .... *sinelinea* p. 39

Euxoa (*Longivesica*) *messoria* (Harris) (Darksided Cutworm*; Ver-gris Moissonneur, m., Fr.)

PL. 1, FIGS. 29, 30; PL. B, FIG. 2; PL. Q, FIG. 1 (RWH 10705).

Agrotis messoria Harris, 1841, *A Report on the Insects of Massachusetts, Injurious to Vegetation,* 324.

Type locality: Massachusetts. [MCZ]

NOTE—The male lectotype of *messoria* was designated by Hardwick (1970: 48).

Agrotis spissa Guenée, 1852, *Histoire Naturelle des Insectes. Species Général de Lépidoptères. 5, Noctuélites,* **1**: 261.
Type locality: North America. [lost]

Mamestra inextricata Walker, 1865, *List of the Specimens of Lepidopterous Insects in the Collection of the British Museum,* **32**: 658.
Type locality: West Canada. [USNM]

Mamestra indirecta Walker, 1865, *List of the Specimens of Lepidopterous Insects in the Collection of the British Museum,* **32**: 659.
Type locality: West Canada. [USNM]

Mamestra displiciens Walker, 1865, *List of the Specimens of Lepidopterous Insects in the Collection of the British Museum,* **32**: 660.
Type locality: West Canada. [USNM]

Mamestra expulsa Walker, 1865, *List of the Specimens of Lepidopterous Insects in the Collection of the British Museum,* **32**: 661.
Type locality: Vancouver Island, British Columbia. [BMNH]

Agrotis ordinata Walker, 1865, *List of the Specimens of Lepidopterous Insects in the Collection of the British Museum,* **32**: 691.
Type locality: West Canada. [lost]

Agrotis reticens Walker, 1865, *List of the Specimens of Lepidopterous Insects in the Collection of the British Museum,* **32**: 692.
Type locality: West Canada. [USNM]

Agrotis cochranis Riley, 1867, *Prairie Farmer,* **35**: 414.
Type locality: [Missouri]. [USNM?]

NOTE—The nominal type of *cochranis* in USNM is probably not the true type (Hardwick, 1970: 49).

Agrotis repentis Grote and Robinson, 1868, *Trans. Amer. Ent. Soc.,* **1**: 350.
Type locality: "Atlantic District." [ANSP]

Agrotis friabilis Grote, 1875, *Can. Ent.,* **7**: 187.
Type locality: Canada [Ontario]. [BMNH]

Agrotis atrifera Grote, 1878, *Bull. U. S. Geol. Surv.,* **4**: 173.
Type locality: California. [BMNH]

NOTE—The female lectotype of *atrifera* was designated by Hardwick (1970: 49).

Carneades territorialis Smith, 1900, *Proc. U. S. Natl. Mus.,* **22**: 436.
Type locality: Denver, Colorado. [USNM]

NOTE—The female lectotype of *territorialis* was designated by Todd (1968: 278).

Carneades fulda Smith, 1900, *Proc. U. S. Natl. Mus.,* **22**: 437.
Type locality: Alameda County, California. [USNM]

NOTE—The female lectotype of *fulda* was designated by Todd (1968: 271).

Carneades pindar Smith, 1900, *Proc. U. S. Natl. Mus.,* **22**: 451.
Type locality: Bluff, Utah. [USNM]

NOTE—The male lectotype of *pindar* was designated by Todd (1968: 275).

Agrotis septentrionalis inordita Barnes and Benjamin, 1926, *Can. Ent.,* **58**: 303.
Type locality: Stockton, Utah. [USNM]

Euxoa messoria is one of the most common and widespread species in the genus. Most adults can be distinguished by wing color and markings. The forewing is grayish brown with all lines and spots normally present and defined by narrow black lines. The antemedial and postmedial lines are double, the latter is prominently scalloped between the veins. A wavy, dark-brown median line is present in most specimens; the terminal space is also dark brown. The claviform, reniform, and orbicular spots are essentially concolorous with the remainder of the forewing and outlined in black; the latter is oval and large, about ⅔ the size of the reniform spot, and the black outline is broken on the costa. In about 5 per cent of the specimens the forewing has a black basal dash and a black bar connecting the reniform and orbicular spots (plate 1, figure 29). Forewing length varies from 13 to 18 mm. The male hindwing is white with smoky-brown shading on the marginal third and on the veins. The hindwing of the female is similar but the basal ⅔ of the wing has a dusting of smoky-brown scales. Specimens of *messoria* usually can be distinguished from those of similar species in other subgenera by the gray-brown forewing color with the maculation defined by fine black lines, and by the relatively large, oval orbicular spot. This species is frequently confused with *septentrionalis* (subgenus *Heteroeuxoa*) in western North America; specimens of the two species can be distinguished by characters given under *septentrionalis* (p. 36).

Males of *messoria* can be recognized by the distinctive shape of the vesica, by the mesially constricted juxta, and by the stout saccular extensions (plate B, figure 2). Females can be distinguished from those of all other species by the genital characters given in the key.

The larva of *messoria,* known as the darksided cutworm, is longitudinally streaked dorsally and laterally with a series of obscure dark-brown and pale-brown stripes. The dark-brown stripe on the side, just above the spiracles, is the widest and darkest and gives the species its common name. The larva is a climbing cutworm and may damage young trees as well as herbs. It is a common pest of vegetable crops and garden plants throughout much of its range; in eastern North America it is a major pest of tobacco. The larva prefers broad-leaved plants but will eat a variety of grasses and is occasionally a pest on corn seedlings in the central United States. The species overwinters in the egg; most larval feeding takes place in May and June.

Euxoa messoria occurs from Newfoundland westward to Yukon and southward to Virginia and Missouri in the East and southern New Mexico, Arizona, and California in the West. Adults have been collected from early July until late October; they fly later in the South than in the North.

Euxoa (*Longivesica*) *divergens* (Walker)

PL. 1, FIG. 31; PL. B, FIG. 3; PL. Q, FIG. 2 (RWH 10702).

Agrotis divergens Walker, [1857], *List of the Specimens of Lepidopterous Insects in the Collection of the British Museum,* **10**: 327.
Type locality: Nova Scotia. [BMNH]

NOTE—The male lectotype of *divergens* was designated by Hardwick (1970: 60).

Agrotis versipellis Grote, 1875, *Can. Ent.,* **7**: 172.
Type locality: Orillia, Ontario. [BMNH]

NOTE—The male lectotype of *versipellis* was designated by Hardwick (1970: 60).

Agrotis fusimacula Smith, 1891, *Trans. Amer. Ent. Soc.,* **18**: 105.
Type locality: California. [USNM]

Agrotis abar Strecker, 1899, *Lepidoptera, Rhopaloceres and Heteroceres, Indigenous and Exotic.* Suppl. **2**: 5.
Type locality: Glenwood Springs, Colorado. [FMNH]

Carneades factoris Smith, 1900, *Proc. U. S. Natl. Mus.,* **22**: 456.
Type locality: Glenwood Springs, Colorado. [USNM]

NOTE—The female lectotype of *factoris* was designated by Todd (1968: 271).

Euxoa divergens can be distinguished from most other *Euxoa* species by the combination of dark-brown or gray-brown forewing color with a black basal dash and a pale cubital vein. Forewing length varies from 14 to 16 mm. *Euxoa divergens* can be confused with *ridingsiana* and with some forms of *tessellata.* From the former, they can be distinguished by lack of a pale streak distal to the claviform spot, lack of streaking in the subterminal area, and by genital characters. Males of *divergens* can be distinguished from those of *ridingsiana* by the relatively long saccular extensions; these are longer than the harpes rather than about half as long. Females of *divergens* can be distinguished from those of *ridingsiana* by the absence of sclerotized flange-like projections on the ovipositor lobes and by the bisaccate rather than unisaccate corpus bursae. Males of *divergens* can be distinguished from those of *tessellata* with similar wing markings by the absence of a yellow "shoulder-tuft" at the base of each forewing and by the more symmetrical saccular extensions and the cucullus less expanded apically in the male genitalia. Females of *divergens* can be recognized by the rounded, finely setose ovipositor lobes and by the bisaccate rather than unisaccate corpus bursae; the ovipositor lobes of *tessellata* females are truncate and are covered with short, conical setae apically.

Specimens of *divergens* from southwestern Colorado, southern Utah, and northern Arizona tend to have a more charcoal-colored forewing with the cubital vein only slightly paler.

The immature stages of *divergens* are known only from laboratory reared material.

Euxoa divergens is a widespread species; it occurs from Newfoundland westward to Alaska and southward to southern New York, central Ontario, and central Michigan in the East. In the West it occurs in montane areas as far south as central New Mexico, central Arizona, and southern California. Adults have been collected from late May until late September; most were collected between mid-June and mid-August.

Euxoa (*Longivesica*) *sinelinea* Hardwick

PL. 1, FIGS. 32, 33 (RWH 10703).

Euxoa sinelinea Hardwick 1965, *Can. Ent.,* **97**: 824.
Type locality: Goose Bay, Labrador. [CNC]

This species is similar to *divergens* but can be recognized by its small size and absence of pale shading on the cubital vein on the forewing. Forewing length varies from 12 to 14 mm. Within its range, *sinelinea* is most likely to be confused with dark specimens of *servita* (subgenus *Euxoa*) but can be distinguished from them by absence of pale streaking in the subterminal area. The male and female genitalia of *sinelinea* differ slightly from those of *divergens,* but the two species cannot safely be distinguished by genital characters.

The immature stages are unknown.

Euxoa sinelinea is a species of the eastern boreal forest zone. It occurs from Newfoundland and Labrador westward to western Ontario and northern Michigan; its range overlaps the eastern range of *divergens* in this area. Adults have been collected from late June until mid-August.

Euxoa (*Longivesica*) *edictalis* (Smith)

PL. 1, FIGS. 34, 35; PL. B, FIG. 4; PL. Q, FIG. 3 (RWH 10704).

Carneades edictalis Smith, 1893, *Ent. News,* **4**: 99.

Type locality: Colorado. [USNM]

NOTE—The male lectotype of *edictalis* was designated by Todd (1968: 269).

Specimens of *edictalis* can be distinguished from those of other species of *Euxoa* by antennal structure, wing color, and genital characters. The antenna is markedly biserrate, almost plumose; the antennal width is about five times that of the central shaft. The forewing is pale brown or gray in most populations; orange brown in Colorado populations. Maculation is present but obscure due to the powdery appearance of the forewing. Forewing length varies from 15 to 17 mm. The hindwing is pale smoky brown. The male and female genitalia of *edictalis* (plate B, figure 4, plate Q, figure 3) cannot be confused with those of any other species. The male genitalia can be recognized by the short harpe and saccular extension, the former lying on the inner surface of the harpe, and by the long vesica. The female genitalia can be recognized by the very small corpus bursae in comparison with the very large appendix bursae.

The immature stages are unknown.

Euxoa edictalis is a rare species known from less than 20 localities. It occurs from south-central British Columbia southward in the Cascades and Sierra Nevada Mountains to Mono County, California and southern Nevada. East of this, it has been collected in east-central Montana, central Utah, and in the western half of Colorado. Adults have a very early flight period; they have been collected from mid-April until late June.

SUBGENUS
Pleonectopoda Grote

Pleonectopoda Grote, 1873.

Pleonectopoda is a moderately large subgenus that contains 23 species in the Nearctic Region. One species is holarctic; the range of *hyperborea* extends from Yukon and Alaska into eastern Siberia. A number of species in the subgenus occur in the Palearctic Region, but the exact number can not be determined until the palearctic fauna has been revised (A. Moberg, personnal communication).

Males of *Pleonectopoda* can be recognized most easily by the prominent twist or subbasal coil of the vesica. When characters of the vesica are not visible, a combination of characters given in the alternate key to subgenera (page 21) must be used. Most species in the *westermanni* group can be recognized by occurrence in arctic and alpine areas and short saccular extensions, but *scandens* and its relatives must be recognized by short saccular extensions and lack of a frontal tubercle. Most members of the *atomaris* and *pleuritica* groups can be recognized by the asymmetrical harpes and cylindrical saccular extensions. The *tristicula, fuscigera,* and *vetusta* groups each contain a single species that can be recognized by external characters.

Members of the subgenus cannot be recognized by characters of the female. Females of *Pleonectopoda* must be associated with the subgenus by characters of the species.

The subgenus *Pleonectopoda* was revised by Hardwick (1970); my treatment of the subgenus is based largely on his revision.

KEY TO SPECIES-GROUPS AND SPECIES OF SUBGENUS *PLEONECTOPODA*

1. Males 2
– Females 28

2. Right saccular extension ⅓, or less, length of right valve 3
– Right saccular extension ⅖, or more, length of right valve 24

3. Frontal tubercle absent or vestigial (*westermanni* group, in part) 4

— Frontal tubercle present 7

4. Antenna markedly biserrate, about three times as wide as central shaft .... *aurulenta* p. 48

— Antenna moderately biserrate, about twice as wide as central shaft 5

5. Right saccular extension short, less than ⅕ length of right valve; hindwing dirty white, buff on marginal quarter .... *scandens* p. 47

— Right saccular extension longer, more than ⅕ length of right valve; hindwing gray brown or with gray-brown shading on marginal half of wing 6

6. Forewing with reniform spot broad, kidney shaped; occurring in boreal forest and western montane areas northward to southern Yukon .... *quebecensis* p. 47

— Forewing with reniform spot narrow, barlike; occurring in northern Alaska and eastern Siberia .... *hyperborea* p. 46

7. Hindwing pale buff, darker on marginal third of wing; vesica with median diverticulum near apex of vesica; saccular extensions as long as harpes and bent away from valve near apex (plate D, figure 2) (*tristicula* group) ...... *tristicula* p. 54

— Hindwing gray brown, slightly paler near base; vesica with median diverticulum near subbasal coil; saccular extensions shorter than harpe and not bent away from valve near apex in most species 8

8. Right saccular extension about as long as right harpe and sicklelike, curved toward valve (plate D, figure 3); subbasal diverticulum of vesica unilobed (*vetusta* group) .... *vetusta* p. 54

— Right saccular extension shorter than harpe in most species and not bent toward valve; subbasal diverticulum of vesica bilobed (*westermanni* group) 9

9. Basal diameter of right saccular extension at least twice that of right harpe 10

— Basal diameter of right saccular extension less than twice that of right harpe 15

10. Right saccular extension less than ⅔ length of right harpe (*lewisi*) 11

— Right saccular extension more than ⅔ length of right harpe 12

11. Orbicular spot about as large as reniform spot and not suffused with white in most specimens; occurring in Cascades and Rocky Mountain regions .... *lewisi lewisi* p. 51

— Orbicular spot about half size of reniform spot and suffused with white in most specimens; occurring in central and southern Sierra Nevada Mountains of California .... *lewisi juliae* p. 51

12. Right saccular extension as long as, or longer than, right harpe; left harpe shorter and stouter than right harpe (plate D, figure 1) ...... *extranea* p. 53

— Right saccular extension shorter than right harpe; left harpe similar in length and diameter to right harpe 13

13. Forewing dark blackish gray with contrasted pale-gray costa and reniform and orbicular spots; occurring in subarctic Canada, southward in west to southern Alberta .... *westermanni* p. 44

— Forewing dark gray brown, without pale costal area; occurring in mountains of central and southern California 14

14. Right saccular extension about ⅓ length of valve, cylindrical, pointed apically; distributed in Sierra Nevada Mountains, California .. *cryptica* p. 52

— Right saccular extension less than ¼ length of valve, flattened, triangular apically; distributed in southwestern California and Baja California, Mexico .... *leuschneri* p. 52

15. Right saccular extension less than ¼ length of right harpe 16

— Right saccular extension more than ⅓ length of right harpe 17

16. Forewing dark brown; widespread in montane regions of western United States .... *altens* p. 51

— Forewing pale buff; known only from San Gabriel Mountains of southern California ... *austrina* p. 52

17. Eye greatly reduced, about 1.5× as high as wide .... *churchillensis* p. 45

– Eye round, or slightly reduced, height similar to width 18

18. Occurring in subarctic zone of eastern and northern Canada 19
– Occurring in montane areas of western North America 20

19. Right harpe at least 1.5 × length of right saccular extension . . . . . *macleani chimoensis* p. 50
– Right harpe less than 1.5 × length of right saccular extension . . . . . *dissona* p. 46

20. Juxta no longer than wide; forewing dark brown . . . . . *trifasciata* p. 50
– Juxta longer than wide; forewing gray or yellow gray 21

21. Thorax and forewings dark gray brown; forewing length 13–15 mm . . . . . *macleani macleani* p. 50
– Thorax and forewings light gray buff or gray with yellow shading; forewing length 16–20 mm (*vallus*) 22

22. Forewing light gray buff, antemedial line almost straight; occurring in central Sierra Nevada Mountains of California . . . . . *vallus bivittata* p. 49
– Forewing gray with yellow shading, or with yellow tufting on thorax; antemedial line zigzagged; occurring in Cascades and Rocky Mountain regions 23

23. Forewing with extensive yellow shading; occurring in montane areas of Colorado . . . . . *vallus luteosita* p. 49
– Forewing with, at most, patches of yellow shading; occurring in montane areas of southwestern Canada and northwestern United States . . . . . *vallus vallus* p. 49

24. Right and left harpes similar in length (plate D, figure 4) (*fuscigera* group) . . . . . *fuscigera* p. 55
– Harpes asymmetrical, left harpe markedly shorter than right harpe (plate D, figures 5–8) 25

25. Juxta about 1.5 × as long as wide; saccular extensions very stout, cylindrical, tapered abruptly near apex (plate D, figure 5) (*atomaris* group) . . . . . *atomaris* p. 55
– Juxta about twice as long as wide; saccular extensions thinner, tapered from base to apex, flattened and bladelike near apex (*pleuritica* group) 26

26. Right harpe 0.59 times, or less, length of right saccular extension (plate D, figure 6); forewing with contrasting pale patches in ground color and length less than 18 mm . . . . . *pleuritica* p. 56
– Right harpe more than 0.59 times length of right saccular extension (plate D, figures 7, 8); forewing either with ground color evenly colored, or with length more than 18 mm 27

27. Forewing length more than 18 mm; genitalia with sacculus disproportionately more massive and triangular than that of *pestula* and saccular extensions longer (plate D, figure 8); occurring in intermontane and Rocky Mountain regions, eastward onto high plains in Alberta . . . . . *simona* p. 58
– Forewing length less than 18 mm; genitalia with sacculus smaller and more crescentic in outline than that of *simona* and saccular extensions shorter (plate D, figure 7); occurring in Great Plains region and eastern portion of Rocky Mountain region . . . . . *pestula* p. 57

28. Sclerotized plate in ventral wall of ductus bursae about twice as long as abdominal segment VIII 29
– Sclerotized plate in ventral wall of ductus bursae less than 1.5 times length of abdominal segment VIII (*westermanni, tristicula,* and *atomaris* groups) 33

29. Corpus bursae massive, about twice as long as ductus bursae (*vetusta* group) . . . . . *vetusta* p. 54
– Corpus bursae smaller, about as long as ductus bursae 30

30. Appendix bursae with ductus seminalis arising at posterior end (*fuscigera* group) . . . . . *fuscigera* p. 55
– Appendix bursae with ductus seminalis arising at anterolateral corner (plate S, figures 7, 8) (*pleuritica* group) 31

31. Forewing length more than 18 mm; distributed in intermontane and Rocky Mountain regions,

eastward onto high plains of Alberta ...... *simona* p. 58

— Forewing length less than 18 mm; distributed from Rocky Mountain and Cascades regions eastward 32

32. Forewing with ground color uniform (plate 2, figures 22, 23); distributed in Great Plains region and eastern portion of Rocky Mountain region .... *pestula* p. 57

— Forewing with contrasting pale patches in ground color (plate 2, figures 20, 21); distributed from Atlantic to Pacific coast in southern Canada and northern United States, southward in Rocky Mountain region .... *pleuritica* p. 56

33. Frontal tubercle absent or vestigial 34

— Frontal tubercle present 37

34. Setae on dorsal margin of ovipositor lobe stout .... *aurulenta* p. 48

— Dorsal margin of ovipositor lobe without a row of stout setae 35

35. Hindwing white, buff colored on marginal quarter of wing .... *scandens* p. 47

— Hindwing gray brown 36

36. Forewing with reniform spot broad, kidney shaped; occurring in boreal forest and western montane areas northward to southern Yukon; corpus bursae bisaccate .... *quebecensis* p. 47

— Forewing with reniform spot narrow, barlike; occurring in northern Alaska and eastern Siberia; corpus bursae unisaccate .... *hyperborea* p. 46

37. Forewing with median line dark, broad and contrasting 38

— Forewing with median line obscure or absent 43

38. Sclerotized plate in ventral wall of ductus bursae markedly bent to right through an angle of about 30° (plate S, figure 2) .... *extranea* p. 53

— Sclerotized plate in ventral wall of ductus bursae straight (plate R, figure 6) 38

39. Forewing and thorax brownish red .... *lewisi* p. 51

— Forewing and thorax yellow brown or dark gray brown 40

40. Forewing and thorax dark gray brown; forewing length 13–15 mm .... *macleani* p. 50

— Forewing and thorax light gray buff, or gray with yellow shading; forewing length 16–20 mm (*vallus*) 41

41. Forewing light gray buff, antemedial line almost straight; occurring in central Sierra Nevada Mountains of California ..... *vallus bivittata* p. 49

— Forewing gray with yellow shading or yellow patches, or with yellow tufting on thorax; antemedial line zigzagged; occurring in Cascades and Rocky Mountain regions 42

42. Forewing with extensive yellow shading; occurring in montane areas of Colorado .... *vallus luteosita* p. 49

— Forewing with, at most, patches of yellow shading; occurring in montane areas of southwestern Canada and northwestern United States .... *vallus vallus* p. 49

43. Ovipositor lobe with 4–6 long setae subbasally, these about ⅓ length of lobe; hindwing brown 44

— Ovipositor lobe with subbasal row of 10–12 long setae, these about ½ length of lobe; subbasal setae shorter in some *atomaris* specimens; these have pale-gray hindwings with dark-brown veins 47

44. Appendix bursae narrower and longer than corpus bursae; forewing with black basal dash 45

— Appendix bursae as broad as corpus bursae and shorter; forewing without basal dash 46

45. Distributed in Sierra Nevada Mountains, California .... *cryptica* p. 52

— Distributed in southwestern California and Baja California, Mexico .... *leuschneri* p. 52

46. Forewing dark brown; widespread in western montane areas .... *altens* p. 51

— Forewing pale brown; known only from San Gabriel Mountains of southern California .... *austrina* p. 52

47. Anterior apophyses short, about 1/3 length of abdominal segment VIII; forewing dark blackish gray with pale contrasting costa and reniform and orbicular spots .... *westermanni* this page

— Anterior apophyses longer, at least 1/2 length of abdominal segment VIII; costa and reniform and orbicular spots not pale and contrasting... 48

48. Bursa copulatrix unisaccate 49

— Bursa copulatrix bisaccate 50

49. Eyes greatly reduced, ellipsoid, about 1.5× as high as wide; diurnal species; forewing length 10–13 mm .... *churchillensis* p. 45

— Eyes slightly reduced, oval, about as high as wide; nocturnal species; forewing length 14–16 mm .... *dissona* p. 46

50. Hindwing pale pearly gray with darker veins; if hindwing dark brown, then forewing almost black 51

— Hindwing dark brown; forewing brown or reddish brown 52

51. Ground color of forewing an even pale gray; forewing length 16–19 mm .... *tristicula* p. 54

— Ground color of forewing light brown to blackish brown; paler specimens with central portion of wing much paler than costal and outer portions; forewing length 13–16 mm .... *atomaris* p. 55

52. Orbicular spot round, concolorous with remainder of forewing, outlined by even, black line .... *trifasciata* p. 50

— Orbicular spot oval, or paler than remainder of forewing in most specimens; cell of forewing with black shading proximal and distal to orbicular spot in most specimens (*lewisi*) 53

53. Forewing orange brown with orbicular spot almost as large as reniform spot, without white shading in orbicular and reniform spots; occurring in Cascades and Rocky Mountain regions .... *lewisi lewisi* p. 51

— Forewing purplish brown with orbicular spot about half as large as reniform spot; orbicular and reniform spots shaded with white in most specimens; occurring in central and southern Sierra Nevada Mountains .... *lewisi juliae* p. 51

westermanni GROUP

The *westermanni* group is the largest species-group in the subgenus *Pleonectopoda* and contains 16 species in North America.

In the male genitalia the saccular extensions are short; the right saccular extension is less than 1/3 length of the valve and shorter than the harpe in most specimens. The median diverticulum of the vesica is closer to the subbasal coil than to the apex of the vesica, and the subbasal diverticulum is bilobed. The female genitalia are similar to those of species in other groups.

Most members of the *westermanni* group occur in boreal habitats, either in subarctic Canada and Alaska, or in alpine habitats in montane areas. Four species occur in sandy habitats and have lost the frontal tubercle; two of them, *aurulenta* and *scandens,* are widely distributed in North America.

Euxoa (*Pleonectopoda*) *westermanni* (Staudinger)

PL. 1, FIG. 36; PL. B, FIG. 5; PL. Q, FIG. 4 (RWH 10707).

Noctua westermanni Staudinger, 1857, *Stettiner Ent. Zeit.,* **18**: 303.
Type locality: Greenland. [Zoological Museum, Copenhagen]

Agrotis westermanni var. *polaris* Bang-Haas, 1910, *Deutsche Ent. Zeits., Iris,* **24**: 35.
Type locality: Labrador. [HUMB]

Specimens of *westermanni* can be distinguished from those of all other species in the subgenus *Pleonectopoda* by the distinctive forewing pattern. No other species has a dark gray-brown forewing with pale, contrasting costa, and reniform and orbicular spots. The pale areas of the forewing may be either pale gray or pale buff brown with a rose suffusion. Forewing length varies from 15 to 17 mm. In forewing color and pattern, specimens of *westermanni* closely resemble those of *idahoensis* (subgenus *Euxoa*); the two species occur together in western Canada. In addition to characters of the male vesica, males of *westermanni* can be distinguished from those of *idahoensis* by the short harpes and saccular extensions; females of *westermanni* can be distinguished from all other *Euxoa* species by the very short anterior apophyses.

The immature stages of *westermanni* are unknown.

The range of *westermanni* is poorly known. It has been collected in southern Greenland and eastern

Labrador in the Northeast; Great Bear Lake, Northwest Territories, and Dawson, Yukon in the Northwest; and in Banff National Park, Alberta in the West. Specimens have been collected in mid- and late July.

Euxoa (*Pleonectopoda*) *churchillensis* (McDunnough)

PL. 1, FIGS. 51–53; PL. B, FIG. 6; PL. Q, FIG. 5 (RWH 10711).

Agrotiphila churchillensis McDunnough, 1932, *Can. Ent.*, **64**: 105.
Type locality: Ft. Churchill, Manitoba. [CNC]

Euxoa (*Pleonectopoda*) *churchillensis alpina* Lafontaine, NEW SUBSPECIES.
Type locality: Pennsylvania Mtn., Park County, Colorado. [CNC]

Euxoa churchillensis is the only species in the subgenus *Pleonectopoda* that has reduced, ellipsoid eyes and is strictly diurnal. Northern populations occur with, and may be confused with, *nomas incognita* (subgenus *Orosagrotis*); Colorado specimens can be confused with those of *montana* (subgenus *Orosagrotis*). They can be distinguished from both of these species by the relatively long saccular extensions of the male genitalia and by the lack of sclerotized flangelike projections on the ovipositor lobes of females. The male genitalia differ from those of most other species in the subgenus *Pleonectopoda* in that the harpe extends dorsally beyond the dorsal edge of the cucullus only at the apex, if at all. The female genitalia are similar to those of *dissona* in that the bursa copulatrix is unisaccate. Females of *churchillensis* can be distinguished from those of *dissona* by small size and by the reduced ellipsoid eyes; the eye is about 1.5× as high as wide.

The immature stages are unknown.

This species is distributed in two, widely separated areas; it occurs in northern Canada and in alpine areas in Colorado and northern New Mexico. Populations from these two areas are arranged in two subspecies.

Euxoa churchillensis is the only species of *Euxoa* that is known from fossil material. J. V. Matthews, Geological Survey of Canada, has collected over ten well-preserved adult heads in the Old Crow Basin in northern Yukon. They were found in mid-Wisconsinan (>35,000 BP) full-glacial deposits in the Beringian Refugium area; at that time, the collecting locality was a dry grass and sagebrush covered slope (J. V. Matthews, personal communication).

Euxoa (*Pleonectopoda*) *churchillensis churchillensis* (McDunnough)

PL. 1, FIG. 51; PL. B, FIG. 6; PL. Q, FIG. 5.

Agrotiphila churchillensis McDunnough, 1932.

Specimens of the nominate subspecies vary greatly in size and maculation. The forewing varies from being pale gray with black-lined maculation to blackish gray with poorly defined maculation. Most specimens are dark gray with pale-gray shades on the costa, in the reniform and orbicular spots, and the subterminal area. Specimens from the Hudson Bay region are smaller (forewing length: 11–13 mm) than are those from the western arctic (forewing length; 13–14 mm).

The subspecies occurs from the MacKenzie Delta in northwestern Canada eastward to the west coast of Hudson Bay. It has been collected as far south as the mouth of James Bay in northern Ontario. Adults have been collected from late June until late July.

Euxoa (*Pleonectopoda*) *churchillensis alpina* Lafontaine, NEW SUBSPECIES

PL. 1, FIGS. 52, 53.

Specimens of *alpina* can be distinguished from those of *churchillensis* by the antemedial and postmedial lines of the forewing. In *alpina* these lines are straight, almost parallel and consist of a broad black line and a whitish-gray shade proximal to the antemedial line and distal to the postmedial line. In subspecies *churchillensis* the antemedial and postmedial lines are similar to those of other *Euxoa* species; they are sinuate and scalloped between the veins, being much closer together near the costal and posterior wing margins than in the middle of the wing.

Antenna of male biserrate, slightly bifasciculate; antenna of female filiform. Frontal tubercle prominent. Eye greatly reduced, ellipsoid. Vestiture of head and thorax blackish brown with scattered white scales. Ground color of forewing brownish gray, heavily dusted with black scales in basal and median areas. Basal line black, incomplete. Antemedial line black, diffuse, almost straight with whitish-gray area proximally. Postmedial line black, slightly scalloped between veins with diffuse black shade proximally and whitish-gray shade distally. Subterminal line pale with series of black, sagittate spots proximal to it. Terminal line a series of small black spots on wing margin. Fringe brownish gray. Basal dash and claviform spot black. Reniform and orbicular spots small, outlined in black, pale inside with whitish-gray shading. Area between reniform and orbicular spots and proximal to orbicular spot black. Hindwing brownish gray; fringe pale gray. Length of forewing 12 to 13 mm. Male and female genitalia indistinguishable from those of subspecies *churchillensis.*

TYPES. Holotype: ♂. Pennsylvania Mtn., Park County, Colorado; 26 July 1980; D. Ford; CNC (Type No. 19,507). Paratypes: 5 ♂, 3 ♀. Same locality as for holotype; 22 July 1978 and 2 August 1978; P. Kevan (1 ♂, 1 ♀); Colorado: Boulder Co., Corona Summit, 11,000′; 4 Aug. 1962; Tom W. Davies (3 ♂, 1 ♀). New Mexico: Taos Co., Wheeler Pk. nr. Twining, 12,000′; 3 July 1966; Mike Toliver (1 ♂, 1 ♀). CNC, LACM, USNM.

Euxoa (Pleonectopoda) dissona (Möschler)
PL. 1, FIGS. 37–39; PL. B, FIG. 7; PL. Q, FIG. 6 (RWH 10706).

Agrotis dissona Möschler, 1860, *Weiner Ent. Monatschr.,* **4**: 365.
Type locality: Labrador. [HUMB]

NOTE—The female lectotype of *dissona* was designated by Hardwick (1970: 96).

Agrotis opipara Morrison, 1874, *Proc. Boston Soc. Nat. Hist.,* **17**: 165.
Type locality: Mt. Washington, New Hampshire. [BMNH]

Agrotis islandica var. *labradoriensis* Staudinger, 1881, *Stettiner Ent. Zeit.,* **42**: 419.
Type locality: Labrador. [HUMB].

Agrotis solitaria Smith, [1888], *Proc. U. S. Natl. Mus.,* **10**: 462.
Type locality: Labrador. [lost]

NOTE—The lectotype of *dissona* and holotype of *labradoriensis* were illustrated by Bang-Haas (1922, pl. 14, fig. 20 (*dissona*) and fig. 21 (*labradoriensis*)).

Specimens of *dissona* occur in three basic wing pattern forms. In the most common form (plate 1, figure 37) the maculation is prominent and defined in black; the forewing has a black basal dash and a large, black claviform spot. Also, a conspicuous transverse black line is on the prothoracic collar. In a less common form (plate 1, figures 38, 39) the maculation is present but obscure, outlined partially by fine black lines; the basal dash is absent; the claviform spot is obscure; and the transverse line on the prothorax is barely traceable. In the third, least common form, the maculation is obscure or absent. In all three forms the forewing ground color may be pale gray or reddish brown. Forewing length varies from 14 to 16 mm. The well-marked form can be distinguished from *churchillensis* by larger size and eye size. The unmarked form can be distinguished from *quebecensis* by the presence of a large frontal tubercle on *dissona* specimens; this is absent in *quebecensis.*

The immature stages of *dissona* are unknown.

Euxoa dissona occurs in subarctic regions of Canada from Newfoundland and Labrador, across northern Quebec to the central Northwest Territories. It also occurs in alpine habitats in the higher mountains of the New England states. Adults have been collected from late June until early August.

Euxoa (Pleonectopoda) hyperborea Lafontaine, NEW SPECIES
PL. 1, FIG. 40; PL. B, FIG. 8; PL. Q, FIG. 7.

Euxoa (Pleonectopoda) hyperborea Lafontaine.
Type locality: Gubik Gas Field, Chandler River, 1 mi S Colville River, Alaska. [KWP]

Euxoa hyperborea is most similar to *quebecensis* in terms of wing color and vestigial frontal tubercle but can be distinguished by the narrow barlike reniform spot on the forewing as well as details of the male and female genitalia. It can be distinguished from all other species of *Euxoa* by the combination of dark wing color and vestigial frontal tubercle.

Antenna of male slightly biserrate and slightly bifasciculate; antenna of female filiform. Frontal tubercle vestigial, consisting of a wrinkled area on frons. Eye slightly reduced. Vestiture of head and thorax mouse gray or gray brown. Ground color of forewing mouse gray, brownish gray, or pale reddish brown. Maculation obscure. Antemedial and postmedial lines obscure or absent. Orbicular spot absent or evident only as an obscure pale ring. Reniform spot dark gray, narrow, and barlike. Other lines and spots absent. Fringe concolorous with remainder of forewing. Forewing length: 14–19 mm; larger specimens from Kulu, U.S.S.R. Hindwing smoky brown, paler toward wing base. Discal spot dark gray brown. Fringe of hindwing yellow buff. Male genitalia similar to those of *quebecensis* but clavus less heavily sclerotized, subbasal diverticulum of vesica lacking cornuti and vesica with an additional, small dorsal diverticulum on subbasal coil. Saccular extensions short; right saccular extension 0.5–0.7 × as long as right harpe, 0.2–0.3 × as long as right valve. Vesica projecting at about 90° to axis of aedoeagus. Female genitalia similar to those of *dissona,* corpus bursae unisaccate.

TYPES. Holotype: ♂. Gubik Gas Field, Chandler River,

1 mi S Colville River, Alaska, 200′; 30 June 1970; J. Hok. Paratypes: 8 ♂, 1 ♀. Same data as for holotype (2 ♂). Magadanskaya Oblast', 1 km E Ust' Chaun, U.S.S.R.; 68°47′N, 170°35′E; 26 July 1978; K.W. Philip (1 ♂). Magadanskaya Oblast', 18 km NW Kulu, U.S.S.R.; 25 July to 9 August 1979; Martchenko (5 ♂, 1 ♀). The holotype and two paratypes are in the collection of K. W. Philip, to be deposited subsequently in USNM; two paratypes are in CNC; and five paratypes are in the collection of Institute of Biology and Pedology, Vladivostok, U.S.S.R.

In addition to the type material, a badly rubbed specimen, probably referable to *hyperborea,* is known from the Ogilvie Mountains, Yukon, Canada (ROM). Although *hyperborea* is most likely to be confused with *quebecensis,* it is possible that *hyperborea* is most closely related to *dissona. Hyperborea* resembles *quebecensis* in that the vesica projects to the right rather than posteriorly, and the frontal tubercle is vestigial; it resembles *dissona* in lacking cornuti in the vesica and the unisaccate corpus bursae.

Euxoa (*Pleonectopoda*) *quebecensis* (Smith)
PL. 1, FIGS. 41, 42; PL. C, FIG. 1; PL. Q, FIG. 8 (RWH 10714).

Setagrotis quebecensis Smith, 1900, *Proc. U. S. Natl. Mus.,* **22**: 417.
Type locality: Quebec. [USNM]

Euxoa moxa Smith, 1907, *Trans. Amer. Ent. Soc.,* **33**: 129.
Type locality: Durango, Colorado. [USNM]
NOTE—The female lectotype of *moxa* was designated by Todd (1968: 274).

Euxoa quinta Smith, 1908, *Ann. New York Acad. Sci.,* **18**: 97.
Type locality: High River, Alberta. [AMNH]
NOTE—The male lectotype of *quinta* was designated by Todd (1968: 277).

Euxoa quebecensis, hyperborea, scandens, and *aurulenta,* differ from other species of the subgenus *Pleonectopoda* in that the frontal tubercle is absent or vestigial. These species are associated with dry, usually sandy, areas; other members of the *westermanni* group occur in coniferous forests or alpine areas.

Euxoa quebecensis can be recognized by the dark wing coloration and broad reniform spot. The frontal tubercle is absent; the male antenna is biserrate with the antennal width about twice that of the central shaft. The forewing is buff gray or mouse gray with the maculation obscure in most specimens. The orbicular spot is obscure or absent. The reniform spot is broad and kidney shaped; a dark spot is present in the lower third in most specimens. The median line is obscure or absent. Forewing length varies from 14 to 17 mm. The hindwing is smoky gray, somewhat paler near the base with a dark-gray discal spot. The male genitalia are similar to those of *scandens,* but the saccular extensions are longer. The right saccular extension is 0.7–0.9× the length of the harpe. The subbasal diverticulum has one or two cornuti in most specimens. In the female genitalia the setae near the apex of the ovipositor lobes are short and conical.

The immature stages are unknown.

Euxoa quebecensis is a species of the boreal forest and western montane regions. The range of *quebecensis* may be disjunct; it has been collected in the boreal forest region of eastern Canada and Maine from Newfoundland westward to western Ontario. In the West it occurs from southern Yukon southward to southern Alberta in the Rocky Mountains and to west-central Oregon in the Cascades; it has also been collected in montane areas in Colorado. Adults have been collected from late June until early September.

Euxoa (*Pleonectopoda*) *scandens* (Riley)
(White Cutworm*; Ver-gris Blanc, m., Fr.)

PL. 1, FIGS. 43, 44; PL. C, FIG. 2; PL. R, FIG. 1 (RWH 10715).

Agrotis scandens Riley, 1869, *First Annual Report of the Noxious, Beneficial and other Insects of the State of Missouri,* 76.
Type locality: Missouri. [USNM]
NOTE—The male lectotype of *scandens* was designated by Hardwick (1970: 115).

Agrotis (*Carneades*) *scandens* form *fulminans* Grote, 1895, *Abhandl. Naturwissenschaftlichen Verein Bremen,* **14**: 67.
Type locality: Buffalo, New York. [lost]

Setagrotis elata Smith, 1898, *Jour. New York Ent. Soc.,* **6**: 106.
Type locality: Colorado. [USNM]
NOTE—The male lectotype of *elata* was designated by Todd (1968: 270).

Euxoa scandens is similar to *quebecensis* in antennal structure and in lacking a frontal tubercle but can be distinguished by the pale coloration and by male genital characters.

The forewing is pale gray or pale buff with a pink suffusion in some specimens. Maculation is similar to that of *quebecensis.* The hindwing is white or very pale buff with a gray-brown discal spot, marginal band, and submarginal band.

Agrotis vetusta Walker frequently is misidentified as a species of *Euxoa* because of the presence of a prominent frontal tubercle and usually as *Euxoa scandens* because of similarity in wing color and markings. The two species can readily be distinguished by the generic characters of the genitalia. Additionally, *scandens* is smaller (forewing length 13 to 18 mm (*scandens*) versus 17 to 21 mm (*vetusta*), lacks a frontal tubercle, and has a dark discal spot on the hindwing. The male genitalia of *scandens* are similar to those of *quebecensis*, but the saccular extensions are shorter. The right saccular extension is 0.3–0.6 × the length of the right harpe. In the vesica, the subbasal diverticulum has one or two cornuti in most specimens, like *quebecensis*, but unlike that species, the vesica has the additional subbasal diverticulum on the basal arc similar to that present in *hyperborea* males. The female genitalia are indistinguishable from those of *quebecensis*.

The larva, commonly known as the white cutworm, is occasionally a pest in areas of dry, sandy soil. The larvae are reported to be climbing cutworms; they usually attack dicotyledonous plants such as sweet clover, young trees and a variety of vegetable crops.

Euxoa scandens occurs from Newfoundland and Nova Scotia, southward to Massachusetts, and westward across southern Canada and northern and central United States to western Alberta, central Montana, central Wyoming, and central Utah. It has also been collected in a dune area near Fort Smith in the southern Northwest Territories. Although adults have been collected from early June until late August, most were collected during July. *Euxoa scandens* is most frequently collected in dry sandy areas.

Euxoa (*Pleonectopoda*) *aurulenta* (Smith)
PL. 1, FIGS. 45, 46; PL. C, FIG. 3; PL. R, FIG. 2 (RWH 10716).

Agrotis aurulenta Smith, 1890, *Bull. U. S. Natl. Mus.*, **38**: 215.
Type locality: Denver, Colorado. [USNM]
NOTE—The male lectotype of *aurulenta* was designated by McDunnough (1950: 371) and Todd (1968: 267).

Agrotis aurulenta ab. *aurulentoides* Strand, [1916], *Archiv für Naturgeschichte*, **81A12**: 145. Infrasubspecific name.

Type locality: United States (Omaha, Nebraska). [BMNH]

Specimens of *aurulenta* are similar to those of *scandens* but can be distinguished by the male antenna, by forewing shape, and by characters of the male and female genitalia.

The male antenna is markedly biserrate with the antennal width about three times that of the central shaft. The frontal tubercle is absent. The ground color of the forewing is highly variable; in most specimens it is pale buff speckled with lighter and darker scales (plate 1, figure 45). In some specimens the forewing is pale yellow buff with the maculation barely visible; at the other extreme the forewing is reddish brown with the costal and posterior margins pale (plate 1, figure 46). *Euxoa aurulenta* usually can be recognized by the relatively narrow forewing. Forewing length varies from 15 to 19 mm. The male genitalia are similar to those of *scandens*, but the saccular extensions are longer; the right saccular extension is 0.6–0.8 × the length of the right harpe. The female genitalia can be distinguished from those of *scandens* by the unisaccate rather than bisaccate corpus bursae and by the presence of a row of stout setae on the dorsal margins of the ovipositor lobes of *aurulenta* females.

The immature stages of *aurulenta* are unknown.

Euxoa aurulenta is widespread in arid regions of central and western North America but is highly localized with rarely more than one or two locations known in each state and province. It has been collected from Ontario westward to Alberta and southward to Illinois, Nebraska, Colorado, Arizona, and Washington. Adults have been collected from early May until late July.

Euxoa (*Pleonectopoda*) *vallus* (Smith)
PL. 1, FIGS. 47–50; PL. C, FIG. 4; PL. R, FIG. 3 (RWH 10709, 10710).

Carneades vallus Smith, 1900, *Proc. U. S. Natl. Mus.*, **22**: 430.
Type locality: Laggan (Lake Louise), Alberta. [USNM]

Carneades luteosita Smith, 1910, *Proc. U. S. Natl. Mus.*, **22**: 433. NEW SYNONYMY, NEW STATUS, SUBSPECIES.
Type locality: Hall Valley, Colorado. [USNM]
NOTE—The male lectotype of *luteosita* was designated by Todd (1968: 272).

Euxoa luteomaculata Hardwick, 1966, *Can. Ent.*, **98**: 764.

Type locality: Okanogan County, Washington. [CNC]

Euxoa (*Pleonectopoda*) *vallus bivittata* Lafontaine. NEW SUBSPECIES.
Type locality: 7 mi WSW Lee Vining, California. [CNC]

Euxoa vallus occurs in three disjunct areas. Although specimens from these three populations cannot be separated by structural characters, they can readily be distinguished by forewing color and pattern and by the color of the thoracic vestiture. The forewing ground color of *vallus* may be slate gray or grayish brown with a scattering of yellow scales, or may be extensively covered with yellow scales. The antemedial line is zigzagged in two subspecies, changing directions at each wing vein; it is almost straight in one subspecies. The postmedial line is scalloped between veins and distinct from the median line for most of its length, except in the Sierra Nevada Mountains where this line is almost straight. In most specimens, the reniform and orbicular spots are outlined in pale yellow. Forewing length varies from 16 to 20 mm. The thoracic vestiture is gray or gray brown, with tufts of yellow scales in two of the subspecies. Specimens of *vallus* are similar to some specimens of *comosa* (subgenus *Euxoa*) but can be distinguished by genital characters. In addition to characters of the vesica, *vallus* has short saccular extensions, about ⅔ the length of the harpe. In *comosa,* the saccular extensions are as long as, or longer than, the harpes. The female of *vallus* has a bisaccate corpus bursae, and the ovipositor lobes lack a sclerotized apical process; that of *comosa* has a unisaccate corpus bursae, and the apex of each ovipositor lobe has a sclerotized process.

The immature stages of *vallus* are unknown.

The range of *vallus* is disjunct: it occurs from southern Alberta and British Columbia southward to northern Washington; it occurs in Colorado; and it occurs in the central Sierra Nevada Mountains of California. Populations in these three areas are arranged in three subspecies. Adults occur from mid-July until early August.

Euxoa (*Pleonectopoda*) *vallus vallus* (Smith)
PL. 1, FIG. 47.

Carneades vallus Smith, 1900.

Euxoa luteomaculata Hardwick, 1966.

Specimens of this subspecies can be recognized by the predominantly gray or grayish-brown color of the forewing. Yellow shading is confined to a scattering of yellow scales, primarily in the basal area and around the reniform and orbicular spots. In some specimens the transverse lines are diffuse and appear somewhat smudged.

This subspecies occurs in southern Alberta and British Columbia and northern Washington. It is found in coniferous forests at or below treeline from 4,500′ to 8,000′.

Euxoa (*Pleonectopoda*) *vallus luteosita* (Smith), NEW STATUS
PL. 1, FIGS. 48, 49.

Carneades luteosita Smith, 1900.

Specimens of *luteosita* can be distinguished from those of subspecies *vallus* by the extensive yellow shading on the forewings.

Euxoa vallus luteosita occurs at high elevations in central Colorado. Specimens have been collected above treeline between the elevations of 11,500′ and 12,500′.

Euxoa (*Pleonectopoda*) *vallus bivittata* Lafontaine, NEW SUBSPECIES
PL. 1, FIG. 50; PL. C, FIG. 4; PL. R, FIG. 3.

This subspecies can be distinguished from the other two subspecies of *vallus* by: lack of yellow shading on forewings and thorax; by the broad, black, almost straight antemedial line; by the indistinct, or lack of, reniform and orbicular spots; and by the fusion of the postmedial line and median line for most of their lengths. *Euxoa vallus bivittata* and *extranea* occur together and are similar to each other in color; however, *extranea* has a wing pattern similar to that of *vallus* and can be distinguished from *bivittata* by the characters listed above. The two species can readily be distinguished by genital characters given in the key.

Antenna of male slightly biserrate, weakly bifasciculate. Frontal tubercle moderately large but frequently obscured by vestiture. Eye slightly reduced, similar in size to that of *vallus vallus.* Vestiture of head and thorax mouse gray. Ground color of forewing buffy gray. Basal line black, incomplete. Antemedial line broad, black, almost straight. Postmedial line black, scalloped between veins, fused proximally with black median line for most of length. Subterminal line made evident only by shading of blackish-brown scales proximal to it. Terminal line a series of small black wedges on wing margin between veins. Fringe concolorous with remainder of forewing. Forewing expanse 16 to 17 mm. Hindwing smoky gray. Fringe yellowish gray. Male and female genitalia indistinguishable from those of *vallus vallus.*

TYPES. Holotype: ♂. Lee Vining, California, 7 mi WSW, 9,600′; 12 August 1967; D. F. Hardwick; CNC (Type No. 19508). Paratypes: 4 ♂, 3 ♀. Same locality and collector as for holotype; 6–14 August 1967. CNC, USNM.

Euxoa (*Pleonectopoda*) *macleani* McDunnough

PL. 1, FIGS. 54, 55; PL. C, FIG. 5; PL. R, FIG. 4 (RWH 10712, 10713).

Euxoa macleani McDunnough, 1927, *Can. Ent.*, **59**: 195.
Type locality: Mt. McLean, British Columbia. [CNC]

Euxoa chimoensis Hardwick, 1966, *Can. Ent.*, **98**: 766. NEW SYNONYMY, NEW STATUS, SUBSPECIES.
Type locality: Ft. Chimo, Quebec. [CNC]

Euxoa macleani usually can be recognized by its relatively small size and dark color. The forewing is dark gray brown with black transverse lines and median line. The reniform and orbicular spots are obscure in most specimens. Forewing length varies from 13 to 15 mm. Specimens of *macleani* could be confused with those of *vallus* or *quebecensis*; they can be distinguished from those of *vallus* by small size and lack of yellow coloration on the wings and thorax and from those of *quebecensis* by presence of a frontal tubercle.

The immature stages of *macleani* are unknown.

Euxoa macleani occurs in two widely separate areas; it has been collected in the Cascade Mountains of southern British Columbia and northern Washington and in northern Quebec and Labrador. Adults have been collected from mid-July until mid-August. Populations of *macleani* are arranged in the following two subspecies.

Euxoa (*Pleonectopoda*) *macleani macleani* McDunnough

PL. 1, FIG. 54; PL. C, FIG. 5; PL. R, FIG. 4.

Euxoa macleani McDunnough, 1927.

Subspecies *macleani* differs from *chimoensis* in having an oblique, almost straight, median line and in minor details of the male genitalia. In the few males examined, the saccular extensions of *macleani* are slightly longer than are those of *chimoensis,* and the subbasal diverticulum of the vesica in *chimoensis* lacks the cornutus present in *macleani.* More specimens are needed to determine the consistency of these differences because *macleani* is known from only five males and one female and *chimoensis* from three males.

Subspecies *macleani* has been collected only in south-central British Columbia and north-central Washington at elevations of 6,000′ to 7,500′.

Euxoa (*Pleonectopoda*) *macleani chimoensis* Hardwick, NEW STATUS

PL. 1, FIG. 55.

Euxoa chimoensis Hardwick, 1966.

Subspecies *chimoensis* is known from three specimens. Two from Fort Chimo, Quebec [CNC] and one from northern Labrador [USNM]. Distinguishing characters of this subspecies are discussed under subspecies *macleani.*

Euxoa (*Pleonectopoda*) *trifasciata* (Smith)

PL. 2, FIG. 1; PL. C, FIG. 6; PL. R, FIG. 5 (RWH 10717).

Agrotis trifasciata Smith, [1888], *Proc. U. S. Natl. Mus.,* **10**: 460.
Type locality: Mt. Hood, Oregon. [MSU]

Specimens of *trifasciata* can be recognized by the even, orange-brown color of the forewings and the prominent black lines defining the maculation. The claviform, orbicular, and reniform spots are essentially the same color as the remainder of the forewing and are outlined in black. The antemedial line is almost straight, only slightly curved between the veins. The postmedial line is slightly scalloped between the veins. The subterminal line is defined by a series of buff spots between the wing veins with small dark sagittate spots proximal to each buff spot. The forewing length varies from 17 to 19 mm. The hindwing is gray brown.

The male genitalia are similar to those of *vallus,* but the juxta is as wide, or wider, than its length in *trifasciata*; in *vallus,* and most other species in the *westermanni* group, it is longer than wide. The female genitalia are indistinguishable from those of *vallus.*

The immature stages of *trifasciata* are unknown.

Euxoa trifasciata occurs at high elevations (6,000′–9,000′) along the Cascades and Sierra Nevada Mountains from south-central Washington southward to the Lake Tahoe area of east-central California. Adults have been collected from late July until late August. Adults usually are collected in open coniferous forests.

Euxoa (*Pleonectopoda*) *lewisi* (Grote)
PL. 2, FIGS. 2, 3; PL. C, FIG. 7; PL. R, FIG. 6 (RWH 10718, 10719).

Pleonectopoda lewisi Grote, 1873, *Bull. Buffalo Soc. Nat. Sci.*, **1**: 137.
Type locality: Colorado. [BMNH]

Agrotis colata Grote, 1881, *Can. Ent.*, **13**: 131.
Type locality: Mt. Hood, Oregon. [BMNH]

Euxoa juliae Hardwick, 1968, *Can. Ent.*, **100**: 272. NEW SYNONYMY, NEW STATUS, SUBSPECIES.
Type locality: Tioga Pass, California. [CNC]

Euxoa lewisi is a highly variable species in terms of wing color and maculation. The forewing varies from pale gray brown with a speckling of reddish-brown scales to dark purplish brown. The transverse lines are usually weakly defined or incompletely defined by darker shading although a dark median line is present in most specimens. The orbicular spot is large in subspecies *lewisi,* often almost as large as the reniform spot, and oval; in subspecies *juliae* the orbicular spot is much smaller than the reniform spot and round. In both subspecies the area between the orbicular and reniform spots, and proximal to the orbicular spot is usually dark blackish brown. The reniform and orbicular spots have an incomplete buffy line inside the black outline. The claviform spot is obscure or absent. The forewing length varies from 14 to 19 mm. The hindwing is smoky brown. *Euxoa lewisi* can be distinguished from other species by the male genitalia. The sacculus is broad and somewhat bowl shaped; the saccular extensions are stouter than the harpes, almost conical, and about ½ the length of the harpes. In other species with similar wing markings the sacculus is narrow and crescentic, and the saccular extensions are about as thin as, or thinner than, the harpes. The female genitalia are similar to those of *trifasciata,* but the ovipositor lobes are more elongate and more acutely angled apically than are those of *trifasciata.*

The immature stages of *lewisi* are unknown.

Euxoa lewisi occurs at middle to high elevations (5,000′–12,000′) in montane coniferous forests from southern Alberta and British Columbia southward in the Cascade Mountains to northern Oregon and southward in the Rocky Mountains to central Colorado; it also occurs in the Uinta and Wasatch Mountains in Utah and as a distinct subspecies in the central Sierra Nevada Mountains of California. Adults have been collected from early July until mid-August.

Populations of *Euxoa lewisi* are arranged in the following two subspecies.

Euxoa (*Pleonectopoda*) *lewisi lewisi* (Grote)
PL. 2, FIG. 2; PL. C, FIG. 7; PL. R, FIG. 6.

Pleonectopoda lewisi Grote, 1873.

Agrotis colata Grote, 1881.

In subspecies *lewisi* the forewing ground color is gray brown, orange brown, or reddish brown with a dark median line in most specimens. The orbicular spot is oval and almost as large as the reniform spot; both spots lack white shading.

It occurs in the Rocky Mountain region and Cascades from southern British Columbia southward to northern Oregon and central Colorado.

Euxoa (*Pleonectopoda*) *lewisi juliae* Hardwick, NEW STATUS
PL. 2, FIG. 3.

Euxoa juliae Hardwick, 1968.

Subspecies *juliae* can be distinguished from subspecies *lewisi* by the purplish-brown forewing, by the small, more rounded shape of the orbicular spot and by the hoary appearance of the forewing due to scattered white scales. The forewing is 14 to 18 mm long. The hindwing is smoky brown. The male and female genitalia are indistinguishable from those of subspecies *lewisi.*

Euxoa lewisi juliae is known only from the central Sierra Nevada Mountains from the Lake Tahoe area southward to Lee Vining in Mono County, California; specimens have been collected between 8,900′ and 9,600′ in early and mid-August.

Euxoa (*Pleonectopoda*) *altens* McDunnough
PL. 2, FIGS. 4, 5; PL. C, FIG. 8; PL. R, FIG. 7 (RWH 10720).

Euxoa altens McDunnough, 1946, *Can. Ent.*, **78**: 30.
Type locality: Mt. Hood, Oregon. [CNC]

Euxoa altens is most similar to species in the subgenus *Euxoa* in wing color and markings, particularly specimens of *terrena, simulata,* and *satis. Euxoa altens* can be distinguished from them, and from all members of the *westermanni* group except *aus-*

trina, by the very short saccular extensions of the male genitalia. In *altens* the forewing ground color is dark brown dusted with gray and black scales. The maculation is outlined in black but is obscure because of the dark wing coloration. Forewing length varies from 15 to 17 mm. The hindwing is smoky brown. In the male genitalia the saccular extension is very short, about ¼ the length of the harpe. The uncus is tapered from the middle to an acutely pointed apex. The subbasal diverticulum of the vesica is somewhat anvil shaped with a curved pouch directed posteriorly. In the female genitalia, the ovipositor lobes are clothed with sparse, short setae; and the subbasal row of longer, hairlike setae is reduced to 4 to 6 setae from the normal 10 to 12 setae.

The immature stages of *altens* are unknown.

Euxoa altens occurs at middle to high elevations in montane areas from southern British Columbia southward in the Cascades and Sierra Nevada Mountains to Inyo County in east-central California and in the Rocky Mountains to central Idaho. It has also been collected in central Utah and central Arizona. At the northern limits of its range specimens have been collected as low as 3,000′ but in most of its range it occurs between the elevations of 7,000′ and 10,500′. Adults have been collected from mid-July until late August.

Euxoa (*Pleonectopoda*) *austrina* Hardwick
PL. 2, FIGS. 6, 7 (RWH 10721).

Euxoa austrina Hardwick, 1968, *Can. Ent.*, **100**: 270.
Type locality: San Gabriel Mts., California. [LACM]

Euxoa austrina is closely related to *altens* but can be distinguished from it by the pale ground color of the forewing and by the larger size. The forewing is buffy brown with a speckling of black scales. The maculation is defined in dark brown. The forewing length varies from 16 to 18 mm. The hindwing is smoky brown. The male and female genitalia are indistinguishable from those of *altens.*

The immature stages of *austrina* are unknown.

Euxoa austrina is known from southern California from the San Gabriel Mountains northward to Mt. Pinos. Adults have been collected from late May until early August at elevations of 6,100′ to 6,500′.

Euxoa (*Pleonectopoda*) *cryptica* Hardwick
PL. 2, FIG. 8; PL. C, FIG. 9; PL. R, FIG. 8 (RWH 10722).

Euxoa cryptica Hardwick, 1968, *Can. Ent.*, **100**: 268.
Type locality: Mosquito Flats near Bishop, California. [CNC]

Euxoa cryptica can be distinguished from other species in the subgenus *Pleonectopoda* by wing color and pattern and by genital characters. The forewing is blackish brown with a dusting of gray and white scales. The maculation is defined in black and bordered by scattered white scales, particularly the reniform and orbicular spots that have a white line inside the black outline. The forewing has a black basal dash bordered by white above. In many specimens the cubital vein is paler than the ground color from the wing base to the reniform spot. The forewing is 14 to 17 mm long. The hindwing is smoky gray. In wing color and pattern *cryptica* is more similar to *tessellata* and *satis* in the subgenus *Euxoa* than it is to other species in *Pleonectopoda.* It can be distinguished from these species by genital characters. In the male genitalia, *cryptica* can be distinguished by the stout, downcurved, relatively long saccular extensions. The right saccular extension is almost as long as the right harpe. The vesica is similar to those of *altens* and *austrina* in that the subbasal diverticulum is curved and anvil shaped but unlike those species, the portion of the vesica posterior to the subbasal coil is disproportionately longer and straighter. The female genitalia are similar to those of *altens* and *austrina* in that the setae on the ovipositor lobes are short and sparse and the subbasal row of longer setae is absent; the female genitalia differ from those of all other species in the subgenus except *leuschneri* in that the appendix bursae is elongate, longer than the corpus bursae.

The immature stages of *cryptica* are unknown.

Euxoa cryptica occurs from Mt. Shasta in northern California southward in the Sierra Nevada Mountains to Inyo County, California at elevations of 8,000′ to 10,000′. Adults of *cryptica* have been collected from late July until mid-August.

Euxoa (*Pleonectopoda*) *leuschneri* Lafontaine, NEW SPECIES
PL. 2, FIG. 9; PL. C, FIG. 10; PL. S, FIG. 1.

Euxoa (*Pleonectopoda*) *leuschneri* Lafontaine.
Type locality: Barton Flats, San Bernardino Mts., San Bernardino Co., California, 6,300′. [LACM]

This species looks like a pale-streaked, well-marked form of *cryptica* but can be distinguished by male genital characters.

Antenna of male moderately biserrate, slightly bifasciculate. Antenna of female filiform. Frontal tubercle present but inconspicuous because of long scales on frons. Eye round. Vestiture of head and thorax a mixture of gray and black scales; prothoracic collar with transverse black line. Ground color of forewing pale gray in basal and subterminal areas, dark gray in median and terminal areas; without extensive dusting of black scales found in *cryptica*. Basal line double, incomplete. Antemedial line double, pale filled, indented opposite basal dash and in some specimens opposite anal vein. Medial line absent. Postmedial line double; basal portion toothed on wing veins, outer portion obscure; area between inner and outer portions of line pale filled. Subterminal line whitish gray, contrasting with dark-gray terminal area and dark-gray wedges adjacent to it in subterminal area. Subterminal line with W-mark opposite veins M_3 and CuA_1. Veins M_3 and CuA_1 pale lined and project into W-mark of subterminal line. Basal dash a black line extending to basal line; a white patch or line above basal dash. Reniform and orbicular spots prominent, outlined in black with white line inside black outline. Claviform spot dark gray with black outline. Forewing with contrasting pale-orange patches in median area distal to claviform and reniform spots. Forewing length 15 to 16 mm. Hindwing pale smoky brown, darker on wing margin, veins and discal spot. Male genitalia similar to those of *cryptica* but saccular extensions shorter, about ¼ length of valve (⅔ length of harpe) rather than ⅓ length of valve (⅘ length of harpe) as in *cryptica*. Saccular extensions broad, flattened, triangular, bending upward toward cucullus apically. Shape of vesica similar to that of *cryptica* but portion of vesica apical to subbasal bend shorter, and median diverticulum curving to project toward apex of aedoeagus rather than parallel to aedoeagus as in *cryptica*. Female genitalia similar to those of *cryptica*.

TYPES. Holotype: ♂. Barton Flats, San Bernardino Mts., San Bernardino Co., California, 6,300′; 15 June 1985; R. H. Leuschner. LACM. Paratypes: 9 ♂, 15 ♀. California, San Bernardino Mts., San Bernardino County: same locality as for holotype; 14, 15 June 1985; R. H. Leuschner (7 ♂, 9 ♀); Camp Norris, Barton Flats, 6,700′; 13–16 June 1985; J. P. Donahue (2 ♀); Green Valley Lake, 7,000′; 6 July 1964 and 27 July 1965; E. Walter (1 ♂, 1 ♀); Barton Flats; 29 June 1970; E. Walter (1 ♀); Upper Santa Ana River; 29 July 1948; G. H. and J. L. Sperry (1 ♀). Riverside County: Pinyon Flats, Santa Rosa Mts.; 4 August 1967; E. Walter (1 ♂). LACM, CNC, AMNH.

I take pleasure in naming the species after Ronald H. Leuschner who recognized the species as being undescribed and sent it to me for description.

Euxoa (*Pleonectopoda*) *extranea* (Smith)
PL. 2, FIG. 10; PL. D, FIG. 1; PL. S, FIG. 2 (RWH 10708).

Agrotis extranea Smith, [1888], *Proc. U. S. Natl. Mus.*, **10**: 459.
Type locality: "Montana" [Mt. Hood, Oregon]. [MSU]

Agrotis quinquelinea Smith, 1890, *Trans. Amer. Ent. Soc.* **17**: 49
Type locality: Sierra Nevada, California. [USNM]

Euxoa extranea usually can be recognized by the color and pattern of the forewing. The ground color varies from pale brownish gray to dark brownish gray. The transverse lines, including the median line, are all present and defined in black. Specimens of *extranea* differ from those of similar species of the *westermanni* group in lacking red or yellow shades on the forewing and thorax. Specimens of *extranea* can be confused with some species in the subgenus *Euxoa* but can usually be distinguished from them by larger size: forewing length varies from 17 to 20 mm. Both sexes of *extranea* can readily be distinguished by characters of the genitalia. In males, the saccular extensions are about as long as the harpes and are flattened apically. The sacculus is oval and disproportionately massive with the dorsal margin convex (plate D, figure 1); in other similar species the sacculus is crescentic with the dorsal margin concave. Females of *extranea* can be recognized by a combination of triangular appendix bursae and sclerotized plate in ventral wall of ductus bursae bent to right (plate S, figure 2).

The immature stages are known only from laboratory reared material. The species overwinters as larvae (Hinks and Byers, 1976).

Euxoa extranea occurs at higher elevations (5,000′–10,000′) in the Cascades and Sierra Nevada Mountains from north-central Washington southward at least as far as Inyo County, in east-central California. Most specimens were collected in subalpine habitats in open stands of pine and fir. Adults have been collected from mid-July until late August.

tristicula GROUP

The *tristicula* group includes only *tristicula*; the characters are given under the species.

Euxoa (*Pleonectopoda*) *tristicula* (Morrison) PL. 2, FIGS. 11, 12; PL. D, FIG. 2; PL. S, FIG. 3 (RWH 10723).

Agrotis tristicula Morrison, (1876), *Proc. Acad. Nat. Sci. Philadelphia,* (**1875**): 429.
Type locality: Orono, Maine. [USNM]

Euxoa nesilens Smith, 1903, *Jour. New York Ent. Soc.,* **11**: 192.
Type locality: Head of Pine Creek near Calgary, Alberta. [AMNH]

Specimens of *tristicula* can be recognized by large size and pale coloration. They occur in two color forms; in both forms the forewing is pale gray with the maculation defined by narrow black lines. The reniform and orbicular spots are slightly paler than the remainder of the forewing. In the most common form (plate 2, figure 11) the forewing has a black basal dash and black shading proximal to the orbicular spot and between the orbicular and reniform spots; there is also a transverse black line on the prothoracic collar. In about ¼ of the specimens these black areas on the forewing and collar are absent (plate 2, figure 12). The forewing length varies from 15 to 20 mm. The hindwing is dirty white with smoky-gray shading on the outer quarter of the wing, on the discal spot, and on the veins. In the male genitalia the sacculus is large and bowl shaped. The saccular extensions are about twice as stout as the harpes and about as long. The apical portion of the saccular extension curves ventrally away from the cucullus. In the vesica, the subbasal diverticulum is massive and somewhat triangular; a small posterior diverticulum is dorsal to the subbasal diverticulum; and the median diverticulum is near the end of the vesica adjacent to an apical diverticulum. In all other species in the subgenus *Pleonectopoda* the median diverticulum is closer to the subbasal coil of the vesica than to the apex. The female genitalia are similar to those of species in the *westermanni* group: the ovipositor lobes are clothed with fine setae; the sclerotized plate in the ventral wall of the ductus bursae is about ⅔ the length of the ductus bursae; and the corpus bursae is bisaccate with the appendix bursae rounded.

The larva of *tristicula* does most feeding in the fall and very early spring, overwintering as an almost mature larva. The larva feeds just below the soil surface, primarily on broad-leaved herbs such as alfalfa, flax, and sugarbeets. It is not a species of economic importance because of the late fall and early spring feeding periods. The life history of *tristicula* was reviewed by Jacobson (1969). The egg was illustrated by Salkeld (1976: 1388, figures 28–31).

Euxoa tristicula occurs throughout the arid regions of western North America from Manitoba and Minnesota westward to British Columbia and California and from the southern Northwest Territories southward to central Arizona and southern California. A disjunct eastern population occurs in coastal dune areas in southwestern Nova Scotia, eastern Maine, and eastern Quebec. Adults have been collected from late May until late August although most were collected during June and July.

vetusta GROUP

The *vetusta* group includes only *vetusta*; the characters are given under the species.

Euxoa (*Pleonectopoda*) *vetusta* (Walker) PL. 2, FIG. 13; PL. D, FIG. 3; PL. S, FIG. 4 (RWH 10724).

Mamestra vetusta Walker, 1865, *List of the Specimens of Lepidopterous Insects in the Collection of the British Museum,* **32**: 662.
Type locality: Vancouver Island, British Columbia. [BMNH]

Lepipolys tetra Walker, 1865, *List of the Specimens of Lepidopterous Insects in the Collection of the British Museum,* **33**: 768.
Type locality: Vancouver Island, British Columbia. [BMNH]
NOTE—The female lectotype of *tetra* was designated by Hardwick (1970: 138).

Agrotis euroides Grote, 1874, *Proc. Acad. Nat. Sci. Philadelphia,* **26**: 202.
Type locality: California. [BMNH]

Agrotis perpura Morrison, 1874, *Proc. Boston Soc. Nat. Hist.,* **17**: 164.
Type locality: "California." [MSU]
NOTE—The nominal type of *perpura,* stated by Morrison to be from California, is labeled "Washington Territory" (Hardwick, 1970: 138).

Specimens of *vetusta* can be distinguished from those of other species in the subgenus by wing color, antennal structure and genital characters. The male antenna is markedly biserrate and bifasciculate, almost plumose; the antenna is about five times as wide as the central shaft. In all other species of the subgenus *Pleonectopoda,* except *fuscigera,* the antenna is only two or three times as wide as the shaft. The ground color of the forewing is pale silver gray, unlike that of any other species in the subgenus. The maculation is defined in black. In most specimens, the space between the orbicular and reniform spots is heavily shaded with black. The fringe of the forewing is pale reddish brown and contrasts with the forewing ground color. The forewing length varies from 16 to 19 mm. The hindwing is smoky brown with a paler fringe. In the male genitalia the saccular extensions are stouter and slightly longer than the harpes, and curve upward toward the cucullus. The vesica has a small, unilobed subbasal diverticulum. The female genitalia differ from those of other species in that the corpus bursae is massive, two or three times the length of the ductus bursae. In other species that have a bisaccate bursae the corpus bursae is less than 1.5× the length of the ductus bursae.

The immature stages are unknown.

Euxoa vetusta occurs along the Pacific Coast from southern British Columbia southward to central California. It has been collected inland at Mt. Rainier, Washington and Mt. Hood, Oregon. Adults have been collected over an unusually long period, for a species of *Euxoa,* from late April until early September.

fuscigera GROUP

The *fuscigera* group includes only *fuscigera*; the characters are given under the species.

Euxoa (*Pleonectopoda*) *fuscigera* (Grote)

PL. 2, FIGS. 14, 15; PL. D, FIG. 4; PL. S, FIG. 5 (RWH 10725).

Agrotis fuscigera Grote, 1874, *Can. Ent.*, **6**: 155.
Type locality: Sausalito, California. [BMNH]

NOTE—The female lectotype of *fuscigera* was designated by Hardwick (1970: 140).

Agrotis feniseca Harvey, 1875, *in* Grote, *Check List of the Noctuidae of America, North of Mexico,* Part **1**: 25.
Type locality: Sausalito, California. [BMNH]

NOTE—The male lectotype of *feniseca* was designated by Hardwick (1970: 140).

Euxoa leucopterota McDunnough, 1946, *Can. Ent.,* **78**: 28.
Type locality: Taft, California. [CNC]

Specimens of *fuscigera* cannot be confused with those of any other species in the subgenus *Pleonectopoda.* They can be confused with some species in other subgenera. The forewing is buffy brown or pale yellow brown with a dusting of black scales. The maculation is defined in black; the transverse lines are double. Forewing length varies from 14 to 18 mm. The hindwing of the male is dirty white with brown shading on the discal spot, on the veins, and on the wing margin. In some specimens the outer quarter of the wing is smoky brown. The hindwing of the female varies from being dirty white, heavily dusted with dark scales, and with a smoky-brown terminal band and median line, to being entirely smoky brown with a contrasting white fringe. This color pattern makes *fuscigera* specimens very similar to some specimens of *difformis* (subgenus *Euxoa*) and to a lesser extent, to *messoria* (subgenus *Longivesica*). Males of *fuscigera* frequently can be distinguished from other similar species by the markedly biserrate antenna; the antennal width is about four times that of the central shaft rather than about twice as wide as in other similar species. Males of *fuscigera* can be distinguished from those of similar species in other subgenera by characters of the vesica. If the vesica has not been everted, the male genitalia are very similar to those of *difformis.* In *fuscigera* the saccular extensions are slightly stouter than the harpes and about 1.25× as long; in *difformis* the saccular extensions are shorter than the harpes in most specimens but may be longer than the harpes in some specimens, particularly in the orange form of *difformis* (plate 7, figure 25) that occurs in southwestern California. Females of *fuscigera* can be distinguished by the bilobed condition of the corpus bursae. The female genitalia are similar to those of many species in the *westermanni* group, but the sclerotized plate in the ventral wall of the ductus bursae is very long, about twice the length of abdominal segment VIII.

The immature stages of *fuscigera* are unknown.

Euxoa fuscigera is a species of arid regions; it occurs in and west of the Coast Ranges of California from Lake County southward at least to San Diego. In southern California it occurs inland as far as the Mojave Desert. Adults have been collected from early September until early December.

atomaris GROUP

The *atomaris* group includes only *atomaris*; the characters are given under the species.

Euxoa (*Pleonectopoda*) *atomaris* (Smith)

PL. 2, FIGS. 16–19; PL. D, FIG. 5; PL. S, FIG. 6 (RWH 10726).

Agrotis atomaris Smith, 1890, *Trans. Amer. Ent. Soc.*, **17**: 47.
Type locality: California. [USNM]

NOTE—The male lectotype of *atomaris* was designated by Todd (1968: 267).

Carneades detesta Smith, 1893, *Bull. U. S. Natl. Mus.*, **44**: 93.
Type locality: Colorado.

NOTE—*Carneades detesta* Smith, 1893, is a replacement name for *Carneades choris* Smith (1890: 158) a misidentification of *Agrotis choris* Harvey (1876: 37). Type material was not selected by Hardwick (1970: 142) or by Todd (1982: 64).

Euxoa esta Smith, 1906, *Can. Ent.*, **38**: 227.
Type locality: Wellington, British Columbia. [AMNH]

NOTE—The female lectotype of *esta* was designated by Todd (1968: 271).

Specimens of *atomaris* usually can be recognized by wing pattern. The forewing ground color varies from pale gray brown to dark blackish brown. The maculation is obscure but is defined by narrow lines of mixed dark and light scales. Thus, in pale specimens the maculation is defined in black; the opposite is true of dark specimens. The distinctive feature of *atomaris* specimens is that the central and posterior portions of the forewing are paler than the costal and outer portions and frequently have a yellow-buff or coppery cast. The extent and shade of this central pale area varies greatly from population to population, but it frequently can be used to distinguish *atomaris* from similar species in the subgenus *Euxoa,* particularly *tessellata* and *satis.* Dark-colored specimens from coastal Washington and British Columbia frequently do not have a pale central area on the forewing (plate 2, figure 19). These specimens can be confused with those of *punctigera* but can usually be distinguished by the paler hindwing. The hindwing of *atomaris* males is white with a variable amount of smoky-brown shading on the wing margin, the veins, and sometimes on the discal spot. The hindwing of the female is similar to that of the male but tends to be darker, particularly on the veins. Forewing length varies from 13 to 17 mm. Males of *atomaris* can readily be recognized by genital characters, particularly by the combination of asymmetrical harpes and stout saccular extensions that taper abruptly near the apex. The left harpe is about ¾ the length of the right. The shape of the vesica of *atomaris* (plate D, figure 5) is also distinctive. The female genitalia are similar to those of many species in the *westermanni* group.

The immature stages of *atomaris* are unknown.

This species is widely distributed in western North America. It occurs from southern Alberta and British Columbia southward to central New Mexico, central Arizona, and southern California. It has been collected as far east as western North Dakota, Nebraska, and Kansas. Most specimens have been collected between mid-August and late October; adults fly in August and September in the North and in September and October in the South. A few specimens have been collected very early in the season, as early as March and April.

Populations of *atomaris* have previously been arranged in three subspecies; typical *atomaris,* with more brown shading in the forewing, occurs in the Coast Ranges of California; subspecies *esta,* with very dark forewing and dusky shading on the hindwing occurs in coastal areas of southern British Columbia and northern Washington; and the inland, pale gray-brown populations are subspecies *detesta.* There seems to be little justification for the retention of these subspecific names because the ground color varies with habitat; dark specimens are collected in the most mesic habitats; the palest specimens are from the driest, Great Basin localities; and intermediate specimens occur in intermediate habitats. Also, many specimens from the Cascade Mountains of Washington and montane areas in Idaho are intermediate between *esta* and *detesta.*

pleuritica GROUP

The *pleuritica* group includes three species. Males can be recognized by the combination of asymmetrical harpes and long, tapered, apically flattened saccular extensions, as well as by the distinctive shape of the vesica. Males of the group also differ from those of other groups in the subgenus in having a long, narrow juxta, which is at least twice as long as wide. Females of the *pleuritica* group can be recognized by the very long sclerotized plate in the ventral wall of the ductus bursae, this over twice as long as abdominal segment VIII and by the shape of the appendix bursae. The appendix bursae is bent anteriorly so that the ductus seminalis arises from the anterolateral corner, rather than from the posterolateral corner as in other *Euxoa* species.

Euxoa (*Pleonectopoda*) *pleuritica* (Grote)

PL. 2, FIGS. 20, 21; PL. D, FIG. 6; PL. S, FIG. 7 (RWH 10727).

Agrotis pleuritica Grote, 1876, *Check List of the*

Noctuidae of America, North of Mexico, pt. **2**: 47.
Type locality: Canada. [BMNH]

Carneades messoria var. *confracta* Smith, 1890, *Bull. U. S. Natl. Mus.,* **38**: 170.
Type locality: unknown. [MSU]
NOTE—Smith (1890: 170) credited the name *confracta* to H. K. Morrison on the basis of a manuscript name on a specimen in the Tepper Collection [MSU].

Specimens of *pleuritica* can be distinguished from those of other species in the *pleuritica* group and most similar species in the subgenus *Euxoa* by relatively small size and the contrasting, mottled appearance of the forewing. The mottled appearance of the forewing is created by patches of pale coppery and gray shading in between areas of dark-brown shading. This combination of colors gives fresh specimens an olive cast. The forewing length varies from 14 to 18 mm. The hindwing is dirty white overlaid with gray-brown scales, these sparse near the wing base but extensive on the veins and on the marginal third of the wing. The hindwing of the female tends to be slightly darker than that of the male. Males of *pleuritica* can be distinguished by: right harpe is about half the length of the right saccular extension; left saccular extension is shorter than the right, about 0.9× as long; right saccular extension is long, about ½ the length of the valve; and sacculus is crescentic or slightly oval. The female genitalia are similar to those of other species in the group.

The immature stages of *pleuritica* are known only from laboratory reared material although one larva was found on Russian thistle (Cook, 1930). It overwinters as a larva and has a long summer aestivation (Hinks and Byers, 1976).

Euxoa pleuritica occurs across southern Canada and northern United States from southern Quebec and eastern Massachusetts westward to south-central British Columbia and central Washington. In the West, *pleuritica* occurs as far south as northern New Mexico, northeastern Arizona, central Idaho, and southern Washington. Adults of *pleuritica* have been collected from late May until mid-October.

In eastern North America, specimens of *pleuritica* are frequently confused with those of *tessellata* in the subgenus *Euxoa*. The two species can be distinguished by brushing away scales at the end of the abdomen and examining the genitalia. In males the saccular extensions are flattened and bladelike apically in *pleuritica* and cylindrical and pointed in *tessellata*. In females, the ovipositor lobes of *pleuritica* are rounded apically and clothed with fine setae; in *tessellata* females they are truncate apically and clothed with short conical setae.

Euxoa (*Pleonectopoda*) *pestula* Smith

PL. 2, FIGS. 22, 23; PL. D, FIG. 7 (RWH 10728).

Euxoa pestula Smith, 1904, *Can. Ent.,* **36**: 150.
Type locality: Calgary, Alberta. [AMNH]
NOTE—The female lectotype of *pestula* was designated by Todd (1968: 275).

Euxoa petruska McDunnough, 1932, *Can. Ent.,* **64**: 235.
Type locality: Calgary, Alberta. [CNC]

Specimens of *pestula* are similar to those of *pleuritica* and *tessellata* (subgenus *Euxoa*). From the former, adults usually can be distinguished by the dark brownish-gray color of the forewing and lack of pale, contrasting, coppery patches in the wing. This gives the forewing of *pestula* a more evenly colored appearance, not the mottled appearance of *pleuritica.* The hindwing tends to be darker than that of *pleuritica.* This species is about the same size as *pleuritica*; forewing length: 15–18 mm. Specimens of *pestula* are most frequently confused with a similarly marked form of *tessellata* that is sympatric. Like *pleuritica, pestula* can readily be distinguished from *tessellata* by examination of characters visible at the end of the abdomen; these are given under *pleuritica.* In the male genitalia, *pestula* can be distinguished from *pleuritica* by the relatively longer harpe and shorter saccular extension (plate D, figure 7). The right harpe is about ⅔ the length of the right saccular extension, and the right saccular extension is noticeably less than half the length of the valve, usually about ⅖ as long. Unlike *pleuritica* males, the right and left saccular extensions are about the same length, and the sacculus is broader with the dorsal margin markedly convex. The female genitalia are similar to those of *pleuritica.*

The immature stages of *pestula* are known only from laboratory reared material.

Euxoa pestula is primarily an inhabitant of the northern Great Plains. It occurs throughout the plains region of Manitoba, Saskatchewan, Alberta, and western Montana. Possibly disjunct populations also occur in montane areas of southern Wyoming and northern Colorado and in open sandy areas in the south-central Northwest Territories. Variation in size and color of some populations of *pestula* in Montana, Wyoming, and Colorado may be the result of hybridization with *simona* in these areas (Hardwick, 1970: 146). Adults of *pestula* occur from mid-June until late August.

Euxoa (Pleonectopoda) simona McDunnough

PL. 2, FIGS. 24, 25; PL. D, FIG. 8; PL. S, FIG. 8 (RWH 10729).

Euxoa simona McDunnough, 1932, *Can. Ent.*, **64**: 234.
Type locality: Yellowstone National Park, Wyoming. [CNC]

Euxoa pleuriticoides Benjamin, (1936), *Bull. So. California Acad. Sci.*, **34**: 203.
Type locality: Crater Lake, Oregon. [USNM]

In wing color, *simona* is similar to *pleuritica* in that the forewing has a mottled appearance with coppery or pale-gray patches on a dark gray-brown ground color. Also, the hindwing is relatively pale, similar to that of *pleuritica*. It can be distinguished from *pleuritica* and *pestula* by large size and by genital characters. Forewing length varies from 18 to 22 mm. The male genitalia are similar to those of *pestula* with respect to the relative lengths of the harpes and saccular extensions but the sacculus is more massive and somewhat triangular (plate D, figure 8).

The immature stages of *simona* are unknown.

Euxoa simona is primarily a species of western montane areas. It occurs from southern Alberta and British Columbia southward to southern Colorado, southern Utah, east-central Nevada, and southwestern California. In southern Washington specimens of *pleuritica* approach *simona* in size and coloration, and the two species may hybridize in this region. Similarly, *pestula* and *simona* may hybridize in some areas in the western Great Plains and southeastern Rocky Mountain region (Hardwick, 1970: 146). Adults of *simona* have been collected from mid-June until mid-September.

SUBGENUS *Euxoa* Hübner

Euxoa Hübner, [1821]

Metaxyja Hübner, [1821]

Exarnis Hübner, [1821]

Brotis Hübner, [1821]

Telmia Hübner, [1821]

Mimetes Hübner, [1821]

Metaxya [sic] Walker [1857]

Carneades Grote, 1883

Paragrotis Pratt, 1902

Metaxyia [sic] Hampson, 1903

Mimetis [sic] Hampson, 1903

The subgenus *Euxoa* is the largest in the genus; it contains 120 species in North America and about 120 species in the Palearctic Region. Members of the subgenus occur throughout North America except the Southeast; unlike species in the subgenus *Pleonectopoda*, few species occur in subarctic or alpine regions.

Males of the subgenus *Euxoa* can most easily be recognized by shape of the vesica: it is from one to two times as long as the aedoeagus; it bends above the apex of the aedoeagus to project dorsally or to the right; and it lacks a prominent twist, coil, or loop subbasally. The valves are bilaterally symmetrical except in respect to the lengths of the right and left saccular extensions. The saccular extensions vary from being about ½ the harpe length to more than twice its length; the saccular extensions are cylindrical, or flattened, saber shaped, and tapered apically, not enlarged and spatulate. If the shape of the vesica cannot be seen, males can be distinguished by characters of the valve except from those of the subgenera *Longivesica* and *Pleonectopoda*. They can be distinguished from males of the subgenera *Chorizagrotis* and *Palaeoeuxoa* by the tapered rather than apically cupped and spatulate saccular extensions, from those of *Heteroeuxoa* by the symmetrically shaped valves, and from those of *Orosagrotis* by the relative shapes of the saccular extensions, harpes, and uncus. They must be distinguished from males of the subgenera *Longivesica* and *Pleonectopoda* on an individual species basis, essentially by eliminating species of these subgenera as possibilities. *Longivesica* contains only four species, and these can be recognized by wing markings and valve shape. *Pleonectopoda* contains 23 species that can be recognized by a combination of characters that are given in the alternate key to subgenera on page 21.

Females of the subgenus *Euxoa* must be associated with the subgenus by characters of individual species. A key to species groups of *Euxoa*, based on female genitalia, starts on page 22.

Species in the subgenus have been arranged in 38 species-groups. The arrangement of species-groups is based on a phylogenetic analysis of the genus (Lafontaine, 1981).

KEY TO THE SPECIES-GROUPS OF THE SUBGENUS *EUXOA* OF AMERICA NORTH OF MEXICO (MALES)

1. Apical third of uncus dorsally flattened, about 6–10× as wide as thick; setae near apex of uncus as stout, or stouter, on dorsal surface as on ventral surface (plate E, figure 8) .. *rufula* group p. 68
– Apical third of uncus cylindrical, or at most slightly compressed dorsally, sometimes spatulate apically, but not more than 2–3× as wide as thick; setae near apex of uncus stouter on ventral surface than on dorsal surface 2

2. Apex of uncus with 2 to 5 stout, heavily sclerotized, conical setae on ventral surface, these conspicuously stouter than remainder of longer setae; right saccular extension about 1.25× length of left saccular extension (plate H, figure 9) .... *tessellata* group (part) p. 87
– Setae on ventral surface of uncus shorter near apex but apical setae not stouter, conical, or more heavily sclerotized than longer setae on ventral surface; saccular extensions symmetrical or asymmetrical with respect to length 3

3. Dorsal surface of apical half of harpe very finely pubescent in addition to much larger scattered setae (most easily observed under magnifications of about 200×) 4
– Dorsal surface of apical half of harpe without pubescence, only larger scattered setae present, these about half as long as diameter of harpe .. 17

4. Aedoeagus expanded laterally and cleft at base; saccus with marked median constriction, anterior portion almost lanceolate (plate F, figure 2) .... *terrena* group p. 71
– Aedoeagus rounded at base; saccus tapered evenly anteriorly, more or less V-shaped 5

5. Right saccular extension shorter than right harpe and stouter, bent toward cucullus near apex 6
– Right saccular extension as long as, or longer than, right harpe, or as thin as harpe and not bent toward cucullus at apex 7

6. Vesica with two small subbasal diverticula on left side in addition to larger footlike subbasal diverticulum; right harpe less than 1.3 mm long (plate H, figure 7) .... *pallipennis* group p. 85
– Vesica without diverticula in addition to footlike subbasal diverticulum; right harpe more than 1.3 mm long (plate H, figure 6) .... *punctigera* group (part) p. 83

7. Right harpe markedly incurved, C-shaped, bent through an angle of about 70° to 90°; dorsal margin of sacculus convex; forewing without a basal dash 8
– Right harpe not markedly incurved, elbowed at middle or near base, bent through an angle of about 30° to 60°; dorsal margin of sacculus concave or straight; forewing with or without a basal dash 10

8. Dorsal margin of sacculus evenly convex from base of saccular extension almost to base of sacculus; sacculus broadening toward base, triangular or teardrop shaped in outline (plate G, figure 6) .... *declarata* group (part) p. 78
– Dorsal margin of sacculus angled or wavy, not evenly convex from base of saccular extension; sacculus broadest near middle, elliptic or rectangular in outline 9

9. Right harpe about 1.5 mm long (plate F, figure 7) .... *serricornis* group p. 73
– Right harpe about 1.0 mm long (plate H, figure 2) .... *simulata* group (part) p. 82

10. Subbasal diverticulum of vesica with large, heavily sclerotized cornutus; cornutus about half as long as apical diameter of aedoeagus (plate K, figure 4) .... *choris* group p. 114
– Subbasal diverticulum of vesica with one to several small, weakly sclerotized cornuti 11

11. Apex of right saccular extension bent around ventral angle of cucullus; right saccular extension sicklelike, compressed near apex, apical portion flattened and bladelike (plate G, figure 9) .... *silens* group (part) p. 80
– Apex of right saccular extension rarely bent around ventral angle of cucullus; if so, then right saccular extension not compressed and bladelike near apex but cylindrical to apex 12

12. Right saccular extension more than 1.25× length of right harpe and sinuate, bent ventrally

away from cucullus near apex (plate I, figure 1) .... *albipennis* group p. 89

– Right saccular extension less than 1.25× length of right sacculus, or evenly incurved and bent toward cucullus near apex 13

13. Right harpe 0.9–1.2 mm long, or forewing with a black basal dash and pale cubital vein 14

– Right harpe 1.3–1.8 mm long; forewing without a basal dash or pale cubital vein, occasionally a trace of basal dash in specimens with forewing almost entirely black 15

14. Prothoracic collar without transverse, black median line; forewing with median line contrasted in most specimens (plate 3, figures 26–32) .... *setonia* group p. 76

– Prothoracic collar with transverse black median line; forewing with median line not contrasted (plate 7, figures 1–6, 32–35) .... *detersa* group (part) p. 118

15. Right saccular extension more than 1.2× length of right sacculus; ventral apex of aedoeagus a raised, fingerlike projection free from vesica at apex (plate F, figure 1); eastern deciduous forest zone .... *scholastica* group p. 70

– Right saccular extension less than 1.1× length of right sacculus; ventral apex of aedoeagus a wedge-shaped plate attached to vesica; Great Plains westward 16

16. Right saccular extension stouter than right harpe, at least toward base, and slightly but evenly incurved throughout its length (plate H, figure 3) .... *punctigera* group (part) p. 83

– Right saccular extension thinner and shorter than harpe (*aequalis*), but if stouter (*conjuncta*), then bent ventrally away from cucullus near apex (plate N, figures 3, 4) ... *aequalis* group (part) p. 138

17. Right saccular extension short, either ¼ or less length of right valve, or less than ¾ length of right harpe, or both 18

– Right saccular extension longer, more than ¼ length of right valve and more than ¾ length of right harpe 25

18. Antenna bipectinate, almost plumose, five or six times as wide as central shaft 19

– Antenna biserrate, less than three times as wide as central shaft 20

19. Right saccular extension less than half length of right harpe (plate D, figure 10) .. *camalpa* group p. 64

– Right saccular extension more than half length of right harpe (plate E, figure 1) ... *serotina* group p. 65

20. Right harpe more than 1.25× length of right sacculus (*melura*, plate N, figure 1) .... *detersa* group (part) p. 118

– Right harpe less than 1.25× length of right sacculus 21

21. Harpes short, less than 0.9 mm long; right saccular extension about ¾ length of right harpe 22

– Harpes longer, more than 1.0 mm long; right saccular extension about ½ length of right harpe 23

22. Apex of uncus dilated; harpe positioned on inner side of cucullus, extended beyond dorsal margin of cucullus only at apex, if at all (plate N, figure 8) .... *cinereopallida* group p. 145

– Apex of uncus tapered, not dilated apically; harpe extended dorsally well beyond dorsal margin of cucullus (plate D, figure 9) .... *hardwicki* group p. 63

23. Base of right saccular extension markedly stouter than base of right harpe; apex of right saccular extension curved inward at apex (plate N, figure 7) .... *misturata* group p. 143

– Base of right saccular extension not stouter than base of right harpe, right saccular extension essentially straight or slightly bent but not curved inward at apex 24

24. Apex of uncus dilated; forewing not streaked, orbicular spot rounded in outline (plate 8, figures 27–29) .... *mitis* group p. 146

– Apex of uncus tapered; forewing longitudinally streaked, orbicular spot oblong in outline (plate 8, figures 19–23) .... *atristrigata* group p. 144

25. Uncus much shorter than harpe, markedly constricted apically into a sharp, hooklike pro-

cess lacking setae (plate E, figure 10) .... *annulipes* group p. 70

– Uncus not shorter than harpe, not constricted apically but tapered evenly near apex; apical portion of uncus slightly curved and with setae present 26

26. Right saccular extension more than twice length of right harpe; harpe with dense cluster of setae near apex (plate I, figure 8) .... *catenula* group p. 95

– Right saccular extension less than twice length of right harpe; harpe with setae scattered over apical third 27

27. Uncus dilated and spatulate near apex; harpe positioned on inner side of cucullus, extended beyond dorsal margin of cucullus only at apex, if at all (plate 0, figure 2) .... *luctuosa* group p. 147

– Uncus tapered, or slightly swollen at apex, not spatulate; harpe usually extended dorsally, well beyond dorsal margin of cucullus 28

28. Apex of right saccular extension curled dorsally around ventral angle of cucullus; right saccular extension compressed near apex, apical portion flattened and bladelike 29

– Apex of right saccular extension usually not curled dorsally around ventral angle of cucullus, if doing so, then right saccular extension not compressed and bladelike apically but cylindrical 30

29. Right sacculus triangular, dorsal margin convex (plate G, figure 6) ...... *declarata* group (part) p. 78

– Right sacculus crescentic, dorsal margin concave (plate G, figure 9) .... *silens* group (part) p. 80

30. Aedoeagus more than 3.2 mm long; right saccular extension at least twice as stout as right harpe at base (plate E, figure 3) .... *bostoniensis* group (part) p. 65

– Aedoeagus less than 3.0 mm long; right saccular extension stout or thin 31

31. Right saccular extension longer than 1.25× length of left saccular extension 32

– Right saccular extension shorter than 1.25× length of left saccular extension 34

32. Subbasal diverticulum of vesica elongate with pouches directed both posteriorly and anteriorly, diverticulum about half length of aedoeagus; left saccular extension about ¾ length of left harpe, stout, length equal to about 6× basal width (plate E, figure 9) .... *intrita* group p. 69

– Subbasal diverticulum of vesica small, usually foot shaped with single pouch directed posteriorly, diverticulum about ⅓ length of aedoeagus; left saccular extension about as long as left harpe, length equal to about 7–10× basal width 33

33. Right saccular extension at least 1.33× length of right harpe; right saccular extension and harpe evenly incurved toward cucullus (plate H, figure 10) .... *tessellata* group (part) p. 87

– Right saccular extension less than 1.33× length of right harpe; either right harpe, or right saccular extension bent away from cucullus near apex 34

34. Forewing with prominent, black basal dash extending well beyond basal line 35

– Forewing without a basal dash, at most with slight thickening of basal line at middle 39

35. Vesica with three pouches in ventral wall between apex of aedoeagus and subbasal bend in vesica in addition to normal subbasal diverticulum (*xasta*, plate I, figure 7) .... *hollemani* group (part) p. 92

– Vesica with at most one pouch in ventral wall between apex of aedoeagus and subbasal bend, in addition to subbasal diverticulum 36

36. Forewing with transverse lines absent; reniform and orbicular spots fused in most specimens (plate 4, figures 44–46) .... *hollemani* group (part) p. 92

– Forewing with transverse lines present, occasionally obscure; reniform and orbicular spots not fused 37

37. Subbasal diverticulum of vesica with pouches directed both anteriorly and posteriorly; saccular extensions twisted (plate K, figures 5, 6) .... *obeliscoides* group p. 115

– Subbasal diverticulum of vesica with either a single pouch directed ventrally, or elbowed at middle and directed posteriorly; saccular extensions straight or curved but not twisted 38

38. Vesica with small subbasal diverticulum on left side in addition to normal subbasal diverticulum; basal dash on forewing beginning at basal line and extending toward claviform spot; ground color of forewing orange or reddish brown (plate 6, figures 4–11) ... *cursoria* group (part) p. 109

– Vesica with only normal subbasal diverticulum; basal dash on forewing beginning at wing base, broken by basal line in some specimens; ground color of forewing variable (plate 6, figures 31–60; plate 7, figures 36–57) .... *detersa* group (part) p. 118

39. Cucullus markedly expanded apically, foot shaped, 1.75–2.00× as wide apically as subapically 40

– Cucullus slightly expanded apically, less than 1.5× as wide apically as subapically 41

40. Right saccular extension longer than right sacculus and twisted (plate I, figure 3); reniform and orbicular spots small and inconspicuous (plate 4, figures 41–43) .... *cincta* group p. 91

– Right saccular extension shorter than right sacculus and straight (plate K, figure 7); reniform and orbicular spots large and contrasting (plate 6, figures 26, 27) .... *lillooet* group p. 116

41. Right saccular extension short, less than 0.85× length of right sacculus 42

– Right saccular extension longer than 0.90× length of right sacculus 45

42. Hindwing of male white with gray-brown shading on wing margin and on veins 43

– Hindwing of male brown 44

43. Right saccular extension about as stout as right harpe for most of its length; longitudinal axis of vesica forming an angle of about 90° with longitudinal axis of aedoeagus (plate K, figure 3); reniform spot filled with black (plate 6, figures 14–16) .... *siccata* group p. 114

– Right saccular extension about twice as stout as right harpe for most of its length in most specimens, only near base in some specimens; vesica bent through angle of about 180°, apex of vesica almost touching aedoeagus (plate E, figure 3); reniform spot of forewing partially filled with black (plate 2, figures 32–39) .... *bostoniensis* group (part) p. 65

44. Cucullus about 1.5× as long as wide; harpe positioned on dorsal margin of cucullus (plate E, figure 2); ground color of forewing evenly colored, pale gray or brown, antemedial and postmedial lines black and contrasting (*bifasciata*, plate 2, figures 40–42) .... *bostoniensis* group (part) p. 65

– Cucullus about 3× as long as wide; harpe extended dorsally, well beyond dorsal margin of cucullus; forewing with transverse lines not particularly contrasting 45

45. Frontal tubercle absent; forewing violet gray with prominent antemedial and postmedial lines, these even, not scalloped between veins (plate 6, figures 1–3) .... *violaris* group p. 109

– Frontal tubercle present, reduced and hidden by overhanging vestiture in some species (e.g., *detersa*); transverse lines, if present, scalloped between veins 46

46. Right harpe markedly incurved throughout its length, C-shaped, bent through an angle of about 90°; clavus barely evident; sacculus broad, almost rectangular in outline (plate H, figure 2) .... *simulata* group (part) p. 82

– Right harpe incurved at middle, bent through an angle of about 45° occasionally more markedly incurved in species with prominent clavus or elongate, crescentic sacculus 47

47. Forewing longitudinally streaked, terminal space with W-mark proximally where pale-lined veins CuA_1 and M_3 project into it 48

– Forewing not longitudinally streaked; veins CuA_1 and M_3 not pale lined 49

48. Forewing with transverse lines absent; forewing pale gray with yellow-orange shading in fold and distal to reniform spot (*pluralis*, plate 3, figures 21, 22) .... *pluralis* group (part) p. 75

– Forewing with transverse lines present; forewing without contrasted yellow-orange shading in fold or distal to reniform spot (plate 6, figures 46–48) .... *detersa* group (part) p. 118

49. Sacculus small and bowl shaped, less than 0.7× length of cucullus and 2.0–2.5× as long as wide in most specimens; saccular extensions very thin, appearing frail (plate J, figures 4–8) ... *infausta* group p. 103

– Sacculus more elongate, more than 0.7 × length of cucullus and 2.5–3.0 × as long as wide in most specimens; saccular extensions usually stout and not appearing frail 50

50. Subbasal diverticulum of vesica with pouches directed both anteriorly and posteriorly, usually to about same degree in each direction (plate G, figure 1) .... *pluralis* group (part) p. 75

– Subbasal diverticulum of vesica either curled in an arc, or elbowed at middle, projected posteriorly 51

51. Juxta markedly constricted at middle (plate K, figure 8) .... *basalis* group p. 117

– Juxta conical or rectangular, without median constriction 52

52. Vesica with small, nipplelike subbasal diverticulum on left side in addition to normal foot-shaped subbasal diverticulum 53

– Vesica with only foot-shaped subbasal diverticulum subbasally 55

53. Forewing black, maculation obscure; hindwing white (*velleripennis,* plate 5, figure 28) .... *comosa* group (part) p. 96

– Forewing gray, brown, or orange; hindwing brown in most species (white in *teleboa*) 54

54. Vesica bent ventrally posterior to subbasal diverticulum for distance equal to twice diameter of aedoeagus before curved in an arc to project dorsally (plate K, figure 1); forewing length: 17–19 mm in most specimens (*ochrogaster, cursoria,* plate 6, figures 4–11) ... *cursoria* group (part) p. 109

– Vesica bent dorsally posterior to subbasal diverticulum (plate L, figures 9, 10); forewing length: 14–16 mm in most specimens (*teleboa, difformis, moerens,* plate 7, figures 13–28) .. .. *detersa* group (part) p. 118

55. Left saccular extension shorter than left harpe and about as stout (*cona, munis,* plate N, figures 5, 6) .... *aequalis* group (part) p. 138

– Left saccular extension longer than left harpe, or at least 1.5 × stouter 56

56. Right saccular extension less than 0.8 × as long as right sacculus; median diverticulum of vesica half way between bend in vesica and apex (*nostra,* plate K, figure 2) .... *cursoria* group (part) p. 109

– Right saccular extension more than 0.8 × as long as right sacculus; median diverticulum of vesica close to subbasal bend in vesica 57

57. Right saccular extension longer than right sacculus; axis of vesica and aedoeagus forming an angle of less than 50° (*citricolor, tronella, latro, murdocki,* plate L, figures 7, 8; plate M, figures 1, 2) .... *detersa* group (part) p. 118

– Right saccular extension usually shorter than right sacculus, slightly longer in some specimens of *comosa ontario* and *occidentalis*; longitudinal axis of vesica and aedoeagus forming an angle of 60° to 90° (plate I, figures 9, 10; plate J, figures 1, 2) .... *comosa* group (part) p. 96

hardwicki GROUP

The *hardwicki* group includes only *hardwicki*; the characters are given under the species.

Euxoa (*Euxoa*) *hardwicki* Lafontaine, NEW SPECIES

PL. 2, FIGS. 26, 27; PL. D, FIG. 9; PL. T, FIG. 1.

Euxoa (*Euxoa*) *hardwicki* Lafontaine.
Type locality: Walla Walla, Washington. [CNC]

This species is most likely to be confused with *Euxoa melura* McDunnough (p. 136), a similarly marked, spring-flying species, but can be distinguished from it by the teardrop-shaped orbicular spot tapered toward the reniform spot and by genital characters. In *melura* the orbicular spot is rounded or oval, tapered away from the reniform spot.

Antenna of male biserrate and bifasciculate. Frontal tubercle moderately large. Eye round. Vestiture of head and thorax a mixture of gray and brown scales. Ground color of forewing brown with light-gray pale areas. Transverse lines present but obscure in most specimens. Basal and antemedial lines dark brown, double when visible. Postmedial line indicated only by change to paler ground color in subterminal area. Subterminal line evident as pale line proximal to dark gray-brown terminal area. A series of dark sagittate spots proximal to subterminal line. Claviform spot obscure or absent. Orbicular spot teardrop shaped, extending toward reniform spot and fused with it in many specimens. Reniform spot relatively narrow and markedly bent. Cubital vein pale silver gray. Veins CuA_1 and M_3 partially pale lined and projected into terminal space. Fringe of forewing dark gray brown with pale basal line. Forewing length 14 to 15 mm. Hindwing gray brown, slightly paler toward base. Fringe white with brown subbasal line. Male genitalia with harpe and saccular extension short, both shorter than sacculus. Saccular extension stouter than harpe and curved inward near apex. Vesica with large subbasal diverticulum. Female genitalia with ovipositor lobes rounded. Corpus bursae bisaccate with left sac as large as, or larger than, right.

TYPES. Holotype: ♂. Walla Walla, Washington; 6 June 1965; Wm. C. Cook; CNC (Type No. 19510). Paratypes: 21 ♂, 8 ♀. Oregon: Cold Spring Junction; 30 May 1960; Wm. C. Cook (1 ♂); Hat Rock State Park; 29 May 1963 (1 ♂). Washington: Prosser; 9–10 June 1955; Klostermeyer (6 ♂, 4 ♀); Richland; 13 May 1949; J. J. Davis (3 ♂); Walla Walla; 27 May 1952, 6 June 1960, 6–19 June 1965; Wm. C. Cook (8 ♂, 4 ♀); Wallula, 23 May and 3 June 1949; Wm. C. Cook (2 ♂). CNC, USNM.

In addition to the type material, I have seen two specimens from Sweet Grass County, Montana, in the collection of John G. Franclemont. These differ from Oregon and Washington specimens in having more gray coloration in the forewing and better defined maculation (plate 2, figure 27).

I take pleasure in naming the species after David F. Hardwick, formerly of the Biosystematics Research Centre, Ottawa. His research on *Euxoa* and extensive field work, over a twelve year period, laid the groundwork that made this revision possible.

camalpa GROUP

This group includes two species, both North American. Males can be recognized by the prominent forewing maculation; prominently biserrate, almost bipectinate, antennae; and occurrence in southwestern United States and Mexico. Males can be recognized by the long harpe that lies on the inner surface of the cucullus running parallel to the dorsal margin, the very short saccular extensions, and by vesica shape. The vesica projects dorsally and is sickle shaped with the posterior half curved in an arc; the subbasal diverticulum is rounded or truncate. In the female genitalia the anterior apophyses are very long. The bursa is bisaccate with a narrow, dumbbell-shaped, corpus bursae and a large, oval, appendix bursae. I revised the group in 1976 (Lafontaine, 1976b).

KEY TO SPECIES OF THE *CAMALPA* GROUP

1. Both fore- and hindwings extensively shaded with brown; male vesica with cluster of cornuti at base of subbasal diverticulum ...... *maderensis* p. 65

— Forewing gray brown or gray, hindwing white with light brown on wing margin; male vesica without cornuti at base of subbasal diverticulum .... *camalpa* this page

Euxoa (*Euxoa*) *camalpa* (Dyar)

PL. 2, FIGS. 28, 29; PL. D, FIG. 10; PL. T, FIG. 2 (RWH 10772).

Porosagrotis camalpa Dyar, 1912, *Proc. U. S. Natl. Mus.*, **42**: 57.
Type locality: Mexico City, Mexico. [USNM]

Euxoa clavigera Dyar, 1922, *Ins. Insc. Mens.*, **10**: 166.
Type locality: Mexico City, Mexico. [USNM]

Euxoa camalpa manca Benjamin, (1936), *Bull. So. California Acad. Sci.*, **34**: 198.
Type locality: Alpine, Texas. [USNM]

This species can be distinguished from other species of *Euxoa* that occur with it by the prominently double antemedial and postmedial lines of the forewing, and by the very large pale, orbicular and reniform spots. Forewing length varies from 13 to 16 mm. The hindwing is white with some pale-brown shading on the wing margin in specimens from north of the Mexican border. Some specimens from central Mexico, such as the holotype of *camalpa,* have more brown shading on the hindwing. Other specimens from the same area, such as the holotype of *clavigera,* are identical to Texas specimens.

The immature stages are unknown.

Euxoa camalpa is widespread but has been collected at relatively few locations. It has been collected in central and western Texas, in central-eastern Arizona, and in central Mexico. Adults fly from late March until late October with no apparent gap in the flight period. It is possible that there are several overlapping generations although no species in *Euxoa* is known to have more than one. Another possibility is that adults may aestivate in montane areas during hot periods and fly only in suitable weather.

Euxoa (*Euxoa*) *maderensis* Lafontaine
PL. 2, FIG. 30 (RWH 10773).

Euxoa maderensis Lafontaine, 1976, *Can. Ent.*, **108**: 665.
Type locality: Madera Canyon, Santa Rita Mts., Santa Cruz County, Arizona. [JGF]

This species is similar to *camalpa* but can be distinguished by larger size, more brown shading on both fore- and hindwings, small size of orbicular and reniform spots, and by male genital characters. The forewing length is 16 to 17 mm. In the male genitalia, the vesica is lightly but conspicuously sclerotized on the anterolateral surface from the subbasal bend almost to the apex. The subbasal diverticulum of the vesica is larger than that of *camalpa* and truncate; in *camalpa* it is rounded. A cluster of 6 to 10 small cornuti near the base of the subbasal diverticulum is absent in *camalpa.*

The immature stages are unknown.

Euxoa maderensis is known only from Madera Canyon in the Santa Rita Mountains of southeastern Arizona. The adult has a brief flight period, compared to that of *camalpa*; adults have been collected from mid-June until early July.

serotina GROUP

The *serotina* group includes only *serotina*; the characters are given under the species.

Euxoa (*Euxoa*) *serotina* Lafontaine
PL. 2, FIG. 31; PL. E, FIG. 1; PL. T, FIG. 3 (RWH 10771).

Euxoa serotina Lafontaine, 1976, *Can. Ent.*, **108**: 665.
Type locality: Welder Wildlife Foundation Refuge, Sinton, San Patricio County, Texas. [USNM]

Specimens of this species can be recognized by the dark brownish-maroon color of the forewing. The maculation is obscure, partially outlined in yellow brown. Forewing length varies from 14 to 15 mm. Like the preceding two species, the male antenna is deeply biserrate, almost bipectinate, about six times as wide as the central shaft. In the male genitalia, the harpe is short, about as long as the sacculus. The saccular extensions are about ⅔ as long as the harpes. The vesica projects to the right; it has an extra diverticulum subbasally on the right side. The female genitalia are similar to those of species in the *camalpa* group but the corpus bursae is very long, unisaccate, and constricted mesially.

The immature stages are unknown.

Euxoa serotina occurs in eastern Texas and west-central Louisiana; it is the only *Euxoa* species in this area. A specimen in the American Museum of Natural History is labeled "Arizona." If this is correct, the species may be more widely distributed than records indicate. *Euxoa serotina* has a very late flight period, it occurs from late October until mid-November.

bostoniensis GROUP

This group includes six species, all North American. They are among the largest species in the genus.

In the male genitalia the sacculus is relatively massive, as long as, or longer than, the saccular extensions. The saccular extensions are much stouter than the harpes. The vesica projects dorsally but is so strongly bent subbasally that its apex is parallel to the apical half of the aedoeagus. In most species the vesica has a large subbasal diverticulum and is swollen apically with two prominent apical diverticula. In the female genitalia, the ovipositor lobes are clothed with fine setae. The corpus bursae is bisaccate.

All species in the group fly in late summer or the fall.

KEY TO SPECIES OF THE *BOSTONIENSIS* GROUP

1. Forewing pale gray brown with contrasting, black antemedial and postmedial lines (plate 2, figures 40–42) .... *bifasciata* p. 67
– Forewing without contrasting antemedial and postmedial lines 2

2. Prothoracic collar without a conspicuous transverse black line; forewing with claviform spot absent; male genitalia with left saccular extension about ⅔ as long as left sacculus 3
– Prothoracic collar with a conspicuous transverse black line; forewing with claviform spot prominent, outlined in black; male genitalia with left saccular extension about as long as left sacculus 5

3. Dark shading in reniform spot forming an inverted comma; left saccular extension not markedly shorter than right 4
– Dark shading in reniform spot not enlarged at

lower end and not comma shaped; left saccular extension about ⅔ as long as right .... *ustulata* p. 68

4. Forewing ground color an even mouse gray; right saccular extension longer than right harpe; occurring from Missouri eastward .... *bostoniensis* this page

– Forewing ground color mottled shades of gray, red, and white; right saccular extension shorter than right harpe; occurring from Wisconsin and Kansas westward .... *medialis* this page

5. Reniform and orbicular spots pale and contrasting, partially filled with yellow buff; median line absent .... *perexcellens* p. 67

– Reniform and orbicular spots concolorous with ground color around spots, outlined in black; median line prominent .... *sculptilis* p. 67

Euxoa (*Euxoa*) *bostoniensis* (Grote)

PL. 2, FIG. 32, 33; PL. E, FIG. 3; PL. T, FIG. 5 (RWH 10812).

Agrotis bostoniensis Grote, 1874, *Proc. Acad. Nat. Sci. Philadelphia,* **26**: 203.
Type locality: Newtonville, Massachusetts. [BMNH]

NOTE—Grote described *bostoniensis* from two specimens, a male (No. 1058) and a female (No. 1057), both collected at Newtonville on 11 September by R. Thaxter. The male specimen in the collection of the British Museum (Natural History) is here designated lectotype. The specimen is in good condition and has been dissected (Brit. Mus. slide No. 6349).

Unlike the next species, specimens of *bostoniensis* show little variation in wing color or maculation. The forewing is mouse gray with obscure maculation. The most prominent wing marking is a dark-gray median line. Worn or faded specimens tend to look gray brown. Forewing length varies from 17 to 22 mm. The male hindwing is dirty white with some smoky-brown shading on the wing margin. The female hindwing is gray brown.

Little is known of the immature stages; the larva has been described from material that was feeding on tobacco (Crumb, 1929: 89).

Euxoa bostoniensis occurs in dry areas from southern Ontario and Massachusetts southward to Virginia, Tennessee, and southern Missouri. Adults fly from early September until mid-October.

Euxoa (*Euxoa*) *medialis* (Smith)

PL. 2, FIGS. 34–39; PL. E, FIG. 4 (RWH 10813).

Agrotis medialis Smith, [1888], *Proc. U. S. Natl. Mus.,* **10**: 459.
Type locality: Texas. [USNM]

NOTE—The female lectotype was designated by Todd (1982: 133).

Carneades kerrvillei Smith, 1900, *Proc. U. S. Natl. Mus.,* **22**: 443.
Type locality: Kerrville, Texas. [USNM]

Euxoa poncha Smith, 1910, *Trans. Amer. Ent. Soc.,* **36**: 258.
Type locality: Colorado. [AMNH]

NOTE—The female lectotype was designated by Todd (1982: 172).

Euxoa truva Smith, 1910, *Trans. Amer. Ent. Soc.,* **36**: 259.
Type locality: Poncha Springs, Colorado. [AMNH]

NOTE—The female lectotype was designated by Todd (1982: 215).

Euxoa discilinea Dyar, 1918, *Proc. U. S. Natl. Mus.,* **54**: 342.
Type locality: Mexico City, Mexico. [USNM]

Euxoa medialis rufosuffusata McDunnough, 1940, *Can. Ent.,* **72**: 195.
Type locality: Maricopa, Arizona. [CNC]

This species is extremely variable, both locally and geographically, in terms of forewing color. Specimens from Arizona and California have a forewing mottled with gray and pinkish orange. One color tends to predominate so that the specimens appear to occur basically in two forms (plate 2, figures 34, 36). A smaller number of specimens are a mottled gray brown or pinkish white (plate 2, figure 38). A pale-gray or yellow-buff form predominates in arid regions in the Great Basin and the Great Plains (plate 2, figure 35). At lower elevations in the Rocky Mountain region from Alberta to Colorado, a form occurs in which the forewing is mottled with light and dark pink (plate 6, figure 37). In all forms the male hindwing is usually white with some gray-brown shading on the wing margin. The hindwing of the female is pale gray brown. Forewing length varies from 15 to 20 mm. *Euxoa medialis* is closely related to *bostoniensis*; they can be distinguished by wing color and by male genital characters given in the key.

The immature stages of *medialis* are unknown.

Euxoa medialis occurs from southern Manitoba and central Wisconsin, westward to southwestern Alberta and west-central California, and southward throughout western United States to south-central Mexico. Adults occur from early September until late October. From limited habitat data, it appears that *medialis* is predominantly a species of open arid lands and dry pine forests.

Euxoa (*Euxoa*) *bifasciata* (Smith)

PL. 2, FIGS. 40–42; PL. E, FIG. 2; PL. T, FIG. 4 (RWH 10796).

Agrotis bifasciata Smith [1888] *Proc. U. S. Natl. Mus.*, **10**: 460.
Type locality: Arizona. [USNM]

Euxoa bifasciata lowensis Benjamin (1936), *Bull. So. California Acad. Sci.*, **34**: 202.
Type locality: Mt. Lowe, California. [USNM]

Euxoa bifasciata bisagittifera Benjamin, (1936), *Bull. So. California Acad. Sci.*, **34**: 202.
Type locality: Glenwood Springs, Colorado. [USNM]

In external appearance, specimens of *bifasciata* are most likely to be confused with those of *infausta* (p. 104) but can be distinguished from them by the black, more contrasting antemedial and postmedial lines. The two species bear little resemblance in genital characters. The forewing length varies from 15 to 17 mm. Males can be distinguished from other similar species by the markedly bent, S-shaped harpe. Females can be recognized by the relatively small size of the appendix bursae.

The immature stages are unknown.

Euxoa bifasciata is widespread but rare. It occurs from western Montana, westward to north-central Washington, and southward to central Colorado, central Arizona, and southern California. Adults have been collected from early July until early September. Most specimens were collected in areas forested with pine and fir.

Euxoa (*Euxoa*) *perexcellens* (Grote)

PL. 2, FIGS. 43–46; PL. E, FIG. 5; PL. T, FIG. 6 (RWH 10816).

Agrotis excellens Grote, 1875, *Trans. Amer. Ent. Soc.*, **5**: 115.
Type locality: Vancouver Island, British Columbia. [BMNH]

NOTE—*Agrotis excellens* Grote, 1875, is preoccupied by *Agrotis excellens* Staudinger, 1867.

NOTE—*Agrotis excellens* was described from a male and a female from Vancouver Island, now in the British Museum (Natural History). The male specimen labeled: Vancouver Is., Grote coll., 81-116/5570/ *Agrotis excellens* male type/Noctuidae Brit. Mus. slide No. 6339/ is here designated lectotype.

Agrotis perexcellens Grote, 1875, *Can. Ent.*, **7**: 139.

NOTE—*Agrotis perexcellens* Grote is a replacement name for *Agrotis excellens* Grote, 1875.

Agrotis infelix Smith, 1890, *Trans. Amer. Ent. Soc.*, **17**: 57.
Type locality: California. [USNM]

NOTE—The male lectotype was designated by Todd (1982: 107).

Euxoa perexcellens differs from other species in the *bostoniensis* group in having a longitudinally streaked forewing. Specimens of *perexcellens* may be confused with those of *ochrogaster* (p. 112) but usually can be distinguished from them by the larger claviform spot and more extensive yellow-buff shading in the reniform and orbicular spots. Males of *perexcellens* can be distinguished from those of *ochrogaster* by the more massive sacculus and longer saccular extension; females can be distinguished by the bisaccate rather than unisaccate corpus bursae. The species is sexually dimorphic in that the forewings of most males are shaded with reddish brown while those of most females are shaded with silver gray. Forewing length varies from 17 to 21 mm.

The immature stages of *perexcellens* are poorly known; the larva is a surface-feeding cutworm; it occasionally is a serious pest of vegetable crops in low-lying areas of southern British Columbia (Beirne, 1971: 14).

Euxoa perexcellens occurs from southern British Columbia southward to southern California. A disjunct population occurs in northern Michigan. Adults occur from mid-July until early October. This species inhabits dry coniferous forests.

Euxoa (*Euxoa*) *sculptilis* (Harvey)

PL. 2, FIG. 47; PL. E, FIG. 6; PL. T, FIG. 7 (RWH 10815).

Agrotis sculptilis Harvey, 1875, *Bull. Buffalo Soc. Nat. Sci.*, **2**: 271.
Type locality: Texas. [BMNH]

NOTE—Harvey listed the type locality as Texas;

however, the nominal type in the British Museum (Natural History) is labeled New Mexico.

Carneades xyliniformis Smith, 1890, *Bull. U. S. Natl. Mus.*, **38**: 221.
Type locality: Las Vegas, New Mexico. [USNM]

Acronycta daedala Druce, 1889, *Biologia Centrali-Americana, Insecta. Lepidoptera Heterocera*, **1**: 260. NEW SYNONYMY.
Type locality: Quezaltenango, Guatemala. [BMNH]

Euxoa sculptilis cannot be confused with any other species. The basal half of the forewing is pale blue gray, the outer half is dark blue gray with the dark median line dividing these two areas. The maculation is defined by fine black lines. Forewing length varies from 15 to 20 mm.

The immature stages are unknown.

This is a southern species. It occurs from western Texas and central Arizona southward to Guatemala. Specimens have been collected from early August until late September.

Euxoa (*Euxoa*) *ustulata* Lafontaine

PL. 2, FIG. 48; PL. E, FIG. 7; PL. T, FIG. 8 (RWH 10814).

Euxoa ustulata Lafontaine, 1976, *Can. Ent.*, **108**: 10.
Type locality: Cedarville, California. [CNC]

Specimens of *ustulata* can be distinguished by their large size and by the burnt reddish-brown color of the forewing. The maculation tends to be obscure because of a dusting of black scales. The forewing length varies from 18 to 21 mm; females are usually larger than males. Unlike other species in the *bostoniensis* group, the saccular extensions in the male genitalia are asymmetrical; the left saccular extension is about ⅔ the length of the right.

This species was misidentified as *Euxoa biformata* for many years.

The immature stages are unknown.

Euxoa ustulata occurs from southeastern Washington southward to central-eastern California. It occurs in coniferous forests. Adults have been collected from early August until late September.

rufula GROUP

The *rufula* group includes only *rufula*; the characters are given under the species.

Euxoa (*Euxoa*) *rufula* (Smith)

PL. 2, FIGS. 49–51; PL. E, FIG. 8; PL. U, FIG. 1 (RWH 10750).

Agrotis rufula Smith, [1888], *Proc. U. S. Natl. Mus.*, **10**: 461.
Type locality: New Mexico. [USNM]

Agrotis basiflava Smith, 1890, *Trans. Amer. Ent. Soc.*, **17**: 52. SUBSPECIES.
Type locality: British Columbia. [USNM]
NOTE—The male lectotype of *basiflava* was designated by Todd (1982: 29).

Agrotis tessellata form *intrusa* Smith, 1890, *Trans. Amer. Ent. Soc.*, **17**: 54.
Type locality: Sierra Nevada, California. [USNM]
NOTE—The male lectotype of *intrusa* was designated by Todd (1982: 113).

Carneades compressipennis Smith, 1900, *Proc. U. S. Natl. Mus.*, **22**: 429.
Type locality: Yosemite, California. [USNM]
NOTE—The male lectotype of *compressipennis* was designated by Todd (1982: 53).

Specimens of *rufula* usually can be recognized by their small size and ruddy coloration. The forewing length varies from 12 to 16 mm. The basal and subterminal areas of the forewing are paler than the medial and terminal areas. The reniform and orbicular spots are outlined in black with a white line inside this. Males can be recognized by the shape and position of the saccular extensions and harpes, but most easily by the uncus. In *rufula* the uncus is very broad and dorsoventrally flattened. Like many *Euxoa* species, the setae near the apex are stout; but unlike all other species, those on the dorsal surface are stouter than the setae near the apex on the ventral surface. The species is similar to a number of others in female genital characters but can be distinguished from species with similar genitalia by wing markings. The corpus bursae is unisaccate and oval and the ovipositor lobes do not have flangelike projections.

The immature stages are unknown.

Euxoa rufula is a species that inhabits dry coniferous forests from southern Yukon southward in the Rocky Mountains to south-central New Mexico and eastern Arizona and southward in the Sierra Nevada Mountains to northeastern California. Adults have been collected from late July until early September.

Populations of *rufula* may be arranged in two subspecies.

Euxoa (*Euxoa*) *rufula rufula* (Smith)
PL. 2, FIG. 49.

Agrotis rufula Smith, [1888].

The nominate subspecies can be recognized by the red-orange coloration of the forewing, the lack of pale streaking, and lack of black, sagittate spots proximal to the subterminal line.

This subspecies occurs from southwestern Montana southward through Wyoming and Colorado to New Mexico and Arizona.

Euxoa (*Euxoa*) *rufula basiflava* (Smith)
PL. 2, FIGS. 50, 51; PL. E, FIG. 8; PL. U, FIG. 1.

Agrotis basiflava Smith, 1890.

Carneades intrusa Smith, 1890.

Carneades compressipennis Smith, 1900.

This subspecies differs from typical *rufula* in having a brownish-red or gray-red forewing; the cubital vein is pale in most specimens; and a series of black, sagittate spots is proximal to the subterminal line.

Subspecies *basiflava* occurs from Yukon southward to southern Alberta and northeastern California.

intrita GROUP

The *intrita* group includes only *intrita*; the characters are given under the species.

Euxoa (*Euxoa*) *intrita* (Morrison)
PL. 2, FIGS. 52–55; PL. E, FIG. 9; PL. U, FIG. 2; PL. EE, FIG. 1 (RWH 10749).

Agrotis intrita Morrison, 1874, *Proc. Boston Soc. Nat. Hist.*, **17**: 164.
Type locality: California. [type lost]

Agrotis strigilis Grote, 1876, *Bull. Buffalo Soc. Nat. Sci.*, **3**: 81.
Type locality: Vancouver Island. [BMNH]

Agrotis alticola Smith, 1890, *Trans. Amer. Ent. Soc.*, **17**: 51.
Type locality: Sierra Nevada, California. [AMNH]
NOTE—The female lectotype of *alticola* was designated by Todd (1982: 13).

Carneades titubatis Smith, 1894, *Trans. Amer. Ent. Soc.*, **21**: 51.
Type locality: Oregon. [USNM]
NOTE—The male lectotype of *titubatis* was designated by Todd (1982: 210).

Agrotis reuda Strecker, 1898, *Lepidoptera, Rhopaloceres and Heteroceres, Indigenous and Exotic,* Suppl., **1**: 6.
Type locality: Seattle, Washington. [FMNH]
NOTE—Strecker described *reuda* from seven specimens, five males and two females. Six of these specimens are in the Strecker collection at Chicago. One of these, a male labeled: "W.T./*Euxoa* slide Strecker n. 6" is here designated lectotype. The specimen is in good condition.

Euxoa intrita is extremely variable in wing color and markings. Specimens from the Great Plains region and the Rocky Mountain region of Alberta and Montana generally have a mottled-brown, or dark, blackish-brown forewing with a dusting of white scales giving the forewing a "frosty" appearance (plate 2, figures 52, 53). Specimens from British Columbia, Washington, and northern Oregon have less white "frosting" on the forewing, and the ground color of many specimens is dark reddish brown (plate 2, figure 54). Specimens from southern Oregon, California, Nevada, and Utah are reddish brown or orange brown (plate 2, figure 55) with obscure maculation. Specimens from southern Montana, Wyoming, and Colorado are a mixture of the more northern dark, maculate forms and the more western, pale, obscurely marked forms; most specimens are intermediate between these two extremes. Forewing length varies from 13 to 16 mm. The hindwing of the male is paler than that of the female. Specimens of *intrita* frequently can be recognized, especially with practice, by the stout, or stubby appearance of the moth created by the relatively broad forewings. Most males have a tuft of orange scales at the base of the forewing similar to that of *tessellata.* Males can be recognized by the disproportionately massive harpes and by the stout, asymmetrical saccular extensions. The female genitalia are similar to those of *rufula* but the corpus bursae is more elongate and slightly constricted mesially.

The immature stages are known only from laboratory reared material. The species is reported to overwinter in the egg and have a long larval aestivation (Hinks and Byers, 1976).

Euxoa intrita is widespread in forested areas of western North America. It occurs from western Ontario westward across central and southern Canada to the west coast. It occurs as far south as southern Manitoba and central Montana in the Great Plains and in the western mountains as far as northern New Mexico, south-central Utah, and south-central California. Adults occur from late July until mid-September.

annulipes GROUP

The *annulipes* group includes only *annulipes*; the characters are given under the species.

Euxoa (*Euxoa*) *annulipes* (Smith)
PL. 3, FIGS. 1–4; PL. E, FIG. 10; PL. U, FIG. 3; PL. EE, FIG. 2 (RWH 10747, 10748).

Agrotis annulipes Smith, 1890, *Trans. Amer. Ent. Soc.,* **17**: 48.
Type locality: Oregon. [USNM]

Agrotis annulipes oncocnemoides Barnes and Benjamin (1927), *Can. Ent.,* **58**: 304. SUBSPECIES.
Type locality: Mt. Lowe, California. [USNM]

Most adults of *annulipes* can be recognized by the combination of pale-gray or white forewing ground color and prominent, dark median line. Forewing length varies from 14 to 17 mm. They can be confused with some specimens of species in the *setonia* group because of similarity in wing markings; however, they can readily be recognized by genital characters. In the male genitalia, the sacculus has a prominent bulge on the dorsal margin giving it a somewhat oval appearance. The saccular extensions are stout and slightly shorter than the harpes. Both the harpes and saccular extensions are incurved so that they look like pincers. The most distinctive feature of the male genitalia is the shape of the uncus: swollen mesially, then constricted apically into a hook. It usually can be observed by brushing away scales from the end of the abdomen. The female genitalia resemble those of members of the *terrena* group in that each ovipositor lobe has a sclerotized rim; the base of the lobe has a dense border of several hundred setae; and the corpus bursae is unisaccate. They differ from those of females in the *terrena* group in that the sclerotized ring of abdominal segment VIII is narrow, about half as long as the anterior apophyses; in the *terrena* group abdominal segment VIII is about as long as the anterior apophyses.

The immature stages of *annulipes* are unknown.

This species occurs from western Colorado westward to southeastern Oregon and southwestern California. Adults have been collected from early June until mid-August. Populations of *annulipes* are arranged in two subspecies; these subspecies have been recognized as distinct species for many years but they differ only in wing color and even this is not diagnostic for all specimens.

Euxoa (*Euxoa*) *annulipes annulipes* (Smith)
PL. 3, FIGS. 1–3; PL. E, FIG. 10; PL. U, FIG. 3.

Agrotis annulipes Smith, 1890.

In subspecies *annulipes* the forewing ground color is pale gray and has a powdery look because of a dusting of black and white scales. The antemedial and postmedial lines may be obscure or prominent. The median line is dark gray and prominent. In most specimens there is dark shading on each side of the subterminal line. The reniform and orbicular spots are partially white filled with a dark-gray central shade. The spots may be partially outlined in black.

Subspecies *annulipes* is primarily an inhabitant of the Great Basin. It occurs from western Colorado westward to southern Oregon, eastern California, and central Arizona. Adults have been collected from early July until mid-August.

Euxoa (*Euxoa*) *annulipes oncocnemoides* (Barnes and Benjamin)
PL. 3, FIG. 4; PL. EE, FIG. 2.

Agrotis annulipes oncocnemoides Barnes and Benjamin, (1927).

Specimens of subspecies *oncocnemoides* differ from those of *annulipes* in the following points: the forewing ground color is creamy white rather than pale gray; the median line is darker and more sharply defined; the reniform and orbicular spots are white filled; and the orbicular spot is round in most specimens.

Subspecies *oncocnemoides* occurs in southwestern California where it has been recorded from Monterey, Los Angeles, and Riverside counties; most records are from the San Gabriel Mountains. Adults have been collected from early June until early August.

scholastica GROUP

The *scholastica* group includes only *scholastica*; the characters are given under the species.

Euxoa (*Euxoa*) *scholastica* McDunnough
PL. 3, FIG. 5; PL. F, FIG. 1; PL. U, FIG. 4; PL. EE, FIG. 3 (RWH 10793).

Euxoa scholastica McDunnough, 1920, *Can. Ent.*, **52**: 161.
Type locality: Meach Lake, Quebec. [CNC]

Specimens of *scholastica* are likely to be confused only with those of *messoria* (subgenus *Longivesica*) in eastern North America. They can be distinguished from *messoria* specimens by the dark brown rather than gray forewing, the smaller and round rather than oval orbicular spot, and the much darker hindwing. In most specimens the orbicular and reniform spots are filled with orange brown. The postmedial line is scalloped between the veins with a series of dark spots on the veins in the subterminal area. Forewing length varies from 15 to 18 mm. *Euxoa scholastica* can readily be recognized by genital characters. In the male genitalia the saccular extensions are long and almost straight. The sacculus is crescentic or slightly oval. The aedoeagus is narrowed apically into a long fingerlike projection that is free from the vesica at the apex. The female genitalia resemble those of the *terrena* group in that each ovipositor lobe has a sclerotized rim, and the corpus bursae is unisaccate. They differ from those of the *terrena* group in having only a small subbasal row of long setae on the ovipositor lobe. *Euxoa scholastica* is the only eastern North American species that has a sclerotized rim on the ovipositor lobe.

The immature stages are unknown.

Euxoa scholastica has a relatively small range. It occurs from Nova Scotia, westward across southern Quebec and Ontario to eastern North Dakota, and southward to central Pennsylvania, northern Illinois, and eastern Nebraska; its range extends farther southward in the Appalachian Mountains to Kentucky and North Carolina. Adults occur from late June until mid-August. The species occurs in dry deciduous forests.

terrena GROUP

This group includes four North American species that can readily be recognized by genital characters. In males the sacculus is triangular with a prominent angle in the dorsal margin; in most species the dorsal margin is irregular and lobed, or sinuate, anterior to this angle and straight posterior to it. The saccular extensions are stout and bladelike. The harpe is C-shaped and densely pubescent. The saccus is constricted mesially and extended anteriorly. The aedoeagus is bilobed. The vesica shape is unlike that found in any other species. The female genitalia resemble those of the *annulipes* group, but abdominal segment VIII is much longer. By examining the end of the abdomen, members of the *terrena* group can be distinguished without dissection from similarly colored members of other species groups by the distinctive harpe shape and by the characters of the female ovipositor lobes.

KEY TO SPECIES OF THE *TERRENA* GROUP

1. Forewing longitudinally streaked; transverse lines obsolescent . . . . *absona* p. 73
— Forewing not longitudinally streaked; transverse lines prominent and double 2
2. Male genitalia with saccular extensions longer than harpes; dorsal margin of sacculus with sinuate and lobed anterior portion about as long as straight-edged posterior portion; occurring north and west of the Great Basin region . . . . *terrena* this page
— Male genitalia with saccular extensions shorter than harpes; dorsal margin of sacculus with sinuate anterior portion about twice as long as straight-edged posterior portion; occurring in Great Basin and southern Rocky Mountain regions 3
3. Forewing brown dusted with gray scales; male genitalia with dorsal margin of sacculus lobed and sinuate (plate F, figure 3) . . . . . . *antica* p. 72
— Forewing gray heavily dusted with black; male genitalia with dorsal margin of sacculus evenly convex (plate F, figure 4) . . . . . . *franclemonti* p. 72

Euxoa (*Euxoa*) *terrena* (Smith)
PL. 3, FIGS. 6–8; PL. F, FIG. 2; PL. U, FIG. 5 (RWH 10742).

Carneades terrenus Smith, 1900, *Proc. U. S. Natl. Mus.*, **22**: 436.
Type locality: Pullman, Washington. [USNM]
NOTE—The female lectotype of *terrena* was designated by Todd (1982: 208).

Carneades lagganae Smith, 1900, *Proc. U. S. Natl. Mus.*, **22**: 439.

Type locality: Laggan [Lake Louise], Alberta. [USNM]

NOTE—Smith listed the type locality as "Laggan, British Columbia;" the locality, now Lake Louise, is in Alberta near the British Columbia border.

In addition to *antica,* specimens of this species are frequently confused with those of *simulata* and *comosa*; they can be distinguished from these latter species by the species-group characters listed above. Specimens of *terrena* cannot safely be distinguished from those of *antica* by external characters; in the narrow zone of overlap of the two species it is necessary to rely on genital characters. Both species have dark-brown forms in mesic habitats such as fir forests (plate 3, figure 6) and a paler form (plate 3, figure 7) in more xeric pine forests. Most specimens of *terrena* are darker in forewing color than are those of *antica.* Forewing length varies from 15 to 19 mm. In the male genitalia, *terrena* differs from *antica* in having longer, twisted saccular extensions; a differently shaped sacculus; a shorter, stouter harpe; and in the vesica: the subbasal diverticulum is somewhat lobed rather than footlike and resembles a rooster comb, and the apex of the vesica is not narrowed as in *antica* males. In the female genitalia of *terrena,* the corpus bursae is shorter and broader than that of *antica*; in *antica* the corpus bursae is constricted mesially, and the appendix bursae leading to the ductus seminalis is longer than that of *terrena.*

The immature stages are unknown.

Euxoa terrena occurs in coniferous forest habitats from southern British Columbia and southwestern Alberta southward to northern Utah, northern Nevada, and in the Sierra Nevada Mountains to the Transverse Ranges of southern California. Adults of *terrena* have been collected from early June until early August.

Euxoa (*Euxoa*) *antica* Lafontaine

PL. 3, FIGS. 9, 10; PL. F, FIG. 3; PL. U, FIG. 6; PL. EE, FIG. 4 (RWH 10743).

Euxoa antica Lafontaine, 1974, *Can. Ent.,* **106**: 651.

Type locality: Cimarron Canyon, Sangre de Cristo Mts., Colfax Co., New Mexico. [CNC]

Euxoa antica replaces *terrena* south and east of the range of that species. In the narrow zone of overlap, specimens must be identified by the genital characters given in the key and discussed under *terrena.* As for *terrena, antica* has both dark and pale color forms (plate 3, figures 9, 10). Forewing length is 16 to 19 mm.

The immature stages of *antica* are unknown.

Euxoa antica occurs from southern Montana southward through Wyoming and Colorado to northern New Mexico and central Arizona; it occurs westward through Nevada to Inyo County, California. Its range overlaps that of *terrena* along the northern and western limits of its range. Adults of *antica* have been collected from early June until mid-August.

Euxoa (*Euxoa*) *franclemonti* Lafontaine, NEW SPECIES

PL. 3, FIGS. 11, 12; PL. F, FIG. 4.

Euxoa (*Euxoa*) *franclemonti* Lafontaine.

Type locality: Walnut Canyon, 6,500′, 6⅓ mi EESE Flagstaff, Coconino Co., Arizona. [JGF]

This species is most likely to be confused with *antica* with which it occurs. Specimens of *franclemonti* can be distinguished from those of *antica* by the dark-gray rather than brown ground color of the forewings and by genital characters discussed below.

Antenna of male moderately biserrate and bifasciculate. Frontal tubercle prominent. Eye round. Vestiture of head and thorax a mixture of gray and black scales. Ground color of forewing dark gray, heavily dusted with black. Basal line double, incomplete. Antemedial line double, consisting of three outward arcs. Postmedial line double, basal portion toothed on veins, outer portion obscure. Median line black, wavy. Subterminal line pale, made evident only by darker shading proximal to it and in terminal space. Orbicular and reniform spots prominent, outlined in black. Basal and subterminal areas and orbicular and reniform spots gray, pale and contrasting because of charcoal shading in median and terminal areas. Forewing length 16 to 18 mm. Hindwing smoky brown in both sexes. Male genitalia similar to those of *antica* but differ in shape of sacculus: dorsal margin of the sacculus even, somewhat crescentic; in male of *antica* dorsal margin with posterior third straight, then bending through an angle of about 30° at prominent, rounded lobe; anterior ⅔ of sacculus with a series of small lobes and bulges. The female genitalia of *franclemonti* are similar to those of *terrena*; corpus bursae shorter and broader than that of *antica.*

TYPES. Holotype: ♂. Walnut Canyon, 6,500′, 6⅓ mi EESE Flagstaff, Coconino Co., Arizona; 5 July 1965; J. G. Franclemont. JGF. Paratypes: 16 ♂, 51 ♀. Arizona: Same location and collector as for holotype; 25 June to 20 July, 1965 (13 ♂, 29 ♀). Fort Valley, 7,350′, 7½ mi NW Flagstaff, Coconino Co.; 4–6 July 1964; J. G. Franclemont (1 ♂, 1 ♀); West Fork, 6,500′, 16 mi SW Flagstaff, Coconino Co.; July 1961, 3 July 1964, 3–21 July 1965; J. G. Franclemont (1 ♂, 11 ♀); Yavapai Co., 5 mi N Prescott, 5,450′; 4–21 June 1963, 9–16 May 1974; L. M. Martin (1 ♂, 6 ♀); Yavapai Co., Prescott; 15 May 1970; L. M. Martin (1 ♀). Colorado: Boulder, 5,400′; 20 July 1961; E. W. Rockburne (1 ♀). New Mexico: McGaffey, Zuni Mts., McKinley Co., 7,500′; 22 July 1962; E. & I. Munroe (1 ♀). Utah: 7 mi W Duchesne, 5,900′; 20 Aug. 1965; D. F. Hardwick (1 ♀). CNC, JGF, USNM. In addition to the type series, I have seen a male in poor condition from Crook County in northeastern Wyoming.

I take pleasure in naming the species after John G. Franclemont who collected most of the type series and has been a constant source of help.

Euxoa (*Euxoa*) *absona* Lafontaine, NEW SPECIES

PL. 3, FIGS. 13, 14; PL. F, FIG. 5.

Euxoa (*Euxoa*) *absona* Lafontaine.
Type locality: Angel Creek, 7,000′, E Humboldt Mts., SSW of Wells, Elko Co., Nevada. [USNM]

Specimens of *absona* can be distinguished from those of the other species in the *terrena* group by the silver-gray ground color and the longitudinally streaked forewing.

Antenna of male moderately biserrate and bifasciculate. Frontal tubercle small but prominent. Vestiture of head and thorax a mixture of light and dark-gray scales; tegulae outlined in black. Ground color of forewing silver gray with extensive pale-brown shading in central part of median area. Transverse lines obscure and incomplete; only posterior portion of antemedial line conspicuous and double. Terminal area dark gray brown, extending proximally into a series of streaks and wedges in subterminal area. A small basal dash and claviform spot present. Reniform and orbicular spots small, outlined in black with pale inner line; orbicular spot elongate. Median area with dark-brown shading proximal to orbicular spot, between orbicular and reniform spots and distal to reniform spot. Cubital vein pale and contrasting. Forewing length 15 to 18 mm. Hindwing smoky brown in both sexes. Male genitalia similar to those of *antica* but sacculus, saccular extension, and harpe disproportionately smaller. Female genitalia indistinguishable from those of *terrena*.

TYPES. Holotype: ♂. Angel Creek, 7,000′, E Humboldt Mts. SSW of Wells, Elko Co., Nevada; 18 July 1971; D. C. Ferguson. USNM. Paratypes: 7 ♀. California: Inyo Co., Westgard Pass, 7,200′; 26 July 1962; C. D. MacNeill, D. C. Rentz & M. R. Lundgren (1 ♀); Inyo Co., White Mts., 5 mi W Deep Springs, 7,000′; 28 June 1973; C. L. and J. N. Hogue (1 ♀); Siskiyou Co., Fowlers Camp, 5 mi E of McCloud; 22 July 1962; D. C. Rentz & C. D. MacNeill (1 ♀). Nevada: Clark Co.; 24–30 June; Barnes Collection (2 ♀); Lincoln Co., Mt. Irish; 19 June 1938; G. H. & J. L. Sperry (1 ♀); Oak Springs; 25 May 1940; G. E. Bohart (1 ♀). AMNH, CAS, CNC, LACM, USNM.

serricornis GROUP

The *serricornis* group contains three North American species. Members of the group usually can be distinguished from other spring-flying species by the dark-filled but pale-outlined orbicular and reniform spots. Species in this group can easily be distinguished from all other *Euxoa* species by genital characters. In males the sacculus is massive with the dorsal margin bulging mesially and convex. The saccular extensions are incurved and somewhat blade-like apically; they are asymmetrical in two species. The harpes are very long, C-shaped, and densely pubescent. The juxta is very large, shield shaped. Female genitalia are similar to those of species in the *terrena* and *annulipes* groups, but the corpus bursae is bisaccate rather than unisaccate. Adults fly in the spring.

KEY TO SPECIES OF THE *SERRICORNIS* GROUP

1. Forewing brownish gray, heavily dusted with pale gray; dorsal margin of sacculus of male genitalia evenly convex (plate F, figure 6) . . . . *scotogrammoides* p. 74
— Forewing orange brown or yellow brown; dorsal margin of succulus with prominent, rounded, central bulge (plate F, figure 7) 2

2. Saccular extensions long, symmetrical, both curved dorsally around ventral angle of cucullus; hindwing smoky brown, basal ⅔ paler in southern California specimens . . . . *tocoyae* p. 74
— Left saccular extension shorter and straighter than right, barely reaching ventral angle of left cucullus in most specimens; hindwing dirty white with pale smoky-brown shading on marginal third of wing . . . . *serricornis* p. 74

Euxoa (*Euxoa*) *serricornis* (Smith)

PL. 3, FIGS. 15, 16; PL. F, FIG. 7; PL. U, FIG. 8; PL. EE, FIG. 5 (RWH 10744).

Agrotis serricornis Smith, [1888], *Proc. U. S. Natl. Mus.*, **10**: 458.
Type locality: Southern California. [MSU]

Carneades itodes Smith, 1900, *Proc. U. S. Natl. Mus.*, **22**: 438.
Type locality: Arizona. [USNM]
NOTE—The female lectotype of *itodes* was designated by Todd (1982: 115).

Euxoa epictata Smith, 1907, *Trans. Amer. Ent. Soc.*, **33**: 130.
Type locality: Southern Arizona. [AMNH]
NOTE—The female lectotype of *epictata* was designated by Todd (1982: 75).

Specimens of *serricornis* can be distinguished from those of other species in the group by their overall paler coloration. Basically two forms occur; the forewing ground color may be either a mottled reddish orange (plate 3, figure 15) or a pale gray buff or yellow buff (plate 3, figure 16). The reddish-orange form predominates in Arizona while in California, the gray-buff form is the more common. The forewing length is 14 to 18 mm.

The immature stages are unknown.

This species inhabits desert regions of southern California, southernmost Nevada and the southern half of Arizona. Adults fly very early, they have been collected from mid-February until early June.

Euxoa (*Euxoa*) *tocoyae* (Smith)

PL. 3, FIGS. 17, 18; PL. F, FIG. 8 (RWH 10745).

Carneades tocoyae Smith, 1900, *Proc. U. S. Natl. Mus.*, **22**: 433.
Type locality: "High Sierras," California. [USNM]
NOTE—Smith stated that the male holotype was collected near North Dome in Yosemite National Park.

Euxoa serricornis sonoma McDunnough, 1941, *Can. Ent.*, **73**: 66.
Type locality: Pine Flat, Sonoma Co., California. [CNC]

This species is like a dark version of *serricornis*, and the two species were confused for many years. Like that species, *tocoyae* has essentially two color forms, a grayish-brown form (plate 3, figure 17) and a reddish-orange form (plate 3, figure 18). The first form predominates in the northern portion of the range while the latter form predominates in the South. Also, the hindwing is darker in specimens from central and northern California than in those from southern California. Males of *tocoyae* can be distinguished from those of *serricornis* by the genital characters given in the key. The symmetrical saccular extensions can be observed by brushing away scales from the end of the abdomen. The species also differs from *serricornis* in details of shape of the vesica. Females of *tocoyae* must be identified by association with males or by their darker coloration. Forewing length is 16 to 17 mm.

The immature stages are unknown.

This species occurs from south-central Washington southward throughout most of California, at least as far as San Diego County. The range of *tocoyae* overlaps that of *serricornis* in southern California. The two species are seldom collected together, however, because *tocoyae* occurs in higher, more heavily forested habitats than does *serricornis*. Adults have been collected from mid-April until late June. The earlier collection dates are at coastal localities while the later dates are from montane localities.

Euxoa (*Euxoa*) *scotogrammoides* McDunnough

PL. 3, FIGS. 19, 20; PL. F, FIG. 6; PL. U, FIG. 7 (RWH 10746).

Euxoa scrotogrammoides McDunnough, 1932, *Can. Ent.*, **64**: 233.
Type locality: Jefferson Co., Montana. [CNC]

This is the most nondescript species of the group. The forewing is gray brown with a dusting of white scales. Forewing length varies from 16 to 17 mm. The maculation is barely traceable in most specimens. Specimens of *scotogrammoides* resemble those of a number of species of other species groups but can readily be recognized by the early flight date. The male genitalia resemble those of *serricornis* males in that the saccular extensions are asymmetrical, but they can be distinguished by the sacculus shape and by the differently shaped vesica. The female genitalia are similar to those of the other two species in the group.

The immature stages are unknown.

This species occupies a range that lies north and east of those of the other two species. It occurs from south-central British Columbia southward to the Wallowa Mountains of Oregon and Modoc County, California, and southeastward through western

Montana to northern Utah and western Colorado. From limited habitat data, it appears that *scotogrammoides* inhabits dry, open coniferous forests. Adults fly from mid-May until early July.

pluralis GROUP

The *pluralis* group includes three North American species that are dissimilar in external appearance but similar in genital characters. In the male genitalia the saccular extensions are slightly longer than the harpes. In the vesica the subbasal diverticulum is bilobed; the median diverticulum is almost half way to the apex from the subbasal bend; the vesica projects dorsally. In the female genitalia the corpus bursae is bisaccate; the ovipositor lobes are broad and rounded and clothed with short, conical setae apically.

KEY TO SPECIES OF THE *PLURALIS* GROUP

1. Forewing brown or orange brown, without contrasting pale areas or longitudinal streaking .... *permixta* p. 76
— Forewing with pale-orange shading and longitudinal streaking 2
2. Forewing with transverse lines obscure or absent, longitudinal pale streaks in fold and distal to reniform spot .... *pluralis* this page
— Forewing with transverse lines prominent, basal and subterminal areas paler orange than median area .... *cinnabarina* this page

Euxoa (*Euxoa*) *pluralis* (Grote)

PL. 3, FIGS. 21, 22; PL. F, FIG. 9; PL. V, FIG. 1; PL. EE, FIG. 6 (RWH 10795).

Agrotis pluralis Grote, 1878, *Bull. Geol. Surv.*, **4**: 174.

Type locality: Nevada. [BMNH]

NOTE—Only one of the two original specimens still exists in the British Museum (Natural History). This specimen, a female in good condition, is labeled "Nevada, Grote Coll., 81–116/*Agrotis pluralis* Grote, Type/Noctuidae, Brit. Mus. slide No. 6311." The specimen is here designated lectotype.

Specimens of *pluralis* can be distinguished from those of other species of *Euxoa* by the combination of pale-orange shading in the fold and distal to the reniform spot of the forewing and pale silver-gray shading along the costal, terminal, and posterior margins of the wing and in the reniform and orbicular spots. Forewing length varies from 14 to 18 mm.

The immature stages are unknown.

This species occurs in dry coniferous forest habitats from southern British Columbia southward to northern New Mexico, central Utah, and southern California. Adults occur from mid-March until mid-August. Early collecting dates are from southwestern California; the latest dates are from high elevations (8,000–10,000′) in the Sierra Nevada Mountains.

Euxoa (*Euxoa*) *cinnabarina* Barnes and McDunnough

PL. 3, FIG. 23; PL. F, FIG. 10; PL. V, FIG. 2 (RWH 10797).

Euxoa cinnabarina Barnes and McDunnough, 1918, *Contrib. Nat. Hist. Lep. N. Am.*, **4** (2): 92.

Type locality: Monachee [sic] Meadows, Tulare Co., California. [USNM]

NOTE—This species was described from three males from Tulare Co., California and one female from San Diego Co., California. One male is in the Canadian National Collection, Ottawa; the other three specimens are in the United States National Museum. A male specimen in the United States National Museum labeled "Aug 8–15/ Monachee Mdws., Tulare Co. 8000 ft, Calif./*Euxoa cinnabarina* B. & McD., Type male/male genitalia on slide Apr. 1966, ELT 2301" is here designated lectotype. The specimen is in good condition.

Specimens of *cinnabarina* can be distinguished from those of other *Euxoa* species by the bright-orange color of the forewing, which is divided into pale-orange basal and subterminal areas and a darker orange median area, and by the white shading on the cubital vein and inside the black outline to the reniform and orbicular spots. Forewing length is 15 to 18 mm. This species has been collected in two areas in California. Specimens from southwestern California (Los Angeles to San Diego counties) differ from typical *cinnabarina* from the southern Sierra Nevada Mountains (Tulare County) in having a series of black, sagittate spots in the subterminal area proximal to the subterminal line.

The immature stages are unknown.

Specimens of *cinnabarina* have been collected from mid-July until early August.

Euxoa (*Euxoa*) *permixta* McDunnough
PL. 3, FIGS. 24, 25; PL. G, FIG. 1; PL. V, FIG. 3 (RWH 10799).

Euxoa brunneigera ab. *coloradensis* Strand, [1916], *Archiv für Naturgeschichte,* **81A12**: 144.
Type locality: Colorado. [BMNH]

NOTE—*Euxoa brunneigera* ab. *coloradensis* Strand is an unavailable, infrasubspecific name.

Euxoa permixta McDunnough, 1940, *Can. Ent.,* **72**: 193
Type locality: Payson Canyon, Utah. [CNC]

Specimens of *permixta* can easily be confused with those of *brunneigera excogita* (p. 107). Both species have a prominent black transverse line on the prothoracic collar and evenly colored orange-brown forewings with obscure maculation. Specimens of *permixta* differ from those of *brunneigera* in having the reniform and orbicular spots outlined in black and concolorous with the remainder of the forewing except for a small, dark spot at the lower end of the reniform spot. The orbicular spot is small and round with a width less than or equal to that of the space between it and the reniform spot. In specimens of *brunneigera excogita* the reniform spot has an irregular black and cream-colored outline and the spot is partially gray filled. The orbicular spot is oval in most specimens and is wider than the width of the space between it and the reniform spot. Males of *permixta* can be distinguished from those of *brunneigera excogita* by the stouter saccular extensions, larger sacculus, and by vesica shape; females of *permixta* can be recognized by the bisaccate rather than unisaccate corpus bursae. In *permixta* forewing length varies from 15 to 17 mm.

The immature stages of *permixta* are unknown.

This species is rare in collections; I have been able to locate less than twenty specimens. It occurs from east-central Oregon and central Washington southward to central New Mexico, southern Utah, and central western Nevada. To the east of the Great Basin, it has been collected only on the Niobrara Escarpment of western Nebraska. On the basis of a limited amount of habitat data, it appears that *permixta* occurs in dry scrubby areas such as piñon-juniper scrub, willow and cottonwood thickets and open pine forests. Adults have been collected from late June until late August.

setonia GROUP

The *setonia* group includes three western North American species. Members of the group can be distinguished from those of most other species-groups by the combination of prominent median line on the forewing and pale reniform and orbicular spots. They may be confused with members of the *annulipes* group (p. 70) but can be distinguished by the pale yellow-brown, or orange-brown, forewing ground color rather than pale gray or white, as in species in the *annulipes* group. In the male genitalia, the right sacculus, saccular extension, and harpe are all similar in length; the harpe is slightly pubescent; the aedoeagus has a rounded, fingerlike apex. The female genitalia are similar to those of members of the *pluralis* group.

KEY TO SPECIES OF THE *SETONIA* GROUP

1. Forewing with median line more prominent than dark shading adjacent to subterminal line in most specimens; species of woodland habitats north and east of Great Basin; male genitalia with saccular extensions directed posteroventrally, bending away from cucullus near apex; subbasal diverticulum of vesica bilobed with lobes projected both anteriorly and posteriorly (plate G, figure 2) .... *setonia* this page

– Forewing with dark shading adjacent to subterminal line as prominent as, or more prominent than, median line in most specimens; desert species of Colorado Plateau, Great Basin, and southern California; male genitalia with saccular extensions directed posteriorly at same angle as longitudinal axis of sacculus, straight or bending toward cucullus near apex; subbasal diverticulum of vesica footlike, projected posteriorly (plate G, figures 3, 4) 2

2. Forewing with reniform spot pale filled; no dark patch in posterior portion of median area extending between transverse lines; male genitalia with right saccular extensions markedly longer than right harpe; occurring in Great Basin and Colorado Plateau .... *pallidimacula* p. 77

– Forewing with dark bar through middle of reniform spot; posterior portion of median area dark filled; male genitalia with right saccular extension about as long as right harpe; occurring in Mojave Desert .... *mojave* p. 77

Euxoa (*Euxoa*) *setonia* McDunnough
PL. 3, FIGS. 26–28; PL. G, FIG. 2; PL. V, FIG. 4; PL. EE, FIG. 7 (RWH 10794).

Euxoa setonia McDunnough, 1927, *Can. Ent.*, **59**: 194.
Type locality: Seton Lake, British Columbia. [CNC]

This is the most common and widely distributed species in the group. Specimens from the southern Rocky Mountain region differ from those from more northerly localities in having the forewing ground color orange brown (plate 3, figure 28) rather than pale yellow brown or buff brown (plate 3, figures 26, 27). Forewing length is 15 to 18 mm. Males can be distinguished from those of other similar species of *Euxoa* by the combination of the saccular extensions curving away from the cucullus and bilobed subbasal diverticulum of the vesica. Females can be recognized by the bisaccate corpus bursae and the broad, rounded ovipositor lobes with short, conical setae apically.

The immature stages are unknown.

Euxoa setonia occurs from central Alberta and southern British Columbia southward to southern Washington, southern Idaho, central Arizona, and central New Mexico; it also occurs in the Black Hills of South Dakota and adjacent Wyoming. Adults have been collected from late May until late July. In most of its range *setonia* occurs in dry, coniferous forests; in some localities in the northern half of its range it also occurs in groves of aspen and poplar.

Euxoa (*Euxoa*) *pallidimacula* Lafontaine, NEW SPECIES

PL. 3, FIGS. 29, 30; PL. G, FIG. 3; PL. V, FIG. 5.

Euxoa (*Euxoa*) *pallidimacula* Lafontaine.
Type locality: Eureka, Utah. [CNC]

Specimens of *pallidimacula* look like pale, diffusely marked specimens of *setonia.* Males can be distinguished from those of *setonia* by the longer, straighter saccular extensions and by the unilobed rather than bilobed subbasal diverticulum of the vesica. Females cannot safely be distinguished from those of *setonia,* but the corpus bursae is less distinctly bisaccate.

Antenna of male biserrate and bifasciculate. Frontal tubercle present. Eye round and globular. Vestiture of head and thorax buffy brown. Ground color of forewing pale buffy brown with a dusting of dark-brown scales. Basal line barely evident. Antemedial and postmedial lines obscure, double, pale filled. Median line prominent, dark gray brown, more diffuse than that in *setonia.* Terminal area dark gray brown, streaked into outer portion of subterminal area; this dark shading as prominent as median line. Subterminal line absent or evident as a pale sinuous line in dark shading along outer portion of wing. Orbicular and reniform spots obscure, partially outlined in black; spots paler than remainder of forewing because of absence of dusting of darker scales. Length of forewing 15 to 18 mm. Hindwing dirty white in basal half, pale smoky brown in outer half. Fringe white with pale-brown subbasal line. Male genitalia similar to those of *setonia* males but saccular extensions longer, projected posteriorly at same angle as sacculus, not bending ventrally away from cucullus as in *setonia*; apex of saccular extensions upcurved toward cucullus in many specimens. Vesica with subbasal diverticulum unilobed, footlike, projected posteriorly, not bilobed as in *setonia.* Female genitalia similar to those of *setonia* but corpus bursae less distinctly bisaccate than in most specimens of *setonia.*

TYPES. Holotype: ♂, Eureka, Utah; 17 July 1911; Tom Spalding; slide No. 6443; CNC (type No. 19512). Paratypes: 20 ♂, 13 ♀. Colorado: Glenwood Springs; 8 June–30 July, 7–10 July 1894; W. Barnes (2 ♂, 5 ♀); Uncompahgre Plateau, 16 mi SW Montrose, Montrose Co., 7,800–8,100′; 25–27 June 1957; F. and P. Rindge (4 ♂). Nevada: Angel Creek, 7,000′, E Humboldt Mts., SSW of Wells, Elko Co.; 15 July 1971; D. C. Ferguson (1 ♂); Indian Creek, Esmeralda Co., 7,700′; 20 July 1966; R. C. Bechtel and P. C. Martinelli (1 ♀). Utah: Dividend; 19 June, 4 July; Tom Spalding (1 ♂, 1 ♀); Eureka; 6–28 July 1911, 1 July 1921; Tom Spalding (6 ♂, 3 ♀); Loop Camp, 13 mi SW Grantsville, Tooele Co., 7,400′; 3–5 July 1960; F., P. and B. Rindge (2 ♂, 1 ♀); Mayfield, 6 mi E, 6,500′; 25 Aug. 1965; D. F. Hardwick (1 ♂); Stockton; 5–12 July 1907 and 14 July 1915 (3 ♂, 2 ♀). AMNH, CNC, USNM.

Euxoa (*Euxoa*) *mojave* Lafontaine, NEW SPECIES

PL. 3, FIGS. 31, 32; PL. G, FIG. 4; PL. V, FIG. 6.

Euxoa (*Euxoa*) *mojave* Lafontaine.
Type locality: Apple Valley, California. [CNC]

Specimens of *mojave* can be distinguished by the mottled yellow-brown and dark-brown color of the forewing, the presence of a transverse dark bar through the middle of the reniform spot, and by the dark shading in the posterior portion of the median area. Males can be distinguished from those of *pallidimacula* by the shorter saccular extensions. Females can be distinguished by the disproportionately large corpus bursae.

Antenna of male biserrate and bifasciculate. Frontal tubercle present. Eye round and globular. Vestiture of head and thorax yellow buff. Ground color of forewing yellow buff dusted with brown. Basal line dark brown, incomplete. Antemedial line double, inner line obscure, outer line prominent, space between lines pale. Postmedial line double, outer element obscure, space between lines pale. Median line irregular, blotchy. Subterminal line pale, made evident by dark-brown streaks in subterminal area proximal to it and gray-brown shading in terminal area. Fringe yellow buff. Reniform and orbicular spots pale, barely visible. Median line forming a brown blotch between reniform and orbicular spots, this extended as a narrow bar through middle of reniform spot to postmedial line. Posterior portion of median area, between transverse lines and between anal vein and edge of wing dark brown. Forewing length 15 to 18 mm. Hindwing pale smoky brown; fringe white. Male genitalia similar to those of *pallidimacula* but: saccular extensions shorter; saccular extensions about as long as, or shorter than, harpes, and vesica with median diverticulum positioned farther toward apex of vesica. Female genitalia similar to those of *pallidimacula* but ductus bursae and sclerotized plates in ductus bursae disproportionately longer and corpus bursae larger.

TYPES. Holotype: ♂. Apple Valley, California; 22 May 1955; W. R. Richards; CNC (type No. 19513). Paratypes: 8 ♂, 4 ♀. California: same locality as for holotype; 13–26 May 1955; W. R. Richards and D. F. Hardwick (7 ♂, 2 ♀); Victorville; 9 May 1936; G. H. and J. L. Sperry (1 ♀); Wildrose Canyon, Panamint Mts.; 29 May 1960; D. S. Verity (1 ♂). Nevada: Clark Co.; 16–23 May (1 ♀). AMNH, CNC, LACM, USNM.

declarata GROUP

The *declarata* group includes four North American species. It can most easily be characterized by the male genitalia. In them, the sacculus is tear-drop shaped, becoming increasingly broad toward the anterior end; the dorsal margin of the sacculus is convex. Both the harpe and saccular extensions are incurved mesially so that they resemble pincers. *Euxoa flavidens* can be distinguished from all other *Euxoa* species by its striking forewing pattern. The other three members of the group frequently are confused with *albipennis* (p. 90) and *lillooet* (p. 116). Males of *albipennis* usually can be distinguished from those of members of the *declarata* group by the white rather than brown hindwing. Specimens of *lillooet* can be recognized by the pale-violet shades in the basal and subterminal areas and in the reniform and orbicular spots. The saccular extensions and harpes are not incurved but are straight or curve away from each other apically in *albipennis* and *lillooet* males. Females of *albipennis* can be distinguished from those of members of the *declarata* group by the unisaccate rather than bisaccate corpus bursae and by the presence of an apical flangelike process on each ovipositor lobe. Females of *lillooet* can be recognized by the disproportionately long corpus bursae (plate Z, figure 8) as compared with those of females in the *declarata* group.

Three species in the group (*declarata, campestris,* and *rockburnei*) have been studied in terms of interfertility and mating behavior (Byers and Hinks, 1978; Byers et al., 1981; Teal and Byers, 1980; Teal et al., 1978), hemocyte morphology (Arnold and Hinks, 1975; Arnold, 1976), thoracic muscle esterases (Hudson and Jui, 1976), egg morphology (Salkeld, 1977), and pheromone structure (Underhill et al., 1981). Although the species differ slightly in many of these aspects, they use essentially the same pheromone and are cross-attractive and completely interfertile. Reproductive isolation in these three species results from differences in female calling time during the night.

A synopsis of the *declarata* group and two similar species (*albipennis* and *lillooet*) was given by Hardwick (1973). A formula for identifying males of species in the group based on a discriminant analysis of genital characters was given by Hardwick and Lefkovitch (1973).

KEY TO SPECIES OF THE *DECLARATA* GROUP

1. Forewing longitudinally streaked; male genitalia with harpe smooth .... *flavidens* p. 80
– Forewing not longitudinally streaked; male genitalia with harpe pubescent 2

2. Male genitalia with sacculus of normal proportions, right sacculus shorter than cucullus and markedly shorter than saccular extensions; right saccular extension and right harpe forming a relatively narrow U, their bases meeting at angle of about 90° (plate G, figure 5) .. *declarata* p. 79
– Male genitalia with sacculus massive, right sacculus similar in length to, or longer than, right cucullus and right saccular extension; right saccular extension and right harpe forming a broad

U, their bases meeting at angle of about 120° (plate G, figure 6) 3

3. Right sacculus similar in length to right saccular extension; right saccular extension markedly longer than right harpe; forewing length 14 to 16 mm; occurring from British Columbia, Utah, eastern Arizona eastward .... *campestris* this page

– Right sacculus longer than right saccular extension; right saccular extension slightly longer than right harpe; forewing length 16 to 19 mm; occurring from southern British Columbia, western Montana, Utah, New Mexico westward .... *rockburnei* p. 80

Euxoa (*Euxoa*) *declarata* (Walker)
PL. 3, FIGS. 33–36; PL. G, FIG. 5; PL. V, FIG. 7; PL. EE, FIG. 8 (RWH 10755).

Mamestra declarata Walker, 1865, *List of the Specimens of Lepidopterous Insects in the Collection of the British Museum,* **32**: 663.
Type locality: Vancouver Island, British Columbia. [BMNH]

NOTE—The two original specimens of *declarata* are in the British Museum (Natural History). The male specimen, labeled "Type/Vancouver Is., Hawkins, 60–13/*Mamestra declarata*/1954–76/Male/Noctuidae, Brit. Mus. slide No. 6365" is here designated lectotype. The specimen is in good condition.

Agrotis decolor Morrison, 1874, *Proc. Boston Soc. Nat. Hist.,* **17**: 162.
Type locality: New York, Massachusetts, Maine. [lost]

NOTE—The type material of *decolor* is apparently lost. A full discussion of the above synonymy was given by McDunnough (1950: 385–386).

Agrotis spectanda Smith, 1890, *Trans. Amer. Ent. Soc.,* **17**: 54.
Type locality: California. [USNM]

Euxoa declarata californica Hardwick, 1973, *Can. Ent.,* **105**: 495. NEW SYNONYMY.
Type locality: Willow Ranch, California, 8 mi NE, 5,600′. [CNC]

This is the most widespread, common, and variable species in the group. Forewing ground color varies from pale violet gray to dark brown. The hindwing varies from dirty white with pale smoky-brown shading on the margin to almost entirely smoky brown. In general, ground color is correlated with habitat. Brown forms occur in forested habitats in eastern North America, western Canada, and in montane areas of western United States. Pale-gray and buff-colored forms occur in open, arid areas throughout the Great Plains and Great Basin. These forms may be highly localized. It is possible to collect in the prairies near an aspen grove and find both pale and dark forms. In most areas, specimens of *declarata* can be distinguished from those of *campestris* by the paler reddish-brown forewing color and paler hindwings. In areas where the two species occur together, *declarata* flies slightly later in the season and tends to be larger. Forewing length varies from 14 to 18 mm. In moist, forested habitats where dark specimens of *declarata* occur, it is necessary to use the genital characters given in the key.

The immature stages are known only from laboratory reared material. The larva has a longer period of aestivation than does that of *campestris* (Hinks and Byers, 1976: 1352).

Euxoa declarata occurs from the Atlantic Provinces of Canada westward to central Alaska and southward to southern Ontario and southern Minnesota in the East and southern New Mexico, Arizona, and California in the West. It has also been collected in the Appalachian Mountains in North Carolina. Specimens from northern California are smaller and darker than those from other areas; they have been described as subspecies *californica.* Adults of *declarata* have been collected from mid-July until late September. Unlike *campestris* and *rockburnei, declarata* occurs in a wide variety of habitats from arid deserts of the Great Basin and Great Plains to dense coniferous forests of the western mountains.

Euxoa (*Euxoa*) *campestris* (Grote)
PL. 3, FIGS. 37, 38; PL. G, FIG. 6 (RWH 10756).

Agrotis campestris Grote, 1875, *Can. Ent.,* **7**: 188.
Type locality: Orillia, Ontario. [BMNH]

NOTE—Grote described this species primarily on the basis of a female specimen from Orillia but also mentions specimens from New York (in BMNH) and Vancouver Island (in AMNH) in the description. McDunnough (1950: 386) considered the Orillia specimen as "the type" and stated his reasons for this. To avoid any possible confusion, I here designated the Orillia specimen lectotype. It is labeled "Type, H.T./Canada, Norman, 75–94/*Agrotis campestris* Grote, Type/1954–74/Female/Noctuidae, Brit. Mus. slide No. 6363."

Euxoa campestris is primarily a species of the boreal forest and Rocky Mountain regions. Specimens usually can be distinguished from those of *declarata* by the dark blackish-brown ground color of the forewings and smaller size. Forewing length varies from 14 to 16 mm. The hindwing is smoky brown in both sexes. Doubtful specimens of *campestris* must be identified by male genital characters given in the key.

The immature stages are known only from laboratory reared material; the larva has little, or no, summer aestivation (Hinks and Byers, 1976: 1352).

Euxoa campestris occurs from Newfoundland westward to Alaska and southward to northern New England and southern Canada from southern Quebec westward to southern British Columbia. In the West it occurs southward in the Rocky Mountain region to southern New Mexico, east-central Arizona, and central Utah. In the East, a disjunct population occurs in the mountains of eastern Kentucky. In any particular area, *campestris* tends to fly about ten days earlier than *declarata.* Adults have been collected from mid-July until mid-September.

Euxoa (*Euxoa*) *rockburnei* Hardwick

PL. 3, FIGS. 39, 40; PL. G, FIG. 7; PL. V, FIG. 8 (RWH 10757).

Euxoa rockburnei Hardwick, 1973, *Can. Ent.*, **105**: 497.
Type locality: Truckee, California. [CNC]

Euxoa rockburnei replaces *campestris* in the western United States. Most specimens can be recognized by their large size and evenly colored wings. Forewing length varies from 16 to 19 mm. The hindwing is smoky brown, paler toward the base. In populations of *declarata* and *campestris* in which the hindwings are this color, the forewings are darker and less evenly colored. Adults of *rockburnei* can be recognized by both male and female genital characters. Males can be recognized by characters given in the key. The male genitalia can be confused only with those of *campestris,* which occurs with *rockburnei* only in the extreme northern and eastern edges of the range of *rockburnei.* Females of *rockburnei* can be distinguished from those of the other two species by the sclerotized plate in the ventral wall of the ductus bursae. In *rockburnei* the margins of this plate are rough and irregular and bulge prominently to the left; in the other species of the *declarata* group the plate is even and symmetrical.

The immature stages are known only from laboratory reared material (Hinks and Byers, 1976).

Euxoa rockburnei occurs from south-central British Columbia southward to northern New Mexico, southern Utah, and east-central California. Adults have been collected from late July until mid-October. Like *campestris, rockburnei* is found in coniferous forest habitats.

Euxoa (*Euxoa*) *flavidens* (Smith)

PL. 3, FIGS. 41, 42; PL. G, FIG. 8; PL. W, FIG. 1 (RWH 10758).

Agrotis flavidens Smith, [1888], *Proc. U. S. Natl. Mus.,* **10**: 455.
Type locality: Colorado. [USNM]

NOTE—The female lectotype was designated by Todd (1982: 87).

Mesembreuxoa intricata Dyar, 1920, *Ins. Insc. Mens.,* **8**: 188.
Type locality: Mexico City, Mexico. [USNM]

Euxoa flavidens is not likely to be confused with any other *Euxoa* species. Adults can be recognized by the streaked pattern of the forewing, the pale yellow-buff color of the costa and distal to the claviform spot, and the very large claviform spot. The forewing length varies from 14 to 20 mm. The male genitalia are intermediate in shape between those of *declarata* and *campestris*; but the saccular extension is very stout; the harpe lacks pubescence; and the vesica is very strongly bent basally. The female genitalia are similar to those of *declarata* and *campestris.*

The immature stages are unknown.

This is a species of the southwestern mountains. It occurs from central Colorado and northern Utah southward through New Mexico and Arizona to south-central Mexico. Adults occur from late July until late September.

silens GROUP

The *silens* group contains three North American species that can most easily be recognized by the genitalia. In the male genitalia the saccular extensions are asymmetrical; the right saccular extension is long and incurved apically with the apical portion flattened and bladelike; the left saccular extension is shorter, straighter and more cylindrical throughout. The shape of the saccular extensions is similar to those of species in the *serricornis* group but unlike those species, the sacculus is crescentic rather than elliptical; the harpe is shorter and with only sparse pubescence; the juxta is smaller; and the vesica shape is different. Females can be distinguished from those of other similar species by the broad, somewhat truncate ovipositor lobes clothed with short setae and by the shape of the corpus bursae.

All three species in the group have basically two forms; one with a black, basal dash and black around the reniform and orbicular spots and the second without these dark areas. In *silens,* the pale form is very rare; in *pimensis* the two forms are equally common; in *immixta* the form with a basal dash is very rare.

Adults fly mainly in the spring and early summer.

KEY TO SPECIES OF THE *SILENS* GROUP

1. Forewing with a series of small black sagittate spots proximal to subterminal line, forewing longitudinally streaked with pale silver gray and dark gray in most specimens (plate 3, figures 43, 44) .... *silens* this page
– Forewing without dark sagittate spots proximal to subterminal line, forewing not longitudinally streaked 2

2. Forewing predominantly gray, hindwing white with pale smoky-brown shading on veins and wing margin; male vesica with subbasal diverticulum as long as width of vesica; occurring from central Colorado and western Texas westward .... *pimensis* this page
– Forewing predominantly brown; hindwing smoky brown, paler toward base; male vesica with subbasal diverticulum about half as long as width of vesica; occurring from western Texas and central Great Plains eastward ..... *immixta* this page

Euxoa (*Euxoa*) *silens* (Grote)

PL. 3, FIGS. 43, 44; PL. G, FIG. 9; PL. W, FIG. 2; PL. EE, FIG. 9 (RWH 10751).

Agrotis silens Grote, 1875, *Can. Ent.,* **7**: 67.
Type locality: Nevada. [BMNH]

Specimens of *silens* usually can be recognized by the streaked forewing (plate 3, figure 43). The rare unstreaked form (plate 3, figure 44) can be recognized by the small, black, sagittate spots in the subterminal area and the silver-gray shading of the forewing. Forewing length varies from 16 to 19 mm. The hindwing is darker than that of most *pimensis* specimens. The male genitalia of *silens* differ from those of other species in the group in having a shorter right saccular extension; in most males this barely reaches the ventral angle of the cucullus. The female genitalia are indistinguishable from those of the other two species in the group.

The immature stages of *silens* are unknown.

Euxoa silens is primarily a species of deserts in the intermontane region where it occurs from southern Washington eastward to western Colorado and southward to central Arizona and southern California. East of the intermontane region, it has been collected in southwestern Montana and southern Alberta. Adults have been collected from late April until early August; they fly later at high elevations in the Sierra Nevada Mountains than at low elevations.

Euxoa (*Euxoa*) *pimensis* Barnes and McDunnough

PL. 3, FIGS. 45–47; PL. G, FIG. 10 (RWH 10752).

Euxoa pimensis Barnes and McDunnough, 1910, *Jour. New York Ent. Soc.,* **18**: 150.
Type locality: Babaquivera [sic] Mts., Pima Co., Arizona. [USNM]

Most specimens of *pimensis* can be recognized by the pale hindwings although those of specimens from central Colorado are as dark as those of *silens.* The pale and black-marked forms are equally common. Forewing length varies from 15 to 19 mm. The pale form of *pimensis* could be confused with specimens of *immixta* but can be recognized by the gray forewing, the paler hindwing, and by the vesica: the subbasal diverticulum is large, its length equal to the width of the vesica.

The immature stages are unknown.

Euxoa pimensis is primarily a species of deserts of the American Southwest. It occurs from central Colorado westward to southern California and southward through New Mexico and Arizona at least to the Mexican border. Adults have been collected from late April until mid-July.

Euxoa (*Euxoa*) *immixta* (Grote)

PL. 3, FIGS. 48, 49; PL. H, FIG. 1 (RWH 10753).

Agrotis immixta Grote, 1880, *Bull. U. S. Geol. Surv.,* **6**: 259.
Type locality: Texas [BMNH]

Agrotis tetrica Smith, [1888], *Proc. U. S. Natl. Mus.* **10**: 458.
Type locality: [Texas]. [lost]

NOTE—Smith described *tetrica* from a male and female specimen from the Bailey collection. He also mentioned a female specimen in the Tepper collection, but it must be excluded from the type series because it is described as atypical. Smith lists the "habitat" as Nevada. Three years later, Smith (1890: 161) redescribed the species and stated that the Bailey specimens were from Texas; the Tepper specimen was from Nevada. The two syntypes in the Bailey collection apparently have been destroyed because they are not in the surviving portion of the collection at the New York State Museum at Albany. I have placed *tetrica* in the synonymy of *immixta* for two reasons: 1) the description of *tetrica* matches Texas specimens of *immixta* in every detail; the evenly colored reddish-luteous forewing with the maculation barely contrasting and the punctiform outer line of the postmedial line are particularly suggestive of *immixta*; 2) Smith (1890, figure 82) illustrated a left valve of *tetrica*; this genitalia preparation, which may be from the original male because Smith had seen only one male, is in the United States National Museum and is a left valve of *immixta*.

Euxoa knoxvillea McDunnough, 1937, *Can. Ent.*, **69**: 153.
Type locality: Knoxville, Tennessee. [CNC]

Specimens of *immixta* can be distinguished from those of *pimensis* by the buff or brown rather than gray forewing, the more punctiform outer element of the postmedial line, and by the darker hindwing. Forewing length varies from 15 to 18 mm. In *immixta* the form with a black basal dash is very rare. *Euxoa immixta* occurs east of the range of *pimensis*. Within its range, *immixta* is most likely to be confused with *messoria* (subgenus *Longivesica*, p. 37) and *scholastica* (p. 71). Males of *immixta* can be distinguished from those of *messoria* and *scholastica* by the asymmetrical and markedly incurved saccular extensions and by the shape of the vesica; females of *immixta* can be recognized by the broad, rounded ovipositor lobes clothed with short, stout setae and by the shape of the corpus bursae.

The immature stages of *immixta* are unknown.

This is a species of the central and eastern Great Plains and relict prairie habitat to the west of the Appalachian Mountains. It occurs from southern Michigan westward to western Montana and southward to eastern Tennessee, northern Arkansas, and southern Texas. Specimens from Texas are pale buffy brown; those from the eastern deciduous forest region are dark brown; specimens from the northern Great Plains are intermediate between these two extremes in color. Adults have been collected from late April until early September; they fly later in the North than in the South.

simulata GROUP

The *simulata* group includes only *simulata*; the characters are given under the species.

Euxoa (*Euxoa*) *simulata* McDunnough
PL. 3, FIGS. 50, 51; PL. H, FIG. 2; PL. W, FIG. 3; PL. EE, FIG. 10 (RWH 10754).

Euxoa simulata McDunnough, 1946, *Can. Ent.*, **78**: 28.
Type locality: Nelson Cr., Plumas Co., California. [CNC]

In wing color and maculation, specimens of *simulata* resemble those of *terrena* (p. 71) and to a lesser extent, those of *declarata* (p. 79) and *satis* (p. 105). The forewing is brown or orange brown. Forewing length varies from 14 to 16 mm. Males of *simulata* can be distinguished by genital characters, most easily by the stout, relatively straight, asymmetrical saccular extensions that can be observed in most males by brushing away scales at the end of the abdomen. Other distinctive characters of the male genitalia are the broad, somewhat rectangular sacculus, the markedly incurved, C-shaped harpes and the shape of the vesica. Females of *simulata* can be recognized by the bullet-shaped ovipositor lobes that are sparsely clothed with short setae and lack a distinctive subbasal row of long setae. Dissected females can also be distinguished by the long, stout setae on the eighth abdominal segment that give it a bristly appearance.

The immature stages of *simulata* are unknown.

Euxoa simulata occurs in pine and fir forests in montane areas of the western United States. Specimens are rare in collections; although the species is widespread, it has been collected at less than two dozen localities and only one or two specimens have been collected at most of them. Also, specimens are usually confused with those of *terrena* in the North and West and with *antica* in the southeastern portion of its range. *Euxoa simulata* resembles these species in external appearance, and they outnumber it at most localities by about ten to one. *Euxoa simulata* has been collected from south-central Washington southward through Utah and western New Mexico; in the West it occurs southward to southern California. Adults have been collected from mid-July until early September.

punctigera GROUP

The *punctigera* group includes six western North American species. The adults are dull-colored moths with obscure wing markings. In dark-colored species, the reniform and orbicular spots have a white line inside the black outline. Members of the group can most easily be recognized by female genital characters; it is the only group that combines the characters of bisaccate corpus bursae and fused dorsal margins of the two ovipositor lobes. The ovipositor lobes have fused, sclerotized, flangelike projections apically in four of the six species. Males can be recognized by the long, pubescent harpe, the relatively massive sacculus with a slightly convex dorsal margin and by vesica shape. Three species can be recognized by their short, stout, incurved saccular extensions.

KEY TO SPECIES OF THE *PUNCTIGERA* GROUP

1. Males 2
– Females 7

2. Right saccular extension as long as, or longer than, right harpe 3
– Right saccular extension much shorter than right harpe 5

3. Forewing very dark brown, appearing black; occurring in mountains of southeastern Arizona . . . . . *stygialis* p. 84
– Forewing dark brown, or paler; occurring north of Mogollon Rim from central Arizona and New Mexico northward 4

4. Forewing dark brown; occurring in Rocky Mountain and Sierra Nevada mountain systems . . . . . *punctigera* this page
– Forewing brownish orange, orange gray, or gray with a dusting of white scales; occurring in intermontane region and eastern Great Plains region . . . . . *aurantiaca* p. 84

5. Right saccular extension more than ¾ length of harpe; forewing fawn gray; occurring in Great Basin from central Utah northward . . . . . *cana* p. 84
– Right saccular extension less than ¾ length of harpe 6

6. Forewing dark brown; occurring in southern Great Basin, Colorado Plateau, southern Rocky Mountains, and Black Hills of South Dakota . . . . . *stigmatalis* p. 85
– Forewing light gray brown to buffy yellow; occurring in northern Great Plains of southern Canada and Montana . . . . . *spumata* p. 85

7. Ovipositor lobe with apical flangelike projection 8
– Ovipositor lobe without apical flangelike projection 11

8. Ventral plate in ductus bursae lanceolate, 8 times as long as wide . . . . . *aurantiaca* p. 84
– Ventral plate in ductus bursae shorter, about 4 times as long as wide 9

9. Forewing light gray brown to buffy yellow; occurring in northern Great Plains of southern Canada and Montana . . . . . *spumata* p. 85
– Forewing dark brown or black; occurring south of Montana 10

10. Forewing very dark brown, appearing black; occurring in mountains of southeastern Arizona . . . . . *stygialis* p. 84
– Forewing dark brown; occurring north of Mogollon Rim from central Arizona and New Mexico northward . . . . . *stigmatalis* p. 85

11. Hindwing white to buff, veins dark brown; occurring in northern Great Basin and Columbia Plateau from central Utah northward to southeastern Washington . . . . . *cana* p. 84
– Hindwing brown; occurring in mountains surrounding Great Basin and Columbia Plateau . . . . . *punctigera* this page

Euxoa (*Euxoa*) *punctigera* (Walker)

PL. 3, FIGS. 52-55; PL. H, FIG. 3; PL. W, FIG. 4 (RWH 10759).

Mamestra punctigera Walker, 1865, *List of the Specimens of Lepidopterous Insects in the Collection of the British Museum,* **32**: 661.

Type locality: Vancouver Island, British Columbia. [BMNH]

Agrotis pastoralis Grote, 1875, *Can. Ent.*, 7: 68.
Type locality: Vancouver Island, British Columbia. [BMNH]

This is the most northerly and westerly occurring species in the group. It is the only species that occurs in the northern Rocky Mountains and in the Cascades and Sierra Nevada Mountains. In most of its range it can be distinguished by the dark-brown color of both fore- and hindwings in both sexes. Specimens of *punctigera* could be confused with those of *stigmatalis* in Utah and Colorado where their ranges overlap. Males of *punctigera* can be recognized by the darker hindwing and longer saccular extensions; females can be distinguished from those of *stigmatalis* by the absence of sclerotized flangelike projections on the ovipositor lobes. Forewing length varies from 16 to 21 mm.

The immature stages of *punctigera* are unknown.

Euxoa punctigera occurs in coniferous forests of spruce, pine, and fir from southern British Columbia southward in the Cascades and Sierra Nevada Mountains to southern California. In the Rocky Mountain system it occurs southward to southern Montana. Apparently disjunct populations occur in the Wasatch Mountains of Utah and the San Juan Mountains of southwestern Colorado. Adults have been collected from mid-July until mid-September.

Euxoa (*Euxoa*) *cana* Lafontaine
PL. 4, FIGS. 1, 2; PL. H, FIG. 4; PL. EE, FIG. 11 (RWH 10762).

Euxoa cana Lafontaine, (1975), *Can. Ent.*, **106**: 1236.
Type locality: Sumpter, Oregon, 28 miles W Baker, 4,430′. [CNC]

Specimens of *cana* can be distinguished from those of other species that occur in the intermontane region by the pale buffy-gray forewing and the white or pale pearly-buff hindwing. Forewing length varies from 13 to 17 mm. In the male genitalia the saccular extensions are shorter than those of *punctigera* and *aurantiaca* but longer than those of *stigmatalis*. The female genitalia are indistinguishable from those of *punctigera*.

The immature stages are essentially unknown; however, adults were reared from larvae collected near Holbrook, Idaho in soil samples.

This is a species of open, sagebrush habitats. It has been found with *punctigera* at a few localities where sagebrush habitats occur in open coniferous forests. *Euxoa cana* occurs from southeastern Washington southward to central Nevada and north-central Utah. Adults have been collected from early August until mid-September.

Euxoa (*Euxoa*) *aurantiaca* Lafontaine
PL. 4, FIGS. 3–5; PL. H, FIG. 5; PL. W, FIG. 5 (RWH 10760).

Euxoa aurantiaca Lafontaine, (1975), *Can. Ent.*, **106**: 1236.
Type locality: Alcova, Wyoming, 4 mi SW, 5,500′. [CNC]

This is the most variable species in the group in terms of wing color. Specimens from Wyoming and western Colorado have bright brownish-orange forewings; specimens from Montana, Utah, and Nevada are gray with a dusting of white scales; specimens from Arizona and New Mexico are orange gray or orange brown with a dusting of white scales. Forewing length varies from 15 to 19 mm. Specimens are most likely to be confused with those of *stigmatalis* and *spumata*. The much longer saccular extensions of *aurantiaca* males can be seen by brushing away the scales at the end of the abdomen. Females can be distinguished from those of *stigmatalis* and *spumata* by the longer sclerotized plate in the ventral wall of the ductus bursae.

The immature stages are unknown.

Euxoa aurantiaca occurs in western Montana and from central Wyoming southward to east-central Nevada, north-central Arizona, and central New Mexico. Adults have been collected from mid-July until early September. They are usually collected in sagebrush habitats in open areas or open pine forests.

Euxoa (*Euxoa*) *stygialis* (Barnes and McDunnough)
PL. 4, FIGS. 6, 7 (RWH 10761).

Feltia stygialis Barnes and McDunnough, 1912, *Contrib. Nat. Hist. Lep. N. Am.*, **1** (5): 8.
Type locality: Santa Catalina Mts., Arizona. [USNM]

Until recently, this species was confused with *stigmatalis*. Although specimens of it are larger and darker than those of *stigmatalis* from most areas, specimens of *stigmatalis* from the White Mountains of Arizona are indistinguishable from those of *stygialis* in external appearance and in female genital characters. Because *stygialis* was known only from specimens of *stigmatalis* from the White Mountains as *stygialis*; the two names were later placed in synonymy (Lafontaine, 1974: 1238). The validity of *stygialis* as a distinct species was demonstrated when John G. Franclemont, Ithaca, New York, collected males of *stygialis* in the Chiricahua Mountains of Arizona. The male genitalia of *stygialis* resemble those of *punctigera,* not those of *stigmatalis.* Forewing length varies from 18 to 20 mm.

The immature stages are unknown.

Euxoa stygialis is known only from the Santa Catalina and Chiricahua Mountains in southeastern Arizona; *stigmatalis* is not known to occur farther south than the White Mountains and east-central Arizona. Adults have been collected from early to mid-July.

Euxoa (*Euxoa*) *stigmatalis* (Smith)
PL. 4, FIGS. 8–10; PL. EE, FIG. 12 (RWH 10764).

Carneades stigmatalis Smith, 1900, *Proc. U. S. Natl. Mus.,* **22**: 425.
Type locality: Glenwood Springs, Colorado. [USNM]

NOTE—The male lectotype was designated by Todd (1982: 200).

Specimens of *stigmatalis* are most likely to be confused with those of *punctigera* and *stygialis* but can be distinguished by characters given in the key. Additionally, the male hindwing of *stigmatalis* is paler than that of *punctigera*; it is dirty white with darker shading on the wing margin and veins rather than smoky brown as in *punctigera.* The forewing ground color in specimens from the mountains of Colorado, New Mexico, and the eastern Great Basin are dark brown; the forewings of specimens from the mountains of Arizona are dark blackish brown. Forewing length varies from 16 to 20 mm.

The immature stages are unknown.

Euxoa stigmatalis occurs in coniferous forests of pine and fir from northern Colorado westward to Lake Tahoe, California and southward to central New Mexico and central Arizona. A disjunct population occurs in the Black Hills of South Dakota and adjacent southeastern Montana. Adults have been collected from mid-July until mid-September.

Euxoa (*Euxoa*) *spumata* McDunnough
PL. 4, FIGS. 11–13; PL. H, FIG. 6; PL. W, FIG. 6 (RWH 10763).

Euxoa spumata McDunnough, 1940, *Can. Ent.,* **72**: 192.
Type locality: Three Forks, Montana. [CNC]

This species and *stigmatalis* are the only two species in the *punctigera* group that are indistinguishable by genital characters in both sexes. Specimens of *spumata* can be distinguished by the pale coloration of the forewing. The forewing is pale gray buff or yellow buff; the forewing is dusted with darker gray in specimens from the foothills of Alberta and Montana. *Euxoa spumata* tends to be smaller than *stigmatalis*; forewing length varies from 14 to 18 mm. The hindwing is pearly white with buffy shading on the wing margin and veins.

The immature stages are unknown.

This species inhabits dry open prairie from south-central Saskatchewan westward to southwestern Alberta and southward to west-central North Dakota, and southern Montana. Its range does not overlap that of any other species in the group. Adults have been collected from early August until early September.

pallipennis GROUP

The *pallipennis* group includes two desert-loving, North American species. Males can be recognized by brushing away the scales from the end of the abdomen and observing the length and shape of the saccular extensions. These are short and bladelike and have a slight twist at about two-thirds the length from the base so that the apex projects dorsally toward the cucullus and slightly to the outside. In other species with short saccular extensions, they bend toward the cucullus or inward toward the aedoeagus apically and lack a definite subapical twist. The most distinctive feature of males in the group is the shape of the vesica, particularly that of the subbasal diverticulum and its accessory pouches. Females are most likely to be confused with those of *satiens* (p. 108), *tronella* (p. 129), *misturata* (p. 143), or *cona* (p. 142); they can be distinguished from those of the first three species by the lack of sclerotized flangelike projections on the ovipositor lobes in *pallipennis.* They can be distinguished from females of *cona* by the fact that the dorsal margins of the ovipositor lobes are not fused and by the smaller and more rounded shape of the orbicular spot on the forewing in *pallipennis* and *baja.*

KEY TO SPECIES OF THE *PALLIPENNIS* GROUP

1. Forewing pale buff or ash gray with slight dusting of dark scales; male genitalia with saccular extensions short, less than ½ length of sacculus and about ⅔ length of harpe; vesica with a

series of pouches in wall of subbasal arc; widely distributed in Great Plains and intermontane region as far southwest as Mojave Desert in south-central California .... *pallipennis* this page

— Forewing buff or gray, heavily dusted with darker scales so that veins appear pale; male genitalia with saccular extension longer, more than ⅗ as long as sacculus and about ¾ length of harpe; vesica without pouches in wall of subbasal arc; occurring only on coast of southwestern California and Baja California from Los Angeles County southward .... *baja* this page

Euxoa (*Euxoa*) *pallipennis* (Smith)

PL. 4, FIGS. 14, 15; PL. H, FIG. 7; PL. W, FIG. 7 (RWH 10765).

Agrotis pallipennis Smith, [1888], *Proc. U. S. Natl. Mus.,* **10**: 461.
Type locality: Colorado. [USNM]

NOTE—The male lectotype of *pallipennis* was designated by Todd (1982: 164).

Euxoa alcesta Smith, 1905, *Jour. New York Ent. Soc.,* **13**: 196.
Type locality: Stockton, Utah. [AMNH]

NOTE—The female lectotype of *alcesta* was designated by Todd (1982: 12).

Euxoa pallipennis and *misturata* are the two most common cutworm species in western aridlands, and specimens of the two species are frequently confused. They can be distinguished by genital characters given in the key and in the discussion of the species-group; these differences can usually be observed without dissecting the specimens. In addition to the genital characters, specimens of *pallipennis* usually can be distinguished from those of *misturata* by the less mottled ground color of the forewing and the less streaked forewing markings. Unlike *misturata, pallipennis* does not show marked geographical variation in ground color. Specimens of *pallipennis* tend to be smaller than those of *misturata*; forewing length varies from 12 to 17 mm.

The immature stages have not been described; the preferred foodplant of the larva in Montana is Russian thistle (*Salsola kali* Linnaeus), and the species has occasionally been reported as a pest on wheat Cook (1930: 262).

Euxoa pallipennis occurs throughout the intermontane region and the western Great Plains from southern Saskatchewan southward to central New Mexico and westward to south-central British Columbia, central Washington and Oregon, and eastern California. Its range extends westward to the Cascades-Sierra Nevada axis; in southern California it occurs westward in the Mojave Desert to the Transverse Ranges. Adults have been collected from mid-August until early October. *Euxoa pallipennis* inhabits arid regions of shortgrass prairie and sagebrush (*Artemisia* spp.); in the Great Basin it is most abundant in areas where shadscale (*Atriplex confertifolia* (Torrey and Frémont) Watson) is common.

Euxoa (*Euxoa*) *baja* Lafontaine, NEW SPECIES

PL. 4, FIG. 16; PL. H, FIG. 8.

Euxoa (*Euxoa*) *baja* Lafontaine.
Type locality: Long Beach, California. [USNM]

This species is closely related to *pallipennis* but differs from it in wing markings and in male genital characters. Specimens of *baja* can be recognized by the darker forewing color with pale, contrasting venation; by the longer saccular extension and smaller sacculus; and by differences in the vesica shape.

Antenna of male biserrate and bifasciculate. Frontal tubercle present. Eye round. Vestiture of head and thorax a mixture of gray and brown scales. Ground color of forewing gray or brown heavily dusted with darker gray or darker brown scales. Forewing veins tend to be pale and contrasting. Transverse lines black, double, appearing slightly smudged. Terminal area dark gray or dark brown. Subterminal area with a series of dark sagittate spots proximal to pale subterminal line. Reniform and orbicular spots concolorous with remainder of forewing, or with some white shading; spots partially outlined in black. Claviform spot obscure or absent; basal line absent. Forewing length 14 to 16 mm. Hindwing of male white with pale smoky-brown shading on wing margin and veins. Hindwing of female similar to that of male but with more dark shading on wing margin, this occupying as much as outer third of wing. Male genitalia similar to those of *pallipennis* but sacculus shorter and narrower with dorsal margin slightly concave rather than convex. Saccular extensions longer than those of *pallipennis,* more than ⅗ as long as sacculus and about ¾ as long as harpe. Vesica projecting at angle of 90° to aedoeagus; in *pallipennis* this angle about 150°. Subbasal arc in vesica even, lacking pouches in wall present in *pallipennis*; subbasal diverticulum elbowed and footlike in *baja* but straighter and more truncate in *pallipennis*. Female genitalia of *baja* are indistinguishable from those of *pallipennis*.

TYPES. Holotype: ♂. Long Beach, California; 26 Sept. 1937. USNM. Paratypes: 11 ♂, 3 ♀. Del Mar, San Diego County, California; 30 Sept. 1957, 21–29 Oct. 1957 and 24 Oct. 1955; A. A. Lee (10 ♂, 1 ♀); San Diego, California; 16 Oct. 1920; Barnes collection (1 ♂); Long Beach, California; 26 Sept. 1937 (1 ♀); Baja California, Mexico, 17 mi N Colonia Guerrero; 3 Nov. 1953; R. E. Ryckman, C. C. Lindt, and C. P. Christianson (1 ♀). AMNH, CNC, SDNH, USNM.

The range of *baja* appears to be separated from that of *pallipennis* in southern California by the Transverse Ranges.

tessellata GROUP

The *tessellata* group includes two North American species. They are dissimilar in forewing markings and are more likely to be confused with species in other groups than with each other. In the male genitalia of both species of saccular extensions are asymmetrical with the right saccular extension about 1.25× the length of the left. In the vesica there is a ventral bulge in the subbasal arc, the subbasal diverticulum is small and the apical third of the vesica is expanded. In the female genitalia the ductus bursae is long; the corpus bursae is T-shaped with both dorsal and ventral bulges at its posterior end.

Sterile hybrids between the two species have been produced in the laboratory. They look like *tessellata* specimens, but genital characters are intermediate.

KEY TO SPECIES OF THE *TESSELLATA* GROUP

1. Males 2
— Females 3

2. Forewing not longitudinally streaked, orbicular spot round or oval; saccular extensions thin, only slightly stouter than harpes, and curved downward near apex away from cucullus . . *tessellata* this page
— Forewing longitudinally streaked, orbicular spot elongate, almost barlike in many specimens; saccular extensions about twice as stout as harpes and curved upward toward cucullus near apex .... *plagigera* p. 89

3. Ovipositor lobe without sclerotized flangelike process at apex but truncate apically .... *tessellata* this page
— Ovipositor lobe tapered to apex with sclerotized flangelike process apically .... *plagigera* p. 89

Euxoa (*Euxoa*) *tessellata* (Harris) (Striped Cutworm*, Ver-gris Rayé, m., Fr.)

PL. 4, FIGS. 17–28; PL. H, FIG. 9; PL. W, FIG. 8; PL. EE, FIG. 13 (RWH 10805).

Agrotis tessellata Harris, 1841, *A Report on the Insects of Massachusetts, Injurious to Vegetation,* 324.
Type locality: Massachusetts. [MCZ]

Mamestra insulsa Walker, 1856, *List of the Specimens of Lepidopterous Insects in the Collection of the British Museum,* **9**: 234.
Type locality: Orilla [sic], West Canada [Ontario]. [BMNH]

Agrotis perlentans Walker, [1857], *List of the Specimens of Lepidopterous Insects in the Collection of the British Museum,* **10**: 332.
Type locality: New York. [BMNH]

NOTE—The original description gives the type locality as New York. The nominal male type, in the British Museum (Natural History), is labeled "U.S." Also, the "blackish streak between the wing base and the orbicular spot, and between the latter and the reniform spot" mentioned in the description is absent in the specimen.

Agrotis insignata Walker, [1857], *List of the Specimens of Lepidopterous Insects in the Collection of the British Museum.* **10**: 353.
Type locality: ?[New York]. [BMNH]

NOTE—The original description of *insignata* indicates that the type locality is unknown. The nominal female type in the British Museum (Natural History) is labeled New York.

NOTE—*Agrotis insignata* Walker, 1857: 353 is a junior homonym of *Agrotis insignata* Walker, 1857: 330, a synonym of *Euxoa ochrogaster.*

Agrotis illata Walker, 1857, *List of the Specimens of Lepidopterous Insects in the Collection of the British Museum,* **11**: 742.

NOTE—*Agrotis illata* is a replacement name for *Agrotis insignata* Walker, 1857: 353.

Agrotis nigricans var. *maizi* Fitch, 1864, *Ninth Report on the Noxious and Other Insects of the State of New York,* 237.
Type locality: New York. [lost]

Agrotis subsignata Walker, 1865, *List of the*

Specimens of Lepidopterous Insects in the Collection of the British Museum, **32**: 706.

NOTE—*Agrotis subsignata* is another replacement name for *Agrotis insignata* Walker, 1857: 353.

Agrotis atropurpurea Grote, 1877, *Bull. U. S. Geol. Surv.,* **3**: 228.
Type locality: Denver, Colorado. [lost]

NOTE—Grote proposed the name *atropurpurea* for the dark form of *tessellata.* His description is based on a Denver specimen although he also mentions having seen specimens from Albany, New York, and elsewhere in "the East." Apparently type material was not selected; the Denver specimen is not in the British Museum (Natural History).

Agrotis tesselloides Grote, 1880, *Bull. U. S. Geol. Surv.,* **6**: 566.
Type locality: Havilah, California. [BMNH]

Agrotis finis Smith, [1888], *Proc. U. S. Natl. Mus.,* **10**: 457.
Type locality: Black Hills [South Dakota]. [USNM]

NOTE—The male lectotype of *finis* was designated by Todd (1982: 83).

Agrotis remota Smith, 1890, *Trans. Amer. Ent. Soc.,* **17**: 48.
Type locality: Sierra Nevada [Truckee]. [USNM]

NOTE—The male lectotype of *remota* was designated by Todd (1982: 183).

Carneades flaviscapula Smith, 1900, *Proc. U. S. Natl. Mus.,* **22**: 422.
Type locality: New Mexico. [USNM]

NOTE—The male lectotype of *flaviscapula* was designated by Todd (1982: 86).

Carneades noctuiformis Smith, 1900, *Proc. U. S. Natl. Mus.,* **22**: 445.
Type locality: Siskiyou, California. [USNM]

NOTE—The male lectotype of *noctuiformis* was designated by Todd (1982: 150).

Carneades neotelis Smith, 1900, *Proc. U. S. Natl. Mus.,* **22**: 446.
Type locality: Pullman, Washington. [USNM]

NOTE—The male lectotype of *neotelis* was designated by Todd (1982: 146).

Carneades atrofusca Smith, 1900, *Proc. U. S. Natl. Mus.,* **22**: 447.
Type locality: Glenwood Springs, Colorado. [USNM]

NOTE—The female lectotype of *atrofusca* was designated by Todd (1982: 26).

Carneades objurgata Smith, 1900, *Proc. U. S. Natl. Mus.,* **22**: 448.
Type locality: Pullman, Washington. [USNM]

NOTE—The male lectotype of *objurgata* was designated by Todd (1982: 153).

Carneades cariosus Smith, 1900, *Proc. U. S. Natl. Mus.,* **22**: 449.
Type locality: Pullman, Washington. [USNM]

NOTE—The male lectotype of *cariosus* was designated by Todd (1982: 41).

Carneades nordica Smith, 1900, *Proc. U. S. Natl. Mus.,* **22**: 455.
Type locality: Calgary, Alberta. [USNM]

NOTE—The male lectotype of *nordica* was designated by Todd (1982: 151).

Carneades caesius Smith, 1900, *Proc. U. S. Natl. Mus.,* **22**: 457.
Type locality: Middle California. [USNM]

NOTE—The female lectotype of *caesius* was designated by Todd (1982: 38).

Carneades acutifrons Smith, 1900, *Proc. U. S. Natl. Mus.,* **22**: 458.
Type locality: California. [USNM]

NOTE—The male lectotype of *acutifrons* was designated by Todd (1982: 7).

Carneades laminis Smith, 1900, *Proc. U. S. Natl. Mus.,* **22**: 460.
Type locality: Middle California. [USNM]

Carneades focinus Smith, 1903, *Jour. New York Ent. Soc.,* **11**: 7.
Type locality: Glenwood Springs, Colorado. [AMNH]

NOTE—The male lectotype of *focinus* was designated by Todd (1982: 88).

Euxoa marinensis McDunnough, 1941, *Can. Ent.,* **73**: 66.
Type locality: Inverness, Marin Co., California. [CNC]

Euxoa tessellata is one of the most common, widespread, and variable species of *Euxoa.* Its many color forms can easily be confused with a variety of species that they resemble. In the East there are forms that resemble *messoria, declarata,* and *redimicula.* Despite this variability, the individual forms, once learned, can usually be recognized without the aid of genital characters. The most common forms in the East (plate 4, figures 17–21) may be reddish brown and resemble *declarata* or gray and resemble *redimicula.* These forms are less common in the prairies and rare in western North America. A less common form in the East lacks black shading near the orbicular spot (plate 4, figures 22–24); it is the most common form west of the Rocky Mountain region; and it is frequently confused with *messoria* and *pleuritica* in the East and with *pestula* in the West. A third, obscurely marked form (plate 4, figures 25, 26) is uncommon in the East and the Far West but is the predominant form in the Rocky Mountain region; it may be confused with *satis* and *atomaris.* Some specimens from coastal California are pale colored and have a pale-lined cubital vein (plate 4, figures 27, 28). They have been described as *caesius* and *marinensis* and have been confused with *divergens.* A good "rule-of-thumb" character for distinguishing males of *tessellata* from those of other similar species is the presence of a tuft of yellowish-orange scales on the thorax at the base of the costa of the forewing. This tuft is present in other species (e.g., *perexcellens, intrita*) but not in those with which *tessellata* is likely to be confused. Fortunately, the many forms of *tessellata* can be distinguished from specimens of other *Euxoa* species by genital characters, usually without having to dissect the specimen. The thin, asymmetrical saccular extensions and the expanded, footlike apex of the cucullus can be observed by removing the scales at the end of the abdomen with a small brush. Females can be recognized by the flat, apically truncate ovipositor lobes that are clothed with short, conical setae apically. The truncate apex and flat sides of the ovipositor lobes, and the lack of fine setae can often be observed with the naked eye because the ovipositor lobes usually protrude from the abdomen. This shape is unique within the genus. Specimens of *tessellata* vary greatly in size; forewing length varies from 13 to 18 mm.

The immature stages have been described briefly by several authors. The egg was illustrated by Salkeld (1975: 1149, figs. 40–42). The larva, commonly called the striped cutworm, was described by Crumb (1932: 83). It is a climbing cutworm that has caused damage to herbaceous plants such as tobacco and garden crops, and woody shrubs including apple, cherry, and pear. The species is reported to overwinter in the egg stage (Hinks and Byers, 1976: 1349).

Euxoa tessellata occurs from Newfoundland westward to Alaska and southward to North Carolina, Kansas, central New Mexico, Arizona, and southern California. Adults have been collected from early June until late September; most records are for the period between late June and early August. Adults occasionally have been collected in open habitats in the Great Plains and Great Basin; however, most records are from forested habitats.

Euxoa (*Euxoa*) *plagigera* (Morrison)
PL. 4, FIGS. 29–31; PL. H, FIG. 10; PL. X, FIG. 1 (RWH 10804).

Agrotis plagigera Morrison, 1874, *Proc. Boston Soc. Nat. Hist.,* **17**: 163.
Type locality: Colorado. [lost]

Specimens of *plagigera* can readily be distinguished from those of *tessellata* by the longitudinally streaked forewing and by the genital characters given in the key. They are more likely to be confused with specimens of the similarly marked *olivalis* (p. 127). Specimens of *plagigera* tend to have less evenly streaked, more mottled, forewings than do those of *olivalis,* and the cubital vein is not pale as in that species. Forewing length varies from 14 to 16 mm. Males of *plagigera* can be distinguished from those of other similarly marked species such as *olivalis, inscripta,* and *melura* by the asymmetrical rather than symmetrical saccular extensions. Females of *plagigera* can be distinguished from those of *melura* by the flangelike projection on the ovipositor lobe rather than stout apical setae, and from females of all three species by the T-shaped rather than oval corpus bursae.

The immature stages are known only from laboratory reared material. The species overwinters in the egg stage and has a brief prepupal aestivation period (Hinks and Byers, 1976: 1349, 1352).

Euxoa plagigera occurs from south-central Saskatchewan westward to southwestern British Columbia and southward to northern New Mexico, Arizona, and southern California. An apparently disjunct population occurs on sagebrush (*Artemisia* spp.) covered slopes in southern Yukon. In the main portion of its range, *plagigera* inhabits both open aridland habitats and open coniferous forests.

albipennis GROUP

This group includes two North American species, *albipennis* and *henrietta.* In wing markings the two species bear little resemblance to each other and are more frequently confused with species in other groups than they are to each other. The male genital characters are very similar. Males can be recognized by the combination of pubescent harpes and long twisted saccular extensions that bend away from the cucullus apically. The right saccular extension is 1.25–1.50× the length of the right harpe; the left saccular extension is slightly shorter. The sacculus is oval, slightly longer than ⅓ the length of the valve. The vesica is also distinctive, particularly with respect to the shape of the subbasal bend and the subbasal diverticulum. The female genital characters are similar to those of several other species groups. The corpus bursae shape differs markedly between the two species. In *albipennis* the corpus bursae is oval, tapered posteriorly and slightly constricted mesially; in those of *henrietta,* it is triangular, tapered anteriorly. The ovipositor lobe may have a sclerotized, apical process (*albipennis*), or it may lack this (*henrietta*). In both species, there is a conspicuous row of long, subbasal setae on the ovipositor lobes and the sclerotized plates in the dorsal and ventral walls of the ductus bursae are very long and narrow.

KEY TO SPECIES OF THE *ALBIPENNIS* GROUP

1. Forewing usually with basal area paler than costal portion of median area; orbicular spot large, about twice as wide as area between it and reniform spot; harpe incurved at base, straight for most of its length; ovipositor lobe with sclerotized projection at apex; widespread .... *albipennis* this page

— Forewing with basal area similar in color, or darker than costal portion of median area; orbicular spot smaller, about as wide as area between it and reniform spot; basal half of harpe incurved; ovipositor lobe without sclerotized projection at apex; occurring in Pacific Coast states .... *henrietta* p. 91

Euxoa (*Euxoa*) *albipennis* (Grote)

PL. 4, FIGS. 32–36; PL. I, FIG. 1; PL. X, FIG. 2; PL. EE, FIG. 14 (RWH 10807).

Agrotis albipennis Grote, 1876, *Bull. Buffalo Soc. Nat. Sci.* **3**: 80.
Type locality: California. [BMNH]

NOTE—It is not clear from Grote's description of *albipennis* how many specimens were in the type series. He does give California (No. 5611) and Canada as localities. Two specimens labeled "type" are in the British Museum (Natural History), a male and a female. Only the male is labeled No. 5611, so this specimen is here designated lectotype. It is labeled "California, Grote Coll. 81–116/5611/*Agrotis albipennis* Grote Male type/Noctuidae Brit. Mus. slide No. 6345." It is generally in good condition although the forewings are slightly worn.

Agrotis verticalis Grote, 1880, *Bull. Brooklyn Ent. Soc.,* **3**: 29.
Type locality: Colorado. [BMNH]

NOTE—Grote described *verticalis* from two specimens, a male and a female collected in Colorado. Both specimens are in the British Museum (Natural History). The male specimen, which is in better condition than the female, is here designated lectotype. It is labeled "Type, H.T./Colorado, Snow, Grote Coll. 81–116/*Agrotis verticalis* Grote Male type/839/Noctuidae, Brit. Mus. slide No. 6355." The specimen is generally in good condition although the right wings are slightly frayed.

Agrotis albipennis female var. *nigripennis* Grote, 1881, *Bull. U. S. Geol. Surv.,* **6**: 159.
Type locality: New York. [? lost]

NOTE—Although Grote refers to "females from N.Y.," the only specimen in the British Museum (Natural History) labeled as type is a specimen labeled "California, Grote Coll. 81–116/*Agrotis nigripennis* Grote, type/Noctuidae, Brit. Mus. slide No. 6309." The first label is a British Museum accession label, and the California locality may have been added incorrectly. The specimen, however, is virtually identical to material of *albipennis* from California in the Canadian National Collection and differs from eastern females of *albipennis* in having paler brown forewings without darker blackish-brown shading on the costa. Considering the uncertainty regarding the place of origin of the specimen, it seems advisable to consider the type material of *nigripennis* to be lost.

Carneades malis Smith, 1900, *Proc. U. S. Natl. Mus.,* **22**: 450.
Type locality: Brandon, Manitoba. [USNM]

Euxoa bialba Smith, 1905, *Jour. New York Ent. Soc.,* **13**: 197.
Type locality: Stockton, Utah. [AMNH]

NOTE—The female lectotype of *bialba* was designated by Todd (1982: 31).

Euxoa indensa Smith, 1910, *Trans. Amer. Ent. Soc.,* **36**: 263.
Type locality: Volga, South Dakota. [AMNH]

NOTE—The female lectotype of *indensa* was designated by Todd (1982: 106).

Specimens of *albipennis* are frequently confused with those of species in the *declarata* and *lillooet* species-groups (pp. 78, 116). Each has a similar forewing pattern, and the ground color varies from dark brown, through various shades of reddish brown, copper brown, and purplish brown to pale gray or pale brown. Most specimens of *albipennis* can be recognized by the combination of a large orbicular spot situated close to the reniform spot and the relatively pale coloration of the basal area of the forewing. Males can be distinguished from those of similar species by the predominantly white rather than brown hindwings, by the straight rather than C-shaped or S-shaped harpe, and by the shape of the saccular extension; this bends away from the cucullus apically in *albipennis* but is straight or bends toward the cucullus apically in the other species. Females can be distinguished from those of similar species by the presence of a narrow, fingerlike sclerotized process at the apex of each ovipositor lobe and by the unisaccate corpus bursae. In other similarly marked species, the ovipositor lobe lacks a sclerotized apical process, and the corpus bursae is bisaccate. Specimens of *albipennis* tend to be smaller than those of other similar species; forewing length varies from 13 to 17 mm. Specimens of *albipennis* can be distinguished from those of *henrietta* by the characters given in the key.

The larvae of *albipennis* have been reported to feed on a wide variety of plants including potato, corn, sweet clover, and sunflower (Cook, 1930: 266). The species overwinters in the egg stage; the larva has a long prepupal aestivation period (Hinks and Byers, 1976). The egg was illustrated by Salkeld (1975: 1148, figures 31–33).

Euxoa albipennis is widely distributed in western North America where it occurs from central Saskatchewan, central Alberta, and southern British Columbia southward to southern Nebraska, southern New Mexico, Arizona, and the Mexican border of California. Its range extends eastward in southern Canada and northern United States to southern Quebec and New York. Adults have been collected from mid-August until early October.

Euxoa (*Euxoa*) *henrietta* (Smith)
PL. 4, FIGS. 37–40; PL. I, FIG. 2; PL. X, FIG. 3 (RWH 10806).

Carneades henrietta Smith, 1900, *Proc. U. S. Natl. Mus.*, **22**: 452.
Type locality: Alameda Co., California. [USNM]
NOTE—The male lectotype of *henrietta* was designated by Todd (1982: 97).

Euxoa delicata Barnes and McDunnough, 1912, *Contrib. Nat. Hist. Lep. N. Am.*, **1** (5): 5.
Type locality: Witch Creek, San Diego Co., California. [USNM]

Euxoa adusta Barnes and McDunnough, 1912, *Contrib. Nat. Hist. Lep. N. Am.*, **1** (5): 6.
Type locality: Plumas Co., California. [USNM]
NOTE—*Euxoa adusta* was described from two females, both in the United States National Museum. The specimen labeled "Sept. 16–23/Plumas Co. Calif./*Euxoa adusta* B. & McD., type female/Female genitalia on slide Apr. 1966 E.L.T. 2305" is here designated lectotype. The specimen is in good condition.

Specimens of *henrietta* can be arranged in three basic forms on the basis of forewing color and pattern. The forewing may be reddish brown with pale yellow-buff or pale reddish-yellow shading on the costa, in the subterminal area, and sometimes in the basal area; the maculation is prominent and contrasting; and the forewing has a black basal dash (plate 4, figures 37, 38). In a second form, the forewing is reddish brown, without areas of pale shading. Like the first form, the maculation is contrasting and there is a basal dash (plate 4, figure 39). In a third form the forewing is reddish brown with obscure maculation except for some yellow-buff shading in the reniform and orbicular spots (plate 4, figure 40). Forewing length varies from 13 to 16 mm. Specimens of *henrietta* can be recognized by wing pattern when the appearance of the three forms has been learned. The two well-marked forms tend to have a "blotchy" appearance that is reminiscent of the much larger *perexcellens* and *wilsoni*. Males can be distinguished from those of all other *Euxoa* species by the characters given for the group and in the key. Females are most similar to those of *tessellata* in shape of the corpus bursae, but the ovipositor lobes in *henrietta* are tapered to a point apically and clothed with fine setae rather than truncate apically and clothed with short, conical setae.

The immature stages of *henrietta* are unknown.

Euxoa henrietta occurs in the Pacific Coast states where it ranges from central Washington southward to southern California. It probably also occurs in western Nevada because it has been collected within a few miles of the Nevada state line in northern and central California. Adults have been collected from late August until early October.

cincta GROUP

This group includes two southwestern species. Both species can be distinguished from most other southwestern species by the combination of prominent, dark median line and dark hindwing. They can readily be distinguished from all other *Euxoa* species by genital characters: markedly constricted cucullus, disproportionately long harpe, and shape of the vesica in males and presence of short, conical setae scattered over the apical third of the ovipositor lobes in females. In other species that have conical setae on the ovipositor lobe, the setae are crowded at the apex of the lobe.

KEY TO SPECIES OF THE *CINCTA* GROUP

1. Ground color of forewing gray; occurring in mountains of southeastern Arizona .... *cincta* this page

— Ground color of forewing orange; known only from Coconino Plateau of northwestern Arizona .... *coconino* this page

Euxoa (*Euxoa*) *cincta* Barnes and Benjamin
PL. 4, FIGS. 41, 42; PL. I, FIG. 3; PL. X, FIG. 4; PL. EE, FIG. 15 (RWH 10823).

Euxoa cincta Barnes and Benjamin, 1924, *Contrib. Nat. Hist. Lep. N. Am.,* **5** (3): 108.
Type locality: Paradise, Cochise Co., Arizona. [USNM]

Specimens of *cincta* can be distinguished from those of most other *Euxoa* species that occur in Arizona by the even gray ground color of the forewing with a prominent, sharply defined median line and the dark hindwing. They are most likely to be confused with specimens of *setonia* and *lucida*; but in these species the forewing has some brown shading, and the median line is less sharply defined. In many specimens of *cincta* the reniform spot is orange and contrasts with the remainder of the forewing color. Forewing length varies from 15 to 16 mm. This species and the next can be distinguished from all other *Euxoa* species by the genital characters given above for the group. The subapically constricted cucullus of the male and the narrow, tapered ovipositor lobes of the female with their distinctive setal pattern can be observed without dissection.

The immature stages of *cincta* are unknown.

Euxoa cincta has been found only in southeastern Arizona, in the region in and around the Huachuca and Chiricahua Mountains. Adults have been collected from late June until late September. Little is known of the habitat preferences of *cincta,* but it has been collected from desert scrub to pine forests.

Euxoa (*Euxoa*) *coconino* Lafontaine, NEW SPECIES
PL. 4, FIG. 43; PL. I, FIG. 4.

Euxoa (*Euxoa*) *coconino* Lafontaine.
Type locality: Kaibab National Forest, Coconino Plateau, Coconino County, Arizona. [CNC]

This species is similar to *cincta* in forewing pattern and structural characters but differs from it in having an orange rather than gray forewing and in details of the male genitalia.

Antenna of male biserrate and bifasciculate. Frontal tubercle present. Eye round. Vestiture of head and thorax orange; prothoracic collar without transverse black line. Forewing ground color orange. Maculation generally obscure except for prominent, sharply defined, dark-brown median line and dark-brown shading in terminal area. Basal, antemedial and postmedial lines double, dark brown but thin and obscure. Reniform and orbicular spots barely evident, defined by narrow pale outline. Forewing length 14 to 15 mm. Hindwing smoky brown, slightly paler toward base. Male genitalia similar to those of *cincta* but harpe disproportionately shorter and thinner and subbasal diverticulum of vesica smaller. Female genitalia indistinguishable from those of *cincta.*

The immature stages of *coconino* are unknown.

TYPES. Holotype: ♂. Kaibab National Forest, Coconino Plateau, Arizona, ca. 7,000′; 20 Aug. 1960; E. G. Munroe; slide No. JDL 10,315; CNC (Type No. 19517). CNC. Paratypes: 1 ♂, 1 ♀. Same data as for holotype. CNC.

hollemani GROUP

This group includes three North American species. Specimens usually can be recognized by a combination of forewing characters including: small size; longitudinal streaking; lacking transverse lines; and teardrop-shaped orbicular spot. In two species the orbicular spot is fused with the reniform spot; they could be confused with members of the *atristrigata* group but can be distinguished from them by smaller size, pale shading on costa and reniform and orbicular spots, long rather than short saccular extensions in the male genitalia and by ovipositor lobe and bursa shape in the female genitalia.

The group is unusual in that the three species show a marked dissimilarity in shape of the vesica and shape of the corpus bursae. The main feature of the vesica shared by the species is the highly modified shape of the subbasal diverticulum and subbasal bend in the vesica. The saccular extensions are as long as, or longer than, the harpes; the harpe is not pubescent as in males of *oblongistigma,* which may be similar in wing markings. The shape of the corpus bursae may be similar to that of females in the *cincta* group (*hollemani*), or the *tessellata* group (*subandera, xasta*) but they can readily be distinguished from species in these groups by wing markings.

KEY TO THE SPECIES OF THE *HOLLEMANI* GROUP

1. Males 2
— Females 4

2. Hindwing dark smoky brown; cucullus markedly constricted subapically; vesica with single ventral bulge in subbasal bend, its surface covered with spinules .... *subandera* p. 94
— Hindwing white or pale smoky brown except on wing margin; cucullus slightly constricted subapically, vesica without ventral spinule-covered bulge in subbasal bend 3

3. Orbicular and reniform spots of forewing fused in most specimens; harpe evenly incurved, C-shaped; vesica with very large subbasal diverticulum but no other subbasal pouches in vesica .... *hollemani* this page
— Orbicular spot not fused with reniform spot; harpe outcurved toward apex, S-shaped; vesica with three pouches in wall subbasally in addition to normal-sized subbasal diverticulum .... *xasta* p. 94

4. Ovipositor lobe with apical flangelike projection .... *xasta* p. 94
— Ovipositor lobe without apical flangelike projection 5

5. Hindwing dark brown; corpus bursae elongate, constricted mesially, enlarged at each end .. .. *subandera* p. 94
— Hindwing pale smoky brown; corpus bursae short and broad .... *hollemani* this page

Euxoa (*Euxoa*) *hollemani* (Grote)

PL. 4, FIGS. 44, 45; PL. I, FIG. 5; PL. X, FIG. 5 (RWH 10820, 10822).

Agrotis hollemani Grote, 1874, *Can. Ent.,* **6**: 156.

Type locality: Siskiyou County, California. [MCZ]

NOTE—The type material of *hollemani* was lost for many years, so the species was determined from the original description. In the description of *hollemani* the hindwings are described as "blackish fuscous." This led Smith to assume that the dark-winged species was *hollemani*; he described the paler species as *andera.* This confusion resulted from two factors: *hollemani* was described from females although this is not stated by Grote; and specimens from the Siskiyou Mountains of California, the type locality of *hollemani,* have darker hindwings than do those from other areas. Now that the type material of *hollemani* has been located, it is clear that the names *hollemani* and *andera* refer to the same species. The type material of *hollemani* consists of two females, both labeled "Cal., Siskiyou Co., Col. Holleman." Only one specimen bears a second label "Type *Agrotis Hollemani* Grote" so it is possible that Grote did not examine both specimens when he described the species. To avoid further confusion the specimen bearing the type label is here designated as lectotype. The specimen is slightly worn and lacks antennae. Both specimens are deposited in the Museum of Comparative Zoology, Harvard University, Cambridge, Massachusetts.

Euxoa andera Smith, 1910, *Jour. New York Ent. Soc.,* **18**: 89. NEW SYNONYMY.
Type locality: Redington, Arizona. [AMNH]

NOTE—The female lectotype of *andera* was designated by Todd (1982: 17).

This is the common, widely distributed member of the group that was known as *andera* for many years. Typical specimens of *hollemani,* in which the orbicular and reniform spots are fused into an hourglass-shaped spot, are unlikely to be confused with those of any other species except *subandera.* Specimens of *hollemani* can be distinguished from those of *subandera* by the pale-colored rather than dark hindwings, the more narrow forewing and by the genital characters discussed below. Forewing length varies from 10 to 16 mm. In some specimens of *hollemani* the orbicular and reniform spots are not fused. These specimens could be confused with those of *oblongistigma* in the *detersa* group (p. 128) but are distinguished from them by genital characters. Males of *hollemani* can be distinguished from those of *subandera* by the differently shaped cucullus, the evenly incurved, C-shaped rather than S-shaped harpe, and by the longer saccular extensions. They can be distinguished from males of *oblongistigma* by the much larger cucullus, stouter saccular extensions, and lack of pubescence on the harpes. Males of *hollemani* can be distinguished from those of these two species, and all other *Euxoa* species, by the unique shape of the vesica, particularly the very large subbasal diverticulum. Females of *hollemani* can be distinguished from those of *subandera* by the oval rather than dumbbell-shaped corpus bursae and from those of *oblongistigma* by the lack of a sclerotized flangelike projection at the apex of each ovipositor lobe.

The immature stages of *hollemani* are unknown.

Most of the range of *hollemani* lies west of the Rocky Mountain region. It occurs in open aridlands and open coniferous forest habitats from south-central British Columbia southward to central New Mexico, central Arizona, and southern California. An apparently disjunct population occurs in the northwestern Great Plains in eastern Montana and northeastern Wyoming (Lafontaine, 1982a: figure 26 (as *andera*)). *Euxoa hollemani* is, unlike *subandera,* a rather common species. I have examined specimens from more than 100 localities. Adults have been collected from mid-August until mid-October.

Euxoa (*Euxoa*) *subandera* Lafontaine, NEW SPECIES
PL. 4, FIG. 46; PL. I, FIG. 6; PL. X, FIG. 6.

Euxoa hollemani of authors, not Grote, 1874.

Euxoa (*Euxoa*) *subandera* Lafontaine.
Type locality: La Tuna Canyon, Los Angeles County, California. [CNC]

For many years, this species was confused with the similarly marked, and much more common species *hollemani.* Also, the name *hollemani* was applied to this species because of confusion surrounding the identity of *hollemani* (see note under *hollemani*). The two species can be distinguished by wing color and by the male and female genital differences discussed below.

Antenna of male biserrate and bifasciculate. Antenna of female filiform, pubescent below. Frontal tubercle present. Eye round. Vestiture of head and thorax brown; prothoracic collar with prominent transverse black line. Forewing ground color brown with dusting of gray scales. Transverse lines absent. Reniform and orbicular spots fused into hourglass-shaped stigma, pale brownish gray, outlined in black. Basal dash prominent, black. Claviform spot outlined in black but inconspicuous. Dark blackish-brown shading in terminal area streaked into subterminal area. Forewings disproportionately shorter and broader than those of *hollemani*; forewing length varies from 12 to 16 mm. Hindwing dark smoky brown, slightly paler toward wing base in males. Male genitalia similar to those of *xasta* but harpe more markedly S-shaped, juxta narrower, and cucullus more markedly constricted through basal ⅔. Vesica with prominent subbasal bulge covered with spinules; lacking series of subbasal pouches found in *xasta.* Female genitalia with corpus bursae narrower and more prominently constricted mesially than in other two species in group. Ovipositor lobe lacks sclerotized flangelike process found in *xasta.*

The immature stages of *subandera* are unknown.

TYPES. Holotype: ♀. La Tuna Canyon, Los Angeles Co., California; 1 Sept. 1948; William H. Evans. CNC (Type No. 19518). Paratypes: 14 ♂, 21 ♀. California: Atascadero; 26 July 1935; E. C. Johnston (2 ♀); Badger, 4 mi NE, 4,800′; 16 Sept. 1966; D. F. Hardwick (1 ♂); Big Pine Area, 6,800′, San Gabriel Mts., Los Angeles Co.; 16, 23 July 1963; C. Henne (2 ♂, 1 ♀); Jackson Lake, 1 mi NW, 5,900′, San Gabriel Mts., Los Angeles Co.; 30 July 1967; C. Henne (1 ♂); Jackson Lake, Big Pine Area, 5,800′, San Gabriel Mts., Los Angeles Co.; 23 July 1963; C. Henne (1 ♂, 1 ♀); Juniper Hills, 3,500′, 2 mi S Pearblossom, Los Angeles Co.; 28 July, 6, 12 Aug. 1967; C. Henne (1 ♂, 3 ♀); Juniper Hills, Mojave Desert, 3,500′, Los Angeles Co.; 19 July, 1963, 22, 24 Aug. 1962; C. Henne (1 ♂, 4 ♀); Kelseyville, Lake Co.; 2, 4 Aug. 1933 (1 ♂, 2 ♀); Lee Vining, 7 mi WSW, 9,600′; 13 Aug. 1967; D. F. Hardwick (1 ♂). Nevada: Panaca, 10 mi E, 6,500′; 1 Sept. 1965; D. F. Hardwick (1 ♀); Preston, 12 mi WNW, 6,600′; 8 Sept. 1969; D. F. Hardwick (1 ♂). Oregon: Cove State Park near Madras, 1,500′; 31 Aug. 1962; W. C. Cook (1 ♀). Utah: Logan; 14 Aug. 1939; G. F. Knowlton (1 ♀). Washington: Rimrock, 2 mi E, 3,500′; 17 Aug. 1961; W. C. Cook (1 ♀); Walla Walla, 24 Aug. 1946, 17 Aug. 1947, 17 Aug. 1950, 19 Aug. 1954, 24 July 1955, 15 Aug. 1962, and 19 Aug. 1966; W. C. Cook (4 ♂, 4 ♀). CNC, LACM.

Euxoa subandera is a rare species known from less than 20 localities. Most collections of it were made in coniferous forests in montane areas from southern Washington southward to southern California. East of this, it has been collected in Nevada, Utah, and Arizona. Adults have been collected from late July until mid-September.

Euxoa (*Euxoa*) *xasta* Barnes and McDunnough
PL. 4, FIGS. 47, 48; PL. I, FIG. 7; PL. X, FIG. 7 (RWH 10821).

Euxoa xasta Barnes and McDunnough, 1910, *Can. Ent.,* **42**: 429.
Type locality: Kerrville, Texas. [USNM]

Euxoa xasta is a poorly known species of the American Southwest. It differs from *hollemani* and *subandera* in that the forewing ground color is much paler and that the orbicular spot is teardrop shaped and tapered toward the wing base. Forewing length varies from 14 to 16 mm. The male genitalia resemble those of *subandera,* but the cucullus is much smaller and not expanded apically, the uncus is broader, and the ventral wall of the vesica has a row of three pouches subbasally. By external appearance, specimens of *xasta* are most likely to be confused with those of *atristrigata* (p. 145). Specimens of *xasta* can be recognized by their gray rather than brown forewings, by males having saccular extensions that are about as long as the harpes rather than about half as long as in *atristrigata* males, and by females having small flangelike projections on the ovipositor lobes, not massive as in *atristrigata* females, and the corpus bursae expanded posteriorly, somewhat T-shaped rather than oval.

The immature stages of *xasta* are unknown.

Euxoa xasta ranges from southern Nevada and Utah southward and eastward to southeastern Arizona, southern New Mexico and western and south-central Texas. Like many species in the southwestern region, it has a long adult flight period. Adults have been collected from mid-May until mid-September. From a limited number of collecting sites, it appears that *xasta* occurs in open deserts.

catenula GROUP

The *catenula* group includes only *catenula*; the characters are given under the species.

Euxoa (*Euxoa*) *catenula* (Grote)

PL. 4, FIGS. 49–51; PL. I, FIG. 8; PL. X, FIG. 8; PL. EE, FIG. 16 (RWH 10809).

Agrotis catenula Grote, 1879, *North Amer. Ent.,* **1**: 44.
Type locality: Idaho Springs, Colorado. [BMNH]

Carneades contagionis Smith, 1900, *Proc. U. S. Natl. Mus.,* **22**: 421.
Type locality: Garfield County, Colorado. [USNM]

NOTE—The female lectotype of *contagionis* was designated by Todd (1982: 56).

Euxoa lindseyi Blackmore, 1923, *Can. Ent.,* **55**: 214.
Type locality: Goldstream, British Columbia. [CNC]

This species looks like a small version of *Agrotis vetusta* Walker and is as likely to be confused with it as with any species of *Euxoa.* In addition to characters of the genitalia that distinguish the two genera, specimens of *catenula* can be recognized by their smaller size (forewing length 13 to 17 mm versus 17 to 21 mm in *vetusta*), lack of a series of pale spots along the postmedial line, and presence of dark shading in the orbicular spot. Within *Euxoa,* specimens of *catenula* are most frequently confused with those of *tronella* (Smith) (*detersa* group, p. 129). They can be distinguished from most specimens of *tronella* by the presence of prominent black spots on the forewing costa at the end of the antemedial and postmedial lines, by the more punctiform postmedial line, by the small, dark orbicular spot, and by the less extensive dark shading in the terminal area. *Euxoa catenula* can be distinguished from *tronella* and all other *Euxoa* species by diagnostic characters of the genitalia. The highly distinctive harpe is very short—less than half as long as the saccular extension, very markedly incurved and C-shaped, and slightly expanded apically with a dense cluster of setae at the apex. The shape of the vesica also is unique in that the apical portion has a row of four or five pouches in the wall distal to the median diverticulum. Females have the unique combination of a dense border of setae subbasally on the ovipositor lobe and an apical flangelike projection. The only other species that have a dense border of setae on the ovipositor lobe are members of the *terrena, serricornis,* and *annulipes* species-groups; females in these groups do not resemble those of *catenula* in wing markings, and the ovipositor lobe has a sclerotized rim along the posterior margin rather than an apical flangelike projection.

Most specimens of *catenula* have pale whitish-gray forewings, often with a slight pink suffusion. Specimens from southeastern Vancouver Island, British Columbia, differ from specimens from other areas in that the forewing ground color is smoky gray and heavily dusted with dark gray; this population was described as *lindseyi* by Blackmore. This population may warrant a subspecific name, but I do not formally recognize it as such here because there is tendency in other species to have darker forms in coastal British Columbia; additional material of *catenula* from western British Columbia is needed to determine the status of this form.

The larvae of *catenula* have been collected in Montana (Cook, 1930: 262) and reared in the laboratory (Hinks and Byers, 1976). In Montana the larvae were reported to feed on wheat, sweet clover, lupines, Russian thistle, mustards, violets, and milk-vetch. The species overwinters in the egg stage. The egg was illustrated by Salkeld (1975: 1151, figs. 52–54).

Euxoa catenula is widely distributed in aridlands and open coniferous forests in western North America. It occurs from the prairie region of southern Saskatchewan westward to southwestern British Columbia and southward to western Kansas, central New Mexico, and Arizona and southern California. Adults have been collected from mid-August until early October.

comosa GROUP

The *comosa* group includes seven species in North America. Members of the group also may occur in eastern Asia because *comosa* occurs on the west coast of Alaska and *hypochlora* Boursin (1964: 10, figs. 1: 6 and 5: 8), described from Nepal, appears to be a member of the *comosa* group.

Members of the group can be recognized with certainty only by genital characters although the combination of a drably colored forewing with a prominent median line and pale-outlined reniform and orbicular spots is a good "rule-of-thumb" character for recognizing species in the group.

The sacculus is crescentic; the saccular extensions are about 1.0–1.5× as long as the harpes and are markedly stouter than the harpes. The left and right saccular extensions are similar in length. The vesica bears a small, footlike diverticulum; the vesica projects dorsally at an angle of about 90° and is tapered near the apex. The ovipositor lobes have a sclerotized flangelike process at the apex of each lobe. The corpus bursae is unisaccate, slightly rectangular, and broader posteriorly with a bulge posteriorly on the left side.

Most species can be identified most easily by subtle differences in wing markings that are difficult to describe but can be discerned by comparison with the illustrations. Also, most species can be recognized by structural characters in one sex only.

The *comosa* group was revised by Lafontaine and Byers (1982). Decisions as to whether various populations should be treated as different species, subspecies, or as a single species with considerable geographical variation, are based on behavioral and hybridization experiments done in the laboratory, or by extrapolation from these experiments. The results of these experiments were reported by Byers and Lafontaine (1982).

KEY TO THE SPECIES OF THE *COMOSA* GROUP

1. Forewing charcoal colored, reniform and orbicular spots outlined in black; hindwing of male white; male genitalia with additional, small nipplelike diverticulum in vesica subbasally; occurring from Montana eastward .. .. *velleripennis* p. 102

– Forewing brown or pale colored; reniform and orbicular spots, if present, outlined by pale line; hindwing of male smoky brown, except in specimens with very pale-colored forewings; male genitalia with only normal footlike diverticulum in vesica subbasally; widely distributed 2

2. Forewing brown; occurring in eastern North America (east of 95th meridian) 3

– Forewing not brown or occurring in central or western North America (west of 95th meridian) 4

3. Forewing an even brown color with all markings obscure except for dark, contrasting antemedial and postmedial lines; male genitalia with cluster of small spines in vesica between apex of aedoeagus and subbasal diverticulum; occurring in eastern deciduous forest zone from southern Canada southward to Pennsylvania .... *fumalis* p. 101

– Forewing mottled brown with maculation obscure or contrasting but if contrasting, then median line prominent as well; male genitalia without cluster of small spines in vesica; occurring in boreal forest zone of Canada and northern New England states ..... *comosa ontario* p. 100

4. Ground color of forewing white dusted with black scales, median line prominent, black; forewing fringe yellow; found only in Guadalupe Mountains in western Texas ... *guadalupensis* p. 100

– Forewing not as described above; occurring north of Texas 5

5. Forewing cream colored with transverse lines, particularly median line, dark and contrasting; hindwing with basal two-thirds pale buff, outer third pale smoky gray; occurring in arid valleys within Rocky Mountain region from Idaho southward to Colorado .... *lineifrons* p. 100

– Not as described above 6

6. Sclerotized flangelike projection on ovipositor lobe stout, heavily sclerotized, partially fused together, bent laterally and forming a triangular shovellike apparatus; forewing pale gray or pale orange with dark shading in terminal area, reniform and orbicular spots outlined in pale yellow; hindwing even smoky brown, frequently darker than forewing .... *lucida* p. 101

– Sclerotized flangelike projection on ovipositor lobe more vertically oriented, not fused or forming a plate; forewing color variable but rarely like that described above 7

7. Male genitalia with saccular extensions not evenly tapered from base to apex but as stout, or stouter, ⅔ from base as at ⅓ from base; vesica swollen apically; right sacculus more than 1.5 mm long in most specimens; forewing charcoal colored with dusting of paler gray and brown scales .... *occidentalis* p. 102

– Male genitalia with saccular extensions evenly tapered from base, not abruptly tapered in apical third; vesica not swollen apically; right sacculus less than 1.5 mm long in most specimens; forewing ground color variable, but in dark specimens it is brown with darker brown markings .... *comosa* this page

Euxoa (*Euxoa*) *comosa* (Morrison)

PL. 4, FIGS. 52–60; PL. 5, FIGS. 1–15; PL. I, FIG. 9; PL. Y, FIG. 1; PL. EE, FIG. 17 (RWH 10780).

Agrotis comosa Morrison, 1876, *Proc. Boston Soc. Nat. Hist.*, **18**: 238.
Type locality: Colorado. [MSU]

Agrotis lutulenta Smith, 1890, *Trans. Amer. Ent. Soc.*, **17**: 50. SUBSPECIES.
Type locality: Sierra Nevada, California. [USNM]

NOTE—The male lectotype of *lutulenta* was designated by Todd (1982: 128).

Agrotis incallida Smith, 1890, *Trans. Amer. Ent. Soc.*, **17**: 50.
Type locality: Sierra Nevada, California. [USNM]

NOTE—*Agrotis incallida* Smith, 1890, is a junior primary homonym of *Agrotis incallida* Walker, [1857], a synonym of *Luperina passer* (Guenée, 1852).

Carneades vulpina Smith, 1895, *Ent. News*, **6**: 335.
Type locality: Calgary, Alberta. [USNM]

Agrotis annir Strecker, 1898, *Lepidoptera, Rhopaloceres and Heteroceres, Indigenous and Exotic*, Suppl., **1**: 6. SUBSPECIES.
Type locality: Loveland, Colorado. [FMNH]

NOTE—*Agrotis annir* was described from four specimens, one male and three females, currently in the Field Museum of Natural History at Chicago. The male specimen labeled "Colo./Male/*A. annir* orig. Type/*Euxoa* slide Strecker No. 8" is here designated lectotype. The specimen is in good condition.

Carneades atropulverea Smith, 1900, *Proc. U. S. Natl. Mus.*, **22**: 428.
Type locality: Colorado. [USNM]

NOTE—The female lectotype was designated by Todd (1982: 26).

Carneades ontario Smith, 1900, *Proc. U. S. Natl. Mus.*, **22**: 440. SUBSPECIES.
Type locality: Sudbury, Ontario. [USNM]

Carneades dakota Smith, 1900, *Proc. U. S. Natl. Mus.*, **22**: 442.
Type locality: Dakota. [USNM]

NOTE—The male lectotype of *dakota* was designated by Todd (1982: 60).

Carneades ternarius Smith, 1900, *Proc. U. S. Natl. Mus.*, **22**: 444.
Type locality: Denver, Colorado. [USNM]

NOTE—The female lectotype of *ternarius* was designated by Todd (1982: 208).

Euxoa vestitura Smith, 1905, *Can. Ent.*, **37**: 201.
Type locality: St. John, New Brunswick. [AMNH]

NOTE—The male lectotype of *vestitura* was designated by Todd (1982: 225).

Euxoa brunneigera masoni Cockerell, 1905, *Can. Ent.*, **37**: 361.
Type locality: Glenwood Springs, Colorado. [? lost]

NOTE—The holotype of *masoni*, reported to be in the Mason Collection, has not been located. The information given in the description is, however, sufficiently detailed to allow the taxon to be identified.

Euxoa brunneigera latebra Benjamin, (1936), *Bull. So. California Acad. Sci.*, **34**: 201.
Type locality: Truckee, California. [USNM]

Euxoa luteotincta McDunnough, 1940, *Can. Ent.*, **72**: 195.
Type locality: Seton Lake, British Columbia. [CNC]

Euxoa altera McDunnough, 1940, *Can. Ent.*, **72**: 196. SUBSPECIES.
Type locality: Cartwright, Manitoba. [CNC]

Euxoa johnstoni McDunnough, 1946, *Can. Ent.*, **78**: 29.
Type locality: Mt. Hood, Oregon. [CNC]

Euxoa yildizae Kocak, 1983, *Priamus*, **3**: 42. NEW SYNONYMY.

NOTE—*Euxoa yildizae* is a replacement name for *Agrotis incallida* Smith, 1890, a junior primary homonym of *Agrotis incallida* Walker, [1857].

This is probably the most variable species of *Euxoa* in North America in wing pattern and color. It can best be identified by eliminating the other, more localized species that are more homogeneous in color. Basically, the forewing ground color is dark brown in coniferous forest habitats in eastern North America, and in the West from Alaska southward to southern California. The ground color is pale gray, pale brown, or yellow in arid areas in the Great Plains and Great Basin with very pale forms occurring in the driest areas; orange or reddish-brown forms occur in the southern Rocky Mountain region and southern Great Basin. The median line is prominent on most dark-colored specimens but is less evident or absent in very pale-colored specimens. The orbicular and reniform spots are outlined by a pale line. Forewing length varies greatly from 12 to 18 mm; pale-colored specimens are, on average, smaller than dark-colored specimens. In most areas the hindwing is smoky brown, but specimens from some populations have the dark shading confined to the outer third of the wing. In them there is a dark median shade on the hindwing. The male and female genitalia lack diagnostic characters by which specimens of this species can be distinguished from those of some other species in the group. Fortunately, most specimens of *comosa* can be distinguished by wing color and pattern.

Knowledge of the immature stages is based almost entirely on laboratory reared material although the larva has been reported as a pest on rye (*Secale cereale* Linnaeus) in Saskatchewan (Beirne, 1971: 17). A comparison of life history information among four subspecies was given by Byers and Lafontaine (1982).

Euxoa comosa inhabits a wide variety of habitats from arid desert areas to mesic coniferous forests. It occurs from Newfoundland westward to western Alaska, southward in the East to the New England States, eastern Ontario, northern Michigan, and southern Minnesota and Nebraska, central New Mexico, southern Arizona, and southern California in the West. Adults have been collected from late July until late September.

Populations of *comosa* have been arranged in five subspecies based on differences in wing color and pattern, genital proportions and life history data. These consist of: *ontario*, a dark-brown form that occurs eastward from western Ontario and northern Michigan; *altera*, a large gray form of the eastern Great Plains and aspen parkland of central Canada; *annir*, a small, pale-colored form of the western Great Plains and Great Basin; *lutulenta*, a dark-brown form of the northern Rocky Mountain region and the montane regions west of the Great Basin; *comosa*, a reddish-brown form of the Colorado Plateau and southern Rocky Mountain region. These five subspecies are discussed in alphabetical order below.

Euxoa (*Euxoa*) *comosa altera* McDunnough
PL. 4, FIGS. 52–55.

Euxoa altera McDunnough, 1940.

In this subspecies the forewing is pale gray or pale grayish brown. The hindwing is smoky brown, slightly paler toward the base; a dark median line is evident in most specimens. Populations of *altera* can be arranged in two forms. Specimens from the aspen parkland of central Canada have pale, relatively evenly colored forewings with dark, contrasting transverse lines (plate 4, figures 52–54). Specimens from North and South Dakota and Minnesota (plate 4, figure 55) have forewings that are extensively dusted with dark-brown scales; the median line is darker than the transverse lines in most specimens.

This subspecies occurs from southern South Dakota and Minnesota, northward and westward through western Manitoba and across central Saskatchewan and Alberta to Fort Simpson in the Northwest Territories. Subspecies *altera* intergrades with *annir* in the northern Great Plains although a disjunct population of *altera* in the Cypress Hills of southwestern Saskatchewan occurs with *annir* without evidence of intergradation. There is, at present, no evidence of intergradation between *altera* and *ontario*; however, no material has been collected in southeastern Manitoba, where their ranges probably meet.

Adults have been collected from mid-August until late September; they are usually found in spruce and aspen forests, or along forested river bottoms in the eastern Great Plains.

Euxoa (*Euxoa*) *comosa annir* (Strecker)
PL. 4, FIGS. 56–60; PL. Y, FIG. 1; PL. EE, FIG. 17.

Agrotis annir Strecker, 1898.

Carneades dakota Smith, 1900.

Relatively small, pale-colored specimens of *comosa* from the western Great Plains and Great Basin are referable to subspecies *annir*. Like other subspecies of *comosa*, populations of *annir* vary markedly from locality to locality. Specimens from the northern Great Plains region (plate 4, figures 56, 57) have pale-buff, or yellow-buff forewings; the hindwing color varies from entirely smoky gray to having smoky gray only on the marginal half of the wing with dirty-white shading in the basal half. In specimens from very arid regions in southwestern Montana and Wyoming (plate 4, figure 58) the forewing is creamy white with a slight dusting of dark-brown scales; the hindwing is entirely smoky gray in most specimens. In forested habitats in the foothills of western Montana and in montane areas of Wyoming, Utah, and Nevada, the forewing is pale brown or pale reddish brown (plate 4, figure 59); this form appears to be intermediate between aridland populations of *annir* and montane populations of *lutulenta*. A fourth form of *annir*, which matches the holotype from Loveland, Colorado, differs from those described above in several ways; the specimens are larger, the forewing is yellowish buff with a dusting of dark-brown scales that gives the wing a slight olive cast, the hindwing is dirty white with dark shading on the wing margin and on the median line (plate 4, figure 60). This latter form occurs in the Great Plains region of eastern Colorado and in central Montana. In Montana, this form blends imperceptibly into the other forms of *annir* in the area.

Subspecies *annir* occurs from southern Saskatchewan and Alberta southward to southern Colorado and westward through Wyoming into the Great Basin region. It intergrades with *altera* in the northern Great Plains, with subspecies *lutulenta* in western Alberta and Montana and with subspecies *comosa* in the Colorado Plateau region. Populations of subspecies *annir* in eastern Colorado and *comosa* in central Colorado occur within sight of each other without any evidence of intergradation in this region. Adults of *annir* have been collected from mid-August until late September.

Euxoa (*Euxoa*) *comosa comosa* (Morrison)
PL. 5, FIGS. 1–5; PL. I, FIG. 9.

Agrotis comosa Morrison, 1876.

Carneades atropulverea Smith, 1900.

Carneades ternarius Smith, 1900.

Euxoa brunneigera masoni Cockerell, 1905.

Specimens of the nominate subspecies of *comosa* can be recognized by the reddish-brown or reddish-orange forewing color. The shade of the forewing varies by locality and habitat. Specimens from mesic habitats, such as Douglas-fir forests, have the darkest forewings (plate 5, figures 1, 2); those from drier habitats, such as open pine forests, are paler (plate 5, figures 3, 4); those from the driest habitats (plate 5, figure 5), such as piñon-juniper woodland, are the palest and the smallest.

Subspecies *comosa* occurs in montane areas from north-central Colorado westward to north-central Utah and southward to central New Mexico and southern Arizona. Along the Colorado Front Range, subspecies *comosa* occurs in montane coniferous forests adjacent to populations of *annir* in the nearby aridlands. In this region there is no evidence of intergradation between subspecies. Where habitat gradients are not as abrupt, such as in the Colorado Plateau region of western Colorado and eastern Utah, piñon-juniper woodland provides a habitat intermediate between that normally inhabited by *comosa* and that by *annir*. In this region, most specimens are intermediate in size and color between typical *annir* and typical *comosa*.

Adults of subspecies *comosa* have been collected from late July until mid-September.

Euxoa (*Euxoa*) *comosa lutulenta* (Smith)
PL. 5, FIGS. 6–11.

Agrotis lutulenta Smith, 1890.

Agrotis incallida Smith, 1890.

Carneades vulpina Smith, 1895.

Euxoa brunneigera latebra Benjamin, (1936).

Euxoa luteotincta McDunnough, 1940.

Euxoa johnstoni McDunnough, 1946.

Euxoa yildizae Kocak, 1983.

Subspecies *lutulenta* occurs north and west of the Great Basin region and like other subspecies of *comosa,* specimens vary markedly in wing color and markings. In the most widespread form, the forewing is dark brown or dark grayish brown with the pale areas of the forewing, particularly the pale outline of the orbicular and reniform spots, pale golden brown or pale buff. This form occurs primarily in the northern half of the range of *lutulenta,* as far south as northern California. To the south of this, along the Cascades-Sierra Nevada Mountain axis, there is a general north-south cline in which the forewing ground color is increasingly paler southward. As a result of the paler ground color, the forewing maculation, especially the median line, becomes more and more prominent and contrasting. At the end of this cline, in the Transverse Ranges of southern California, specimens resemble large, prominently marked specimens of subspecies *annir.*

Subspecies *lutulenta* is most commonly found in coniferous forests of spruce, pine, and Douglas-fir.

Euxoa (*Euxoa*) *comosa ontario* (Smith)

PL. 5, FIGS. 12–15.

Carneades ontario Smith, 1900.

Euxoa vestitura Smith, 1905.

The eastern subspecies of *comosa* resembles the western subspecies, *lutulenta,* in wing color and markings. Specimens from Nova Scotia and New Brunswick tend to be slightly paler and more reddish brown than do specimens from other parts of the range. The genitalia cannot be distinguished from those of specimens of other subspecies although the saccular extensions and harpes are, on average, longer in *ontario* males than in other subspecies.

Subspecies *ontario* occurs in boreal forest habitat from Newfoundland westward to western Ontario and southward to northern New England and northern Michigan. Specimens of *ontario* are most likely to be confused with those of *fumalis* but can be distinguished from them by characters given in the key.

Euxoa (*Euxoa*) *lineifrons* (Smith)

PL. 5, FIGS. 16, 17 (RWH 10784).

Carneades lineifrons Smith, 1890, *Bull. U. S. Natl. Mus.,* **38**: 219.
Type locality: Colorado. [USNM]

Carneades audentis Smith, 1894, *Trans. Amer. Ent. Soc.,* **21**: 51.
Type locality: Colorado. [USNM]

NOTE—The male lectotype of *audentis* was designated by Todd (1982: 26).

This species is closely related to *comosa,* or possibly conspecific with it; the male and female genitalia of the two species are indistinguishable. Specimens of *lineifrons* can be recognized by the cream-colored forewing with prominent, dark median line. In wing pattern, *lineifrons* appears to be an extreme form of the Colorado prairie form of *comosa annir* that is disjunct in interior valleys of the Rocky Mountain region. I retain *lineifrons* as a distinct species because it occurs sympatrically with *comosa comosa* in Colorado and with *comosa lutulenta* in Idaho without evidence of intergradation. It is possible that these populations form a "circle of races" in which *lineifrons* and *annir,* and *annir* and *comosa,* are interfertile but *lineifrons* and *comosa* are not. Living material of *lineifrons* has not been obtained to test the fertility of these crosses. The forewing length of *lineifrons* varies from 15 to 17 mm.

The immature stages of *lineifrons* are unknown.

Euxoa lineifrons is known only from central Colorado, northeastern Utah, and east-central Idaho. It occurs in dry valleys in the interior of the Rocky Mountain region. Adults have been collected from late July until mid-September.

Euxoa (*Euxoa*) *guadalupensis* Lafontaine and Byers

PL. 5, FIGS. 18, 19.

Euxoa guadalupensis Lafontaine and Byers, 1982, *Can. Ent.,* **114**: 581.
Type locality: Bear Canyon, Guadalupe Mountains, Texas. [USNM]

If *lineifrons* can be described as being a degree removed from Colorado populations of *comosa annir,* then *guadalupensis* can be considered to be several degrees further removed. The forewing ground color is white with black shading on the median line and proximal to the subterminal line along with a sprinkling of black scales over the entire forewing surface. The effect of the black scales over a white ground color is that the forewing has a very powdery appearance unlike that of any other species in the group. The forewing length varies from 15 to 17 mm. Only a single male of *guadalupensis* is known so any remarks on genital characters must be considered tentative. In this male, the saccular extensions are short (right side 1.18 mm, left side 1.13 mm), similar to those of *lineifrons* (right 1.19 ± 0.07 mm, N = 7, left 1.15 ± 0.10 mm) but shorter than those of *comosa annir* males (right 1.37 ± 0.12, N = 20, left

1.20 ± 0.12 mm) (Lafontaine and Byers, 1982). The female genitalia of the three species are indistinguishable.

The immature stages of *guadalupensis* are unknown.

This species is known only from the Guadalupe Mountains of western Texas where it has been collected in Bear Canyon and MacKittrick Canyon. Specimens range in collection date from 29 August until 2 October.

Euxoa (*Euxoa*) *lucida* Barnes and McDunnough

PL. 5, FIGS. 20–23; PL. I, FIG. 10; PL. Y, FIG. 2 (RWH 10782).

Euxoa lucida Barnes and McDunnough, 1912, *Contrib. Nat. Hist. Lep. N. Am.*, **1** (5): 7.
Type locality: Eureka, Utah. [USNM]

NOTE—*Euxoa lucida* was described from three males, now in the collection of the United States National Museum of Natural History. One of these males, labeled "Eureka, UT.; VIII.26.11; Tom Spalding/*Euxoa lucida* B. & McD., type male/Barnes Collection/male genitalia on slide Apr. 1966; ELT 2265" is here designated lectotype.

Specimens of *lucida* can be recognized with certainty only by the female genitalia. They usually can be distinguished from those of *comosa annir* by subtle differences in wing color and markings. Differences between the two vary from locality to locality because of geographical variation in both species. Where the two occur together, males can be identified by association with females. In general, except for dark smoky-gray shading in the terminal area and proximal to it in the subterminal area, the forewing appears washed out. The forewing ground color is pale yellowish gray in specimens from Montana, Utah, and Nevada, and yellowish orange in those from Colorado and Arizona. The fringe and pale areas of the forewing are pale yellow, and in most specimens the orbicular spot lacks a dark central spot. Forewing length varies from 12 to 16 mm. As mentioned above, the only diagnostic structural character of *lucida* is in the female genitalia: the sclerotized flangelike structures at the apex of the ovipositor lobes are heavily sclerotized, bent sharply to the sides and appear dog eared; they frequently are fused partially so that together they form a flattened, somewhat triangular, spadelike structure. In females of other species in the *comosa* group the processes are vertically oriented although they frequently curve slightly to the sides.

The immature stages of *lucida* are unknown.

Euxoa lucida has a relatively limited range, occurring in only a portion of the range of *comosa annir*. It has been collected in western Montana, and in the intermontane region from northern Utah southward to southeastern California (New York Mountains), central Arizona, and west-central Colorado. Adult collection dates range from mid-August until mid-September.

Euxoa (*Euxoa*) *fumalis* (Grote)

PL. 5, FIGS. 24, 25; PL. J, FIG. 1; PL. Y, FIG. 3 (RWH 10781).

Agrotis fumalis Grote, 1873, *Bull. Buffalo Soc. Nat. Sci.*, **1**: 98.
Type locality: New York. [lost]

NOTE—The types of both *fumalis* and *permunda* are lost; this was discussed in full by McDunnough (1950: 380). Morrison's two specimens, from Massachusetts and Canada, may well have represented both *fumalis* and *comosa ontario*; his description applies equally well to both species. Because *permunda* has been in synonymy with *fumalis* for over 100 years, and to avoid any possible confusion, I here restrict the type locality to Massachusetts, where only *fumalis* occurs.

Agrotis permunda Morrison, 1874, *Proc. Boston Soc. Nat. Hist.*, **17**: 163.
Type locality: Massachusetts; Canada. [lost]

Considerable confusion existed for many years as to the relationship of *fumalis* and *ontario* until the status of the two species was clarified by McDunnough (1950: 380–382). Throughout most of its range, specimens of *fumalis* can readily be recognized by wing markings. Also, its range overlaps that of *ontario* only in southeastern Canada. The forewing ground color is a light chocolate brown, without the dusting of pale-gray and black scales that is present on the forewing in *ontario*. Unlike specimens of *ontario*, the median line on the forewing is barely evident or absent although the antemedial and postmedial lines are dark and contrasting. Specimens of *fumalis* are, on average, larger than are those of *ontario*; forewing length ranges from 16 to 19 mm (14 to 18 in *ontario*). The harpes and saccular extensions are, on average, longer than are those of *ontario* males. The only reliable genital difference between *fumalis* and *ontario* is in the vesica. In *fumalis* there is a cluster of 8 to 10 small spines between the apex of the aedoeagus and the subbasal diverticulum. These spines, absent in *ontario* males, can usually be seen inside the aedoeagus when the vesica is not everted. The corpus bursae of *fumalis* tends to be larger than that of *ontario,* but the size difference is not consistent.

The immature stages of *fumalis* are unknown.

Euxoa fumalis occurs in eastern deciduous forest habitats from Prince Edward Island westward to southern Ontario and southward to southern New York and west-central Pennsylvania. A specimen in the United States National Museum of Natural History is labeled "District of Columbia," but occurrence of *fumalis* in this area needs to be confirmed. In eastern Canada where the range of *fumalis* overlaps that of *comosa ontario, fumalis* occurs in open deciduous forests, *ontario* occurs in forests of spruce and fir. Adults of *fumalis* have been collected from late July until mid-September.

Euxoa (*Euxoa*) *occidentalis* Lafontaine and Byers

PL. 5, FIGS. 26, 27; PL. J, FIG. 2; PL. Y, FIG. 4.

Euxoa occidentalis Lafontaine and Byers, 1982, *Can. Ent.*, **114**: 587.
Type locality: Sumpter, Oregon, 28 mi W of Baker, 4,430′. [CNC]

Euxoa occidentalis has only recently been described as a species distinct from *comosa lutulenta.* In *occidentalis* the forewing is so extensively dusted with black scales that the forewing appears to be pale charcoal colored with black transverse lines. Specimens of *comosa lutulenta* have brown forewings with darker brown markings. In many specimens of *occidentalis* the dark central area in the orbicular spot is reduced so that the orbicular spot appears as a pale blotch or ring on the forewing. Forewing length varies from 14 to 18 mm. Males of *occidentalis* can be distinguished from those of *comosa* by genital characters: 1) *occidentalis* has longer saccular extensions and sacculus (length of right saccular extension 1.65 ± 0.12 mm, N = 20, versus 1.40 ± 0.10 mm, N = 20, in *comosa lutulenta*; length of right sacculus 1.68 ± 0.07 mm versus 1.46 ± 0.07 mm in *comosa lutulenta*); 2) The saccular extensions do not taper evenly from base to apex as they do in *comosa lutulenta* males (either stout for most of their length, then tapered abruptly apically, or the saccular extensions are stouter at ⅔ from the base than at ⅓ from the base); 3) *occidentalis* has the vesica more strongly bent subbasally than that of *comosa lutulenta,* so that it projects dorsoanteriorly rather than dorsally, and the vesica is markedly swollen near the apex. Females of *occidentalis* are indistinguishable from those of *comosa lutulenta* by genital characters; they must be identified by wing marking characters or by association with males.

The immature stages of *occidentalis* are unknown.

Euxoa occidentalis occurs in montane coniferous forests from south-central British Columbia southward in the mountains of Washington and Oregon and the Sierra Nevada of California to the Transverse Ranges of southern California. Two specimens in the Canadian National Collection from western Montana appear to be this species; both are females and thus can not be identified with certainty. Adults have been collected from late August until mid-October. *Euxoa occidentalis* has a later flight season than does *comosa lutulenta*; most specimens of *occidentalis* were collected in September, whereas most specimens of *comosa lutulenta* were collected in August.

Although *occidentalis* is most likely to be confused with *comosa lutulenta,* characters of the male genitalia indicate that *occidentalis* may be more closely related to *hypochlora* (Boursin, 1964), described from Nepal, than to *comosa.*

Euxoa (*Euxoa*) *velleripennis* (Grote)

PL. 5, FIGS. 28, 29; PL. J, FIG. 3; PL. Y, FIG. 5 (RWH 10803).

Agrotis velleripennis Grote, 1874, *Rept. Peabody Acad. Sci.*, **6**: appendix, 25.
Type locality: New York. [BMNH]

Males of *velleripennis* can readily be distinguished from those of all other species within its range by the combination of charcoal-colored forewing with black maculation and white hindwing. The male genitalia are similar to those of *occidentalis,* but the saccular extensions are more evenly tapered and the vesica has a ventral pouch in the subbasal bend. Females of *velleripennis* may be confused with those of *perpolita* (subgenus *Orosagrotis,* p. 151) in that both species have dark-colored fore- and hindwings. Females of *velleripennis* can be recognized by three wing marking characters: 1) the antemedial and postmedial lines are defined in black in *velleripennis,* but these lines are absent in *perpolita*; 2) the reniform spot has some pale scaling inside the black outline, absent in *perpolita*; 3) the hindwing fringe is white with a brown subbasal line in *velleripennis* but pale brown with a darker brown subbasal line in *perpolita.* In the female genitalia, the flangelike projections on the ovipositor lobes are long and narrow and not fused together; those of *perpolita* females are short, rounded and fused together. The corpus bursae is entirely different from that of *per-*

polita (see subgeneric characters); it differs from those of other species in the *comosa* group in being more enlarged posteriorly and T-shaped. Forewing length in *velleripennis* varies from 15 to 18 mm.

The immature stages of *velleripennis* have not been described. A larva collected in Nebraska was reared to adult (Walkden, 1950: 19).

Euxoa velleripennis occurs from Prince Edward Island westward to southwestern Manitoba and southward to southeastern Montana, north-central Wyoming, northern Kansas, and western North Carolina. Adults have been collected from early August until early October. *Euxoa velleripennis* is found most commonly in dry, forested areas.

infausta GROUP

The *infausta* group is probably the most complex and poorly understood species group of *Euxoa* in North America. It contains eight species, all North American, that are distinguished from those of other groups by genital characters.

In the male genitalia, the sacculus is very small in relation to the length and width of the cucullus, being only ⅓ to ⅖ the length of the valve. The saccular extensions are long and thin and appear frail. The subbasal diverticulum of the vesica is large and either lobed both anteriorly and posteriorly, somewhat T-shaped, or elbowed and projected posteriorly with a median or subapical constriction. In all species except *satiens,* the vesica has a nipplelike subapical diverticulum dorsolaterally on the right. In the female genitalia, the ovipositor lobes are broad and rounded apically; the setae are shorter and more numerous toward the ovipositor apex so that apical setae appear short and conical, although they are fine tipped. In *satiens* the dorsal margins of the ovipositor lobes are fused with a single, finlike plate situated between the lobes apically. The ductus bursae is situated at the posterior end of the corpus bursae, or in *satiens,* about one-third of the way from the posterior end. The corpus bursae is curved so that the right side is convex, the left side concave.

Although it is relatively easy to assign specimens to this group by genital characters, it may be extremely difficult to identify them to species. Only *satiens* has diagnostic genital characters; *selenis* and *piniae* can be recognized by distinctive wing color and markings. The remaining five species form a complex of extremely variable, structurally similar species that appear to hybridize at localities throughout western North America. This complex was analyzed by arranging specimens in geographically localized, structurally homogeneous units. About 3,000 specimens were arranged in 15 units, and 400 male genital preparations were analyzed using numerical clustering techniques and discriminant analyses. On the basis of these analyses, and by using external characters, the localized units were combined into the five species recognized in the *infausta* species complex. Genitalia of the three most common and widespread species, *infausta, satis,* and *brunneigera,* can be distinguished by numerical techniques at most localities where the species occur together. They can not, however, be distinguished by these techniques, in general, because of geographical variation. Details of geographical variation, possible hybridization, and problems with distinguishing these taxa are discussed under each species.

Lack of structural characters and extensive geographical variation made preparation of the key to species particularly difficult. It should be considered a guideline only; extensive use of the illustrations will greatly enhance its usefulness.

Rearing and hybridization studies of the members of the group would help to resolve the inter-relationship of the species and their many geographic forms.

KEY TO SPECIES OF THE *INFAUSTA* GROUP

1. Male hindwing white with some dusky shading on wing margin and veins; male vesica without subapical dorsal diverticulum; female with sclerotized, apical finlike structure on ovipositor lobes . . . . *satiens* p. 108

— Male hindwing brown, or at least with smoky-gray shading on outer third of wing (*selenis*); male vesica with small, subapical, dorsal diverticulum; female ovipositor lobes without sclerotized projection apically 2

2. Hindwing white with brown shading on outer third or quarter of wing; occurring in southern California . . . . *selenis* p. 108

— Hindwing brown, pale brown near base in some specimens 3

3. Forewing bright orange red, maculation obscure except for contrasting, black antemedial and postmedial lines; occurring in northern California from Mt. Shasta southward to Yosemite National Park . . . . *piniae* p. 108

— Forewing gray, brown, or dull reddish brown;

maculation obscure or contrasting but transverse lines not particularly so 4

4. Prothoracic collar with two semicircular black patches fused to form a conspicuous transverse black bar 5
— Prothoracic collar with thin, transverse, black line forming an arc on each side of collar 8

5. Ground color of forewing and prothoracic collar reddish brown .... *brunneigera* (part) p. 106
— Ground color of forewing and prothoracic collar brown, gray, or white; if with reddish-brown shading on forewing, then costal area and prothoracic collar heavily dusted with pale-gray scales 6

6. Forewing white dusted with gray scales; known only from Inyo and Mono Counties, California .... *inyoca* p. 108
— Forewing gray or brown; occurring from western slope of Sierra Nevada Mts. westward 7

7. Forewing gray, pale brown, or pale reddish brown; if reddish brown then costal area and prothoracic collar pale gray .... *bicollaris* p. 107
— Forewing and prothoracic collar dark brown .... *brunneigera* (part) p. 106

8. Forewing ground color pale brown or pale gray; forewing length less than 15 mm in most specimens; occurring in Coastal Ranges of western California from Lake County southward; reniform spot outlined by pale-yellow line or with some yellow scales in reniform spot between dark central spot and black outline .... *brunneigera* (part) p. 106
— Populations in Coastal Ranges of California with adults larger, forewing length 15 mm, or more, in most specimens; forewing ground color dark brown or orange brown, if pale colored, then reniform spot inconspicuous, outlined by thin, black line; widely distributed 9

9. Forewing ground color brown, grayish brown or pale reddish brown; reniform spot inconspicuous, outlined by thin, black line, this often incomplete .... *infausta* this page
— Forewing dark brown, appearing almost black in many populations; forewing either dusted with pale scales, or with maculation contrasted and lined by pale scales; reniform and orbicular spot with at least partial pale-yellow line between black outline and dark central area ... *satis* p. 105

Euxoa (*Euxoa*) *infausta* (Walker)

PL. 5, FIGS. 30–35; PL. J, FIG. 4 (RWH 10785).

Hadena infausta Walker, 1865, *List of the Lepidopterous Insects in the Collection of the British Museum.* **33**: 729.
Type locality: Vancouver Island, British Columbia. [BMNH]

Agrotis rubefactalis Grote, 1881, *Bull. U. S. Geol. Surv.*, **6**: 154.
Type locality: Washington. [BMNH]
NOTE—Grote described *rubefactalis* from "two fresh specimens," a male and a female collected in "Washington Territory." There are three specimens from Washington in the British Museum (Natural History) labeled type; these are one male and two females. One female specimen differs from the other two specimens in that it is not labeled as to its sex, so this may be the one added subsequently. The male specimen, labeled "Washington Terr., Grote coll. 81–116/*Agrotis rubefactalis* Grote; Type male/Noctuidae, Brit. Mus. slide No. 6321" is here designated lectotype. The specimen is generally in good condition but is slightly rubbed.

Agrotis sponsa Smith, [1888], *Proc. U. S. Natl. Mus.*, **10**: 457.
Type locality: Washington. [USNM]

Agrotis numa Strecker, 1898, *Lepidoptera, Rhopaloceres and Heteroceres, Indigenous and Exotic.* Suppl. **1**: 5.
Type locality: Seattle, Washington. [FMNH]

Carneades relaxus Smith, 1900, *Proc. U. S. Natl. Mus.*, **22**: 432.
Type locality: San Francisco, California. [USNM]
NOTE—The male lectotype of *relaxus* was designated by Todd (1982: 183).

Carneades holoberba Smith, 1900, *Proc. U. S. Natl. Mus.*, **22**: 443.
Type locality: Calgary, Alberta. [USNM]
NOTE—The female lectotype of *holoberba* was designated by Todd (1982: 98).

The most widespread form of *infausta* (plate 5, figures 30–33) can be recognized by the even buffy-brown or pale grayish-brown ground color of the forewing. This form occurs in dry open habitats in the northern intermontane region and the northern Great Plains. At some localities in northern Utah and southwestern Montana, where *infausta* occurs sympatrically with *brunneigera excogita,* a reddish-brown form of *infausta* occurs (plate 5, figure 34) that has a more prominent, prothoracic line. These specimens appear to be intermediate between typical members of each species and may be of hybrid origin. In forested habitats, in southwestern Alberta, southern British Columbia, and western Washington, a darker brown form occurs that has a more mottled ground color (plate 5, figure 35). Specimens from other areas tend to be intermediate. Throughout the western half of its range, *infausta* occurs sympatrically with *satis* although *infausta* is more common in dry habitats with *satis* more common in forests of fir and spruce. At some localities, populations are intermediate in appearance between *infausta* and *satis* and may be areas of extensive hybridization. *Euxoa infausta* is larger than *satis* in most areas; forewing length varies from 14 to 17 mm.

The immature stages have not been described although Cook (1930: 263) reports having reared the larva on a variety of broad-leaved plants including cabbage, alfalfa, and sweet clover.

Euxoa infausta occurs across southwestern Canada from Saskatchewan westward to Vancouver Island, British Columbia. It occurs southward to central South Dakota, northern Utah, northern Nevada, and southern California. In California it occurs south of Modoc and Siskiyou Counties only in the western half of the state where its range extends southward to San Diego County. An apparently disjunct population of *infausta* occurs at Fort Smith in the Northwest Territories of Canada. Adults of *infausta* have a long flight season that extends from mid-May until late August. In general, adults are on the wing in May and June in the southern portion of its range and in July and August in the northern portion.

Euxoa (*Euxoa*) *satis* (Harvey)
PL. 5, FIGS. 36–40; PL. J, FIG. 5 (RWH 10786).

Agrotis satis Harvey, 1876, *Can. Ent.,* **8**: 36.
Type locality: California. [BMNH]

Agrotis micronyx Grote, 1878, *Bull. U. S. Geol. Surv.,* **4**: 171.
Type locality: California. [BMNH]

Agrotis perfusca Grote, 1883, *Papilio,* **3**: 77.
Type locality: Soda Springs, California. [USNM]

NOTE—The single original type in the United States National Museum is labeled "Ariz.," whereas the type locality, Soda Springs, is northwest of Lake Tahoe, California. This is probably a labeling error because the female specimen matches both the original description and other specimens from this part of California. *Euxoa satis* has not been recorded from Arizona.

Euxoa cocklei Smith, 1908, *Ann. New York Acad. Sci.,* **18**: 96.
Type locality: Kaslo, British Columbia. [AMNH]

NOTE—The female lectotype of *cocklei* was designated by Todd (1982: 49).

Throughout most of the range of *satis,* specimens can be distinguished from those of other species in the group by the dark blackish-brown forewing color (plate 5, figures 36–39). This form of *satis* is more frequently confused with the dark form of *tessellata* that occurs with it, than with other species in the *infausta* group. Males of *satis* can be distinguished from those of *tessellata* by the lack of a tuft of yellow scales on the thorax at the base of the costa of the forewing; the cucullus is less expanded apically, and the saccular extensions are more symmetrical in *satis.* These genital characters can be observed by brushing away the scales from the end of the abdomen. Females of *satis* can be distinguished from those of *tessellata* by the lack of a small white tuft of scales at the base of the forewing costa, and by the rounded, rather than apically truncate and flattened, ovipositor lobes. A second form of *satis* (plate 5, figure 40) is most common in Washington, Oregon, and California. This form of *satis* has a paler ground color and could be confused with dark specimens of *infausta* but can be distinguished from them by the presence of pale shading adjacent to the transverse lines on the forewing and pale shading in the reniform and orbicular spots between the black outline and the dark central area. This species appears to hybridize with *infausta* at some localities. It is possible that *satis* and *infausta* are not completely distinct species but may be habitat forms that are kept distinct by different habitat preferences. The two taxa, however, do occur together at many localities without the presence of intermediate forms that would be expected if they are hybridizing. Laboratory crosses between the dark form of *satis* from British Columbia and the prominently marked form from southern Oregon have been made in the laboratory without any evidence of reduced fertility in the offspring or any dissynchrony in emergence of males and females. *Euxoa satis* varies less in size than does *infausta*; forewing length is from 11 to 17 mm.

The immature stages of *satis* are known only from laboratory reared material.

Euxoa satis occurs in western montane coniferous forests from southwestern Alberta and southern British Columbia southward to central Colorado, northern Utah, northern Nevada, and southern California. In California it occurs in the Sierra Nevada Mountains as far south as Mono County and in the Coast Ranges to Monterey County. Adults have been collected from mid-June until late August.

Euxoa (*Euxoa*) *brunneigera* (Grote), REVISED STATUS

PL. 5, FIGS. 41–46; PL. Y, FIG. 6 (RWH 10785, part, 10787).

Agrotis brunneigera Grote, 1876, *Bull. Buffalo Soc. Nat. Sci.*, **3**: 80.
Type locality: Vancouver Island, British Columbia. [BMNH]

NOTE—Grote stated that *brunneigera* was described from a male and a female, numbers 5654 and 5612, collected on Vancouver Island. There is a male and female in the British Museum (Natural History) labeled type but the male is number 5641 (the female is 5612). Specimen number 5654 is a female in the American Museum of Natural History. It is possible that Grote mistook the female (No. 5654) for a male, or the numbers on the male in the British Museum (Natural History) and the female in the American Museum of Natural History have been interchanged. Because of this discrepancy, I here designate the female in the British Museum (Natural History) as lectotype. The specimen is in good condition and is labeled "Type/Vancouver I., Grote coll. 81–116/ 5612/*Agrotis brunneigera* Grote, Type, 13-3-80/ Noctuidae, Brit. Mus. slide No. 6361."

Carneades loya Smith, 1900, *Proc. U. S. Natl. Mus.*, **22**: 422. NEW SYNONYMY.
Type locality: High Sierras, California. [USNM]

Carneades excogita Smith, 1900, *Proc. U. S. Natl. Mus.*, **22**: 423. NEW SYNONYMY, SUBSPECIES.
Type locality: Glenwood Springs, Colorado. [USNM]

NOTE—The male lectotype of *excogita* was designated by Todd (1982: 77).

Noctua monteclara Smith, 1906, *Jour. New York Ent. Soc.*, **14**: 10. NEW SYNONYMY.
Type locality: Claremont, California. [AMNH]

NOTE—The male lectotype of *monteclara* was designated by Todd (1982: 143).

"*Euxoa*" *obscura* Hill, 1924, *Bull. So. California Acad. Sci.*, **23**: 185. NEW SYNONYMY.
Type locality: San Diego, California. [USNM]

Euxoa brunneigera is the most variable species geographically in the *infausta* group. Several geographic forms are so dissimilar in size and wing markings that they would not be considered to be conspecific except for the presence of broad zones where intermediate forms occur. The variation in this species ranges from large, dark reddish-brown forms with the maculation obscure (plate 5, figure 46) through smaller, reddish-brown forms with well-defined maculation (plate 5, figure 41) to small, pale-buff forms in southern California (plate 5, figure 44). Forewing length may be as small as 11 mm in California to as large as 18 mm in Utah. Details of geographical variation and possible hybridization between *brunneigera* and other species is discussed under the subspecies.

The immature stages of *brunneigera* are unknown.

Euxoa brunneigera occurs from northern Montana and southern British Columbia southward to central Colorado, southern Utah, northeastern Nevada, and southern California. Adults have been collected from mid-May until late August; adults are on the wing in May and June at low elevations in the South and in July and August in the North and at high elevations. Populations of *brunneigera* are arranged in two subspecies.

Euxoa (*Euxoa*) *brunneigera brunneigera* (Grote)

PL. 5, FIGS. 41–44.

Agrotis brunneigera Grote, 1876.

Carneades loya Smith, 1900.

Noctua monteclara Smith, 1906.

"*Euxoa*" *obscura* Hill, 1924.

Subspecies *brunneigera* occurs from southern British Columbia and western Montana southward to southern California. Specimens from the northern portion of its range and from eastern California are smaller (forewing length 13 to 16 mm) than those of subspecies *excogita* and have reddish-brown forewings. Specimens vary little in appearance throughout this entire region (plate 5, figures 41, 42). In western California the situation appears to be much more complex, possibly due to hybridization with *bicollaris.* Adults are similar in size to those from the Sierra Nevada Mountains and northward (forewing length 11 to 15 mm), but in most specimens the transverse, black band on the prothoracic collar is narrow. Specimens vary in forewing color from brown or reddish brown in mesic habitats to pale brown in xeric habitats (plate 5, figures 43, 44). There is also a tendency for dark-colored specimens to have a wider prothoracic transverse line than pale-colored ones. Also, a wide prothoracic line is more common in Transverse Range populations that it is in Coastal Range populations. At some localities in the Coastal Range, many specimens appear to be intermediate between *brunneigera* and *bicollaris*; this may be a region of extensive hybridization. In southern California, and in the Sierra Nevada Mountains, *brunneigera* and *bicollaris* frequently occur together without evidence of hybridization. In general, where the two species occur together, *brunneigera* flies later in the season than does *bicollaris* although there is usually some overlap (see *bicollaris*). In western California *brunneigera* occurs from Sonoma County southward to San Diego. There is no evidence of hybridization between *brunneigera brunneigera* and *infausta* although there is between *brunneigera excogita* and *infausta.* Both subspecies of *brunneigera* are widely sympatric with *infausta.*

Euxoa (*Euxoa*) *brunneigera excogita* (Smith)
PL. 5, FIGS. 45, 46; PL. Y, FIG. 6.

Carneades excogita Smith, 1900.

Subspecies *excogita* occurs in the Rocky Mountain region from Montana southward to Colorado, the Wasatch Range of Utah, and the Ruby Mountains in Nevada. Adults are larger than those of subspecies *brunneigera*; forewing length varies from 15 to 18 mm. Typical specimens are large with obscure forewing markings and a dusting of hoary-gray scales, particularly in the terminal area (plate 5, figure 46). At some localities in Utah and at most localities in Montana, specimens are smaller, have less hoary forewing shading and better defined maculation (plate 5, figure 45); these specimens appear to be intermediate between the large form of subspecies *excogita* and subspecies *brunneigera.* At most localities where large and small specimens of *excogita* occur, the majority of specimens are intermediate. At one locality in Montana the two forms of *excogita* occur together without evidence of intermediate forms. At some localities where *brunneigera excogita* occurs sympatrically with *infausta,* intermediate specimens that appear to be hybrids occur.

Euxoa (*Euxoa*) *bicollaris* (Grote)
PL. 5, FIGS. 47–49 (RWH 10788).

Agrotis bicollaris Grote, 1878, *Bull. U. S. Geol. Surv.,* **4**: 173.
Type locality: Havilah, California. [BMNH]
NOTE—The two original type specimens of *bicollaris* (numbers 6524 and 6517) are in the British Museum (Natural History). Only the female is labeled "type." It is in good condition, labeled "California, Havilah, Grote Coll. 81–116/6524/*Agrotis bicollaris* Grote, type/*Agrotis bicollaris* Grote/Noctuidae, Brit. Mus. slide No. 6314;" it is here designated lectotype.

Agrotis abnormis Smith, 1890, *Trans. Amer. Ent. Soc.,* **17**: 41.
Type locality: Sierra Nevada, California. [AMNH]
NOTE—The female lectotype of *abnormis* was designated by Todd (1982: 4).

Euxoa bicollaris occurs in basically three forms. In the northern portion of its range, from northern California northward, the forewing is reddish gray with extensive gray shading on the costal margin and on the prothoracic collar. In most of the Sierra Nevada Mountain region and in the Transverse Ranges of southern California, specimens have a gray forewing with a slight red tint on the posterior portion of the wing and proximal to the pale-buff subterminal line (plate 5, figures 47, 48). In specimens from coastal localities in southern California, the forewing tends to be a buffy-gray or pale-brown color (plate 5, figure 49). At some localities, this latter form appears to hybridize with the coastal form of *brunneigera.* In general, the coastal form of *bicollaris* differs from *brunneigera* specimens from this area in being larger, paler, and with a wider black transverse band on the prothroacic collar. Forewing length in *bicollaris* varies from 14 to 17 mm.

The immature stages of *bicollaris* are unknown.

Euxoa bicollaris has been collected from extreme western Montana and southeastern Washington southward through California and western Nevada to southern California. The adult flight period of *bicollaris* tends to be earlier than that of *brunneigera* where they occur together. At low elevations in southern California, collecting dates range from late April until late May for *bicollaris* but from mid-May until mid-July for *brunneigera.* At higher elevations in the Transverse Ranges, *bicollaris* flies until mid-July, *brunneigera* until late July. In the northern portion of its range, *bicollaris* occurs in July and early August.

Euxoa (*Euxoa*) *inyoca* Benjamin
PL. 5, FIG. 50 (RWH 10789).

Euxoa inyoca Benjamin, (1936), *Bull. So. California Acad. Sci.*, **34**: 199.
Type locality: Inyo County, California. [USNM]

This species is probably a pale, desert form of *bicollaris* that is isolated in the arid regions in Mono and Inyo Counties east of the Sierra Nevada Mountains. Like *bicollaris,* the prothoracic collar has a prominent, transverse black band. Specimens of *inyoca* are much paler than those of *bicollaris*; the forewing ground color is white with a dusting of dark scales that give it a dirty white appearance. There is also smoky-gray shading in the terminal area and proximal to the pale subterminal line. The forewing length varies from 14 to 16 mm.

The immature stages of *inyoca* are unknown.

Euxoa inyoca is known only from a few localities in Inyo and Mono Counties. Specimens were collected between mid-June and early July. The status of *inyoca,* with respect to *bicollaris,* will only be resolved after more extensive collecting has been done in southeastern California and hybridization and life history data are available. I have seen two rubbed specimens, possibly referable to this species, from Coconino and Mojave Counties, Arizona; these are in the collection of the Natural History Museum, Los Angeles County, California.

Euxoa (*Euxoa*) *selenis* (Smith)
PL. 5, FIGS. 51–53; PL. J, FIG. 6; PL. Y, FIG. 7 (RWH 10790).

Carneades selenis Smith, 1900, *Proc. U. S. Natl. Mus.*, **22**: 459.
Type locality: Los Angeles County, California. [USNM]

Euxoa claromonta Smith, 1906, *Jour. New York Ent. Soc.*, **14**: 11.
Type locality: Claremont, California. [AMNH]
NOTE—The female lectotype of *claromonta* was designated by Todd (1982: 48).

Specimens of *selenis* can be distinguished from those of other species in the *infausta* group by the following characters: hindwing white with smoky-brown shading confined to the outer third or quarter of the wing; forewing pale gray with the reniform spot, and in many specimens the orbicular spot, filled with dark-gray shading; most specimens with some dark-gray shading between the reniform spot and the orbicular spot. The only other species in the group that has a pale hindwing is *satiens,* which has a very different forewing pattern and a much later flight period. In *selenis* forewing length varies from 14 to 17 mm. The male and female genitalia are similar to those of the species in the *infausta* complex, but the corpus bursae of the female is longer.

The immature stages of *selenis* are unknown.

This species has a relatively small range; it has been collected in southwestern California from Kern and Santa Barbara Counties southward to San Diego County. Adults have a very early flight period that extends from late March until mid-April at low elevations and until mid-June in the mountains.

Euxoa (*Euxoa*) *piniae* Buckett and Bauer
PL. 5, FIGS. 54, 55; PL. J, FIG. 7; PL. EE, FIG. 18 (RWH 10791).

Euxoa piniae Buckett and Bauer, 1964, *Can. Ent.*, **96**: 97.
Type locality: Johnsville, Plumas County, California. [UCD]

The combination of small size, bright orange-brown forewing color, and black, contrasting, transverse lines distinguish specimens of *piniae* from those of other species of *Euxoa.* The orbicular spot is defined by a pale circle; the reniform spot is gray with a pale outline. In some specimens there is a trace of some hoary shading distal to each transverse line. Forewing length varies from 13 to 15 mm. The male and female genitalia are similar to those of species in the *infausta* complex.

The immature stages are unknown.

Euxoa piniae is known from a relatively small area in California that extends from Siskiyou and Plumas Counties southward to Tuolomne County. Adults have been collected in coniferous forest habitats from mid-July until mid-September.

Euxoa (*Euxoa*) *satiens* (Smith)
PL. 5, FIGS. 56–60; PL. J, FIG. 8; PL. Y, FIG. 8; PL. EE, FIG. 19 (RWH 10792).

Agrotis satiens Smith, 1890, *Trans. Amer. Ent. Soc.*, **17**: 45.
Type locality: British Columbia. [USNM]
NOTE—The female lectotype of *satiens* was designated by Todd (1982: 191).

Specimens of *satiens* are not likely to be confused with those of any other species in the *infausta* group.

They are, however, frequently confused with specimens of *pallipennis* (p. 86) and *cona* (p. 142). Forewing ground color varies with habitat from white with scattered dark scales in desert areas (plate 5, figures 56, 57) to buff or reddish brown in grassland areas and pine parkland (plate 5, figures 58, 59) and smoky gray or dark gray in more mesic coniferous forests (plate 5, figure 60). In general, the forewing and hindwing veins are lined with dark scales that give the wings a finely streaked appearance that is not found in *pallipennis* or *cona.* The male genitalia of *satiens* differ from those of *pallipennis* and *cona* males in that the sacculus is very small in proportion to the cucullus size and the saccular extensions are long, thin and appear frail in *satiens.* Females of *satiens* can be distinguished from those of all other *Euxoa* species by the presence of a single, sclerotized, finlike plate between the dorsal margins of the ovipositor lobes. All other *Euxoa* species that have sclerotized plates, or projections, on the ovipositor lobes have two structures, one on each lobe. The male vesica in *satiens* differs from those of other species in the *infausta* group in that the vesica is expanded apically rather than subapically, and a dorsolateral apical diverticulum is absent. In *satiens,* forewing length varies from 14 to 17 mm.

The immature stages of *satiens* are unknown.

Euxoa satiens inhabits both open aridlands and coniferous forests. It is primarily an inhabitant of the intermontane region and the surrounding montane areas. It occurs from southern British Columbia southeastward to northeastern Wyoming and southward to central New Mexico, central Arizona, and southern California. Adults have been collected from mid-August until mid-October.

violaris GROUP

The *violaris* group includes only *violaris*; the characters are given under the species.

> *Euxoa* (*Euxoa*) *violaris* (Grote and Robinson)
>
> PL. 6, FIGS. 1–3; PL. J, FIG. 9; PL. Z, FIG. 1 (RWH 10810).
>
> *Agrotis violaris* Grote and Robinson, 1868, *Trans. Amer. Ent. Soc.,* **1**: 353.
> Type locality: Atlantic District. [ANSP]

This is probably the most easily recognized *Euxoa* species in eastern North America. It is one of the few species in the genus that lacks a frontal tubercle; the others are *detersa* in the *detersa* group and the *scandens-aurulenta* complex in the *westermanni* group. Like other species that lack a frontal tubercle, *violaris* inhabits areas of loose, shifting sand such as beach and dune areas. The male antenna is very prominently biserrate and bifasciculate, more so than any other eastern species. Adults of *violaris* can most easily be recognized by their distinctive forewing color and pattern. In most specimens the ground color is pale gray with scattered brown scales that give the wing a blue-gray appearance. The outer half of the median area, between the median line and the postmedial line, is shaded in grayish brown. The antemedial and postmedial lines are unusual in that they are not scalloped between the wing veins; both lines are pale. In most specimens the line is bordered by a prominent brown line that is distal to the pale antemedial line and proximal to the postmedial line. In some specimens the pale transverse lines are bordered on both sides by faint, brown lines. The orbicular spot is obscure, partially outlined either by a faint pale line or a narrow dark line. The reniform spot is dark gray with a pale outline and in many specimens with a partial dark line outside this. A dark form of *violaris* (plate 6, figure 2), about 5 percent of the specimens examined, has a dark brownish-charcoal forewing with the maculation defined by pale lines. Forewing length varies from 14 to 18 mm. The genitalia are similar to those of members of the *comosa* group, but the vesica is differently shaped, the lobe of the corpus bursae bearing the ductus seminalis projects dorsally rather than to the left, and the ovipositor lobe lacks a sclerotized apical projection.

The immature stages of *violaris* are unknown.

Euxoa violaris occurs in sandy areas along the Atlantic coast from Massachusetts southward to North Carolina. I have seen a single specimen from the Gulf of Mexico, this from Ocean City, Okaloosa County in northwestern Florida. This specimen differs from other specimens examined in that it lacks dark shading in the median area (plate 6, figure 3). Adults of *violaris* have been collected from mid-September until late October.

cursoria GROUP

The *cursoria* group is a large palearctic species-group that probably includes at least several dozen species in Asia, including *decora* ([Denis and Schiffermüller], 1775), *recussa* (Hübner, [1818]), and *oberthuri* (Leech, 1900). Only three species in the *cursoria* group occur in North America, and two of them occur in the Old World as well.

The best characters for recognizing members of the group are in the vesica, but the three North

American species can be recognized by other genital characters and by wing markings.

In most species in the *cursoria* group, the vesica bends sharply ventrally just beyond the subbasal diverticulum; it projects ventrally for a distance of about twice the width of the aedoeagus before bending through 180° to project dorsally. The subbasal diverticulum of the vesica is round or oblong, rather than foot shaped as in most other species-groups, and there is a large bulge in the vesica wall on the left side adjacent to the subbasal diverticulum. In most species the saccular extensions are markedly asymmetrical with the left saccular extension much shorter than the right one although in two of the three North American species the left saccular extension is only slightly shorter than the right one. In most species the left saccular extension is much shorter than the left sacculus. In the female genitalia the corpus bursae is elongate and constricted mesially so that it appears somewhat dumbbell shaped. In shape of the bursa, females of species in the *cursoria* group are similar to those of *olivia, serotina, hollemani,* and species in the subgenus *Chorizagrotis*; however, they are not likely to be confused with them because of differences in wing markings and other genital characters.

KEY TO NORTH AMERICAN SPECIES OF THE *CURSORIA* GROUP

1. Male genitalia with right saccular extension long, at least 1.25 × length of left saccular extension, as long as, or longer than right sacculus, and extending beyond ventral angle of cucullus; female genitalia with corpus bursae slightly constricted mesially, and about twice as long as wide; known only from arid areas in Yukon and western Northwest Territories .... *cursoria* this page

— Male genitalia with right saccular extension shorter, less than 1.25 × length of left saccular extension, shorter than right sacculus, and not extending as far posteriorly as ventral angle of cucullus; female genitalia with corpus bursae prominently constricted mesially and more than twice as long as wide; widely distributed 2

2. Forewing shaded with orange buff; reniform spot predominantly buff filled with black outline; male vesica with median diverticulum half way between subbasal bend and apex of vesica; female genitalia with corpus bursae narrow, as wide anteriorly as posteriorly; ovipositor lobes short and rounded apically .... *nostra* p. 113

— Forewing with many color forms; reniform spot predominantly dark gray with conspicuous white line inside inconspicuous black outline; male vesica with median diverticulum close to subbasal bend of vesica; female genitalia with corpus bursae broader, posterior portion much wider than anterior portion; ovipositor lobe tapered and narrow near apex .... *ochrogaster* p. 112

Euxoa (*Euxoa*) *cursoria* (Hufnagel)

PL. 6, FIGS. 4–7; PL. J, FIG. 10; PL. Z, FIG. 2 (RWH 10802).

Phalaena cursoria Hufnagel, 1766, *Berlinisches Magazin,* **3**: 416.
Type locality: Berlin, Germany.

Noctua mixta Fabricius, 1794, *Entomologia Systematica,* **3** (2): 118.
Type locality: Italy. [Zoological Museum, Copenhagen]

Noctua concolor Haworth, 1809, *Lepidoptera Britannica,* **2**: 243.
Type locality: Great Britain.

Agrotis venosa Stephens, 1829, *A Systematic Catalogue of British Insects,* 67.
Type locality: Great Britain.

Agrotis detorta Eversmann, 1851, *Bull. Ent. Soc. Moscow,* **2**: 628.
Type locality: U.S.S.R. [Zoological Institute, Leningrad]

Agrotis cespitis Swinhoe, 1885, *Trans. Ent. Soc. London,* **1885**: 349.
Type locality: Afghanistan. [BMNH]

Agrotis cursoria var. *currens* Staudinger, 1896, *Deutsche Ent. Zeits., Iris,* **9**: 249. SUBSPECIES.
Type locality: Uliassutai. [HUMB]

NOTE—The term variety apparently was used by Staudinger in a geographical race context, not as an infrasubspecific category. The name *currens* was used as a subspecific name for Mongolian populations of *cursoria* by Kováks and Varga (1973: 295).

Euxoa wirima Hardwick, 1965, *Can. Ent.,* **97**: 674. NEW SYNONYMY, NEW STATUS, SUBSPECIES.
Type locality: Fort Smith, Northwest Territories, Canada. [CNC]

NOTE—The following infrasubspecific names have been proposed for local color forms of *cursoria*. They are listed in alphabetical order.

ab. *asiaeminoris* Strand, (1916), *Archiv für Naturgeschichte,* **81A12**: 144. [Asia Minor]

var. *brunnea* Tutt, 1892, *British Noctuae and Their Varieties,* **2**: 43. [Great Britain]

var. *coerulea* Tutt, 1892, Op. cit., **2**: 41. [Great Britain]

var. *costacoerulea* Tutt, 1892, Op. cit., **2**: 41. [Great Britain]

var. *distincta* Tutt, 1892, Op. cit., **2**: 42. [Great Britain]

var. *marginata* Tutt, 1892, Op. cit., **2**: 43. [Great Britain]

form *muelleri* Hanel, 1920, *Internationale Entomologisches Zeits.,* Guben, **13**: 185. [Germany]

form *nigrescens* Hanel, 1920, Op. cit., **13**: 185. [Germany]

form *nigrovittata* Hanel, 1920, Op. cit., **13**: 185. [Germany]

ab. *obscura* Staudinger, 1871, *Catalog der Lepidopteren des Europaeischen Faunengebiets. I. Macrolepidoptera*: 86. [Poland]

var. *obsoleta* Tutt, 1892, Op. cit., **2**: 42. [Great Britain]

var. *ochrea* Tutt, 1892, Op. cit., **2**: 42. [Great Britain]

var. *pallida* Tutt, 1892, Op. cit., **2**: 42. [Great Britain]

ab. *sagittata* Staudinger, 1896, *Deutsche Ent. Zeits., Iris,* **9**: 249. [Poland]

ab. *vaga* Staudinger, 1896, Op. cit., **9**: 250. [Mongolia]

NOTE—The type specimens of Staudinger's names, in the collection of the Zoologisches Museum der Humboldt-Universität zu Berlin, were illustrated by Bang-Haas, 1922, *Deutsche Ent. Zeits., Iris,* **36**: pl. 13, 14.

Within its very limited North American range, specimens of *cursoria* are likely to be confused only with *ochrogaster*. Specimens of *cursoria* tend to look like small, bleached specimens of *ochrogaster*; they occur in color forms that are similar to those of *ochrogaster,* but the various forms are more poorly defined. Both species occur in four basic color forms: a pale-buff, or pale gray-brown, well-marked form (plate 6, figure 4); a reddish-brown, well-marked form (plate 6, figure 5); a reddish-brown obscurely marked form (plate 6, figure 6); and a buff, or pale gray-brown obscurely marked form (plate 6, figure 7). Unlike *ochrogaster,* however, the first three forms are very rare in *cursoria*; the pale, obscurely marked form and a form intermediate between this and the pale well-marked form make up about 90 percent of the individuals. Although most specimens of *cursoria* can be recognized without dissection, positive identification depends on genital characters given in the key.

The immature stages of *cursoria* are unknown in North America but have been described from Great Britain. The species is thought to overwinter as an early instar larva; although the adult flight period is similar to those of species that overwinter in the egg stage. The larva feeds on a wide variety of dune inhabiting plants, both grasses and broad-leaved herbs (Bretherton et al., 1979: 136–137).

Euxoa cursoria is widely distributed in the Palearctic Region where it occurs from Great Britain eastward to eastern Siberia. It has not previously been reported from North America but is now known to occur in southern Yukon and southwestern Northwest Territories. In most of its range, *cursoria* occurs in sandy beach and dune areas. Adults are in flight from July until September.

Populations of *cursoria* are arranged in three subspecies; in most of the Palearctic Region, populations are assigned to the nominate subspecies. The forewing ground color in specimens of subspecies *cursoria* is pale gray or yellow buff in pale forms and reddish brown in dark forms. Specimens from Mongolia and the Altai Mountains of south-central Soviet Union have orange-brown or dark gray-brown forewings; these have been treated as a distinct subspecies, *cursoria currens* (Kovåks and Varga, 1973: 295). Populations of *cursoria* in North America are assigned to subspecies *wirima* as discussed below.

Euxoa cursoria wirima Hardwick, NEW STATUS

PL. 6, FIGS. 4–7; PL. J, FIG. 10; PL. Z, FIG. 2.

Euxoa wirima Hardwick, 1965.

Nearctic populations of *cursoria* are intermediate in color between those of typical *cursoria* and those of subspecies *currens*. Specimens of the three subspecies are indistinguishable by genital characters. I tentatively retain the name *wirima* for the disjunct populations in North America. Life history and hybridization studies between New and Old World populations of *cursoria* may reveal differences between these populations that would not be suspected from studies of adult morphology.

Subspecies *wirima* has been collected in southern Yukon and southwestern Northwest Territories from late July until late August. It occurs on dry hillsides and in dune areas where aridland plants, such as sagebrush (*Artemisia* spp.), are common. It was probably more widely distributed in Alaska and Yukon during the last glacial advance (the Wisconsinan) when aridlands were more widespread in the Beringian refugium area of eastern Siberia and northwestern North America.

Euxoa (*Euxoa*) *ochrogaster* (Guenée) (Red-backed Cutworm*; Ver-gris à dos Rouge, m., Fr.)

PL. 6, FIGS. 8–11; PL. K, FIG. 1; PL. Z, FIG. 3; PL. EE, FIG. 20 (RWH 10801).

Noctua ochrogaster Guenée, 1852, *Histoire Naturelle des Insectes. Species Général de Lépidoptères. 5, Noctuélites,* **1**: 327.
Type locality: North America. [lost]

Agrotis insignata Walker, [1857], *List of the Specimens of Lepidopterous Insects in the Collection of the British Museum,* **10**: 330.
Type locality: Nova Scotia. [BMNH]

NOTE—*Agrotis insignata* Walker, 1857: 330 is a senior homonym of *Agrotis insignata* Walker, 1857: 353, a junior synonym of *Euxoa tessellata.*

NOTE—*Agrotis insignata* Walker, 1857: 330 was described from two specimens. Of these, only a single female has been located in the British Museum (Natural History). This specimen labeled "Type/Nova Scotia, Redman/53 *Agrotis insignata*/*illata*/Noctuidae, Brit. Mus. slide No. 6342" is here designated lectotype. It is generally in good condition although the antennae are missing.

Agrotis islandica Staudinger, 1857, *Stettiner Ent. Zeit.,* **18**: 232. NEW SYNONYMY, NEW STATUS, SUBSPECIES.
Type locality: Iceland. [Syntypes in HUMB]

Agrotis cinereomacula Morrison, 1874, *Proc. Boston Soc. Nat. Hist.,* **17**: 164.
Type locality: St. Louis, Missouri. [USNM]

Agrotis turris Grote, 1875, *Can. Ent.,* **7**: 226.
Type locality: Orillia, Ontario. [BMNH]

NOTE—In the original description of *turris,* Grote refers several times to "specimens" although he does not indicate how many specimens are in the type-series. Only one specimen in the British Museum is labeled "type." This specimen labeled "Grote coll. 81–116, Orillia, Ont./*Agrotis turris* Grote, type/*Agrotis cressida* Morr., type [unpublished]/*Agrotis cinereomacula* Morrison/Female" is here designated lectotype. The specimen is missing the right antenna and the abdomen; otherwise it is in good condition.

Agrotis gularis Grote, 1875, *Proc. Acad. Nat. Sci. Philadelphia,* **27**: 424.
Type locality: Orillia, Ontario. [BMNH]

Agrotis islandica var. *rossica* Staudinger, 1881, *Stettiner Ent. Zeit.,* **42**: 419. NEW SYNONYMY, NEW STATUS, SUBSPECIES.
Type locality: Saisan, Urals, Margelan, U.S.S.R. [Syntypes in HUMB]

NOTE—The remaining palearctic synonymy of *ochrogaster* has been omitted because the palearctic subspecies and synonymy are being revised by A. Moberg and Z. Varga.

Specimens of *ochragaster* occur in four basic forms, two based on wing color and two on maculation. The forewing ground color may be either orange buff or reddish brown; in maculation, the forewing may have a black subbasal dash, a prominent black claviform spot and black shading around the reniform and orbicular spots, or the forewing may lack dark shading. All four forms, and intermediate ones, can be reared from an egg batch from a single female so the forms are frequently collected together. The pale unmarked form of *ochrogaster* (plate 6, figure 8) is most likely to be confused with specimens of *nostra* but can be distinguished by the characters given in the key. Specimens of the pale, well-marked form (plate 6, figure 9), particularly those in which the postmedial line is not prominently toothed and punctiform on the veins, are frequently confused with a similarly marked form of *perexcellens* (plate 2, figure 43). Most specimens of these two species can be distinguished by subtle differences in forewing pattern, but in some instances it is necessary to use genital characters to distinguish them. The reddish-brown forms of *ochrogaster* can be very difficult to distinguish from the corresponding forms of *lidia* (plate 1, figures 2, 3). Fortunately, the reddish-brown color form in *lidia* is almost entirely restricted to males, which can readily be distinguished by removing scales from the end of the abdomen to observe the shape of the saccular extensions. These are tapered and pointed apically in *ochrogaster* and expanded and spatulate apically in *lidia.* Females of *lidia* can usually be recognized, without dissection, by the dorsoventrally flattened abdomen. To a lesser extent, the reddish-brown, unmarked form of *ochrogaster* can be confused with the western red form of *mimallonis.* Specimens of *ochrogaster* can be distinguished from those of this species by the darker hindwing, the dark-filled reniform spot and by genital characters. Overall, *ochrogaster* is most similar to *cursoria*; the latter species occurs in North America only in northwestern Canada. The two species can be distinguished by characters given in the key. Specimens of *ochrogaster* vary markedly in size; forewing length varies from 14 to 19 mm.

The larva of *ochrogaster,* the redbacked cutworm, is among the most economically important cutworm pests in North America. Most crop damage occurs in the northern Great Plains and in British Columbia, Washington, and Oregon. No complete account of the life history of *ochrogaster* has been published, although, much information is available in piecemeal form. Beirne (1971: 15–17) discusses larval habits, reviews the economic importance of the species, and gives a list of references. More recently, aspects of the life history and control of the species in Oregon and Washington have been published in *Environmental Entomology* and *Journal of Economic Entomology* (e.g., Berry, 1975; Tamaki et al., 1975). The larva feeds at night at, or below, the ground surface, primarily on broad-leaved plants although extensive damage to grain crops has been reported. The larva feeds on a wide variety of plants. The species overwinters in the egg stage, so crop damage occurs in May and June. The egg was illustrated by Salkeld (1975: 1144, figures 7–11).

Euxoa ochrogaster is widely distributed across Asia, Canada, and northern United States. In North America it occurs from Newfoundland westward to western Alaska and southward to the northern tier of states in eastern and central North America. In the West it occurs southward to southern Montana and Oregon and in the mountains to north-central New Mexico and east-central Arizona. Adults have been collected from mid-June until late September, although most records extend from mid-July until mid-August.

Populations of *ochrogaster* are arranged in three subspecies. Two subspecies occur in the Palearctic Region in Iceland and Fennoscandia (*islandica*) and central and eastern Siberia (*rossica*). The nominate subspecies, *ochrogaster,* is confined to North America. The three subspecies are indistinguishable in genital characters and differ only slightly in wing markings. The North American subspecies is discussed below.

Euxoa (*Euxoa*) *ochrogaster ochrogaster* (Guenée)

PL. 6, FIGS. 8–11; PL. K, FIG. 1; PL. Z, FIG. 3; PL. EE, FIG. 20.

Noctua ochrogaster Guenée, 1852.

Agrotis insignata Walker, 1857.

Agrotis cinereomacula Morrison, 1874.

Agrotis turris Grote, 1875.

Agrotis gularis Grote, 1875.

North American populations of *ochrogaster* differ from those from the Palearctic Region in that the four color forms of the species tend to be discrete and almost all specimens can readily be assigned to one of the four forms. In the Palearctic Region the forms are more variable and form a gradient from light to dark and unmarked to well marked. Also, the dark forms in North America are dark reddish brown while those from Asia are dark brown.

Euxoa (*Euxoa*) *nostra* (Smith)

PL. 6, FIGS. 12, 13; PL. K, FIG. 2; PL. Z, FIG. 4 (RWH 10800).

Agrotis nostra Smith, 1890, *Trans. Amer. Ent. Soc.,* **17**: 55.

Type locality: Sierra Nevada, California. [USNM]

NOTE—The male lectotype of *nostra* was designated by Todd (1982: 151).

Compared to other members of the *cursoria* group, *nostra* shows only a small amount of variation in forewing color and pattern. Specimens of *nostra* are most similar to those of the pale, obscurely marked form of *ochrogaster* but can be distinguished from them by a combination of forewing characters. In *nostra* the forewing is pale orange brown with a dusting of pale-gray scales that give most specimens a hoary appearance. The hoary shading tends to be most pronounced in the orbicular spot, on the costa, and at the wing apex. The reniform and orbicular spots are obscure; they are concolorous with, or slightly paler than, the remainder of the forewing except for a small amount of gray shading at the posterior end of the reniform spot. The spots are outlined with a narrow black line; in some specimens there is an obscure pale-colored line inside this. Forewing length varies from 16 to 19 mm. In specimens of the pale form of *ochrogaster,* the forewing lacks hoary shading, and the reniform and orbicular spots are dark gray with a conspicuous pale outline; these spots contrast with the remainder of the forewing ground color. The differences between *nostra* and *ochrogaster* are minor, but they combine to give the forewings of the two species a very different appearance so that specimens of the two species are rarely confused. The male genitalia of *nostra* can be distinguished from those of *ochrogaster* with certainty only by vesica shape: the vesica of *nostra* lacks the prominent subbasal crook present in *ochrogaster,* the subbasal diverticulum is differently shaped; and the median diverticulum is half way between the subbasal bend and the apex in *nostra* rather than close to the subbasal bend. The female genitalia of *nostra* differ from those of *ochrogaster* by the characters given in the key.

The immature stages of *nostra* are unknown.

Euxoa nostra occurs in dry coniferous forests and piñon-juniper woodland from southern Alberta and southern British Columbia southward to central Colorado, southern Utah, northeastern Nevada, and southern California. A disjunct population occurs at Fort Smith in south-central Northwest Territories. Adults have been collected from mid-June until late September.

siccata GROUP

The *siccata* group includes only *siccata*; the characters are given under the species.

Euxoa (*Euxoa*) *siccata* (Smith)

PL. 6, FIGS. 14–16; PL. K, FIG. 3; PL. Z, FIG. 5; PL. FF, FIG. 1 (RWH 10811).

Carneades siccata Smith, 1893, *Ent. News,* **4**: 99.
Type locality: Colorado. [USNM]

NOTE—The male lectotype of *siccata* was designated by Todd (1982: 195).

Euxoa placida Barnes and McDunnough, 1912, *Contrib. Nat. Hist. Lep. N. Am.,* **1** (5): 4.
Type locality: Kerrville, Texas. [USNM]

Specimens of *siccata* can be recognized by a combination of characters: the male antenna is very prominently biserrate, with its width about 4 times that of the central shaft; the forewing is dirty white, pinkish white, pale orange or pale yellow with the transverse lines indistinct in most specimens; the reniform and orbicular spots are prominent and charcoal colored; and, the hindwing is white in the male, dirty white in the female. In some specimens there is extensive charcoal shading in the basal, subterminal, and terminal areas of the forewings; in them the median line is frequently well defined. On the basis of forewing color and pattern, specimens of *siccata* are most likely to be confused with those of *teleboa* (p. 130), *medialis* (p. 66), *citricolor* (p. 129), *catenula* (p. 95), and a rare form of *tronella* (plate 7, figure 12). They can be distinguished from specimens of *medialis* by the dark-filled reniform spot that in *medialis* is only partially dark filled and takes the form of an inverted comma, and from specimens of other other four species by larger size, stouter body, more prominently biserrate antennae, and details of forewing maculation. In *siccata,* forewing length varies from 14 to 18 mm. Specimens of *siccata* can be distinguished from all other *Euxoa* species by genital characters: the sacculus is disproportionately long—almost as long as the cucullus and about ⅓ longer than the harpe and saccular extension; the saccular extension is similar in length to the harpe (in similar species the harpe and saccular extension are as long as the sacculus or longer); the subbasal diverticulum of the vesica is rounded or oval rather than elbowed and footlike, and there is an additional, small, bubblelike diverticulum subbasally on the right dorsolateral surface (this additional diverticulum is also present in *teleboa* males); and the vesica projects to the right at 90 degrees with respect to the aedoeagus. The most distinctive feature of the female genitalia of *siccata* is the presence of a sclerotized rim on the apical third of the dorsal margin of the ovipositor lobe. In other *Euxoa* species there is either a sclerotized rim around the entire posterior margin of the lobe, or a sclerotized process projecting from the apex of the lobe. The corpus bursae is somewhat T-shaped with the ductus bursae at the posterior end of the expanded portion.

The immature stages of *siccata* are unknown.

Euxoa siccata is primarily an inhabitant of the western Great Plains region where it occurs from southern Alberta southward to south-central Texas. In the southern portion of its range it occurs as far west as southeastern Arizona. Adults have a late flight period; they have been collected from early September until late October.

choris GROUP

The *choris* group includes only *choris*; the characters are given under the species.

Euxoa (*Euxoa*) *choris* (Harvey)

PL. 6, FIGS. 17, 18; PL. K, FIG. 4; PL. Z, FIG. 6; PL. FF, FIG. 2 (RWH 10819).

Agrotis choris Harvey, 1876, *Can. Ent.,* **8**: 37.
Type locality: Nevada. [BMNH]

Agrotis cogitans Smith, 1890, *Trans. Amer. Ent. Soc.,* **17**: 46.
Type locality: California. [USNM]

NOTE—The male lectotype of *cogitans* was designated by Todd (1982: 50).

Agrotis achor Strecker, 1899, *Lepidoptera, Rhopaloceres and Heteroceres, Indigenous and Exotic,* Suppl., **2**: 5.
Type locality: Colorado. [FMNH]

Euxoa choris is a large, distinctively marked species that has pale-brown or pale gray-brown forewings. Specimens of *choris* are most likely to be confused with those of *messoria* but can be distinguished from them, and from those of other large, pale-brown species, by the single rather than double postmedial line of the forewing and by the presence of a narrow, black basal dash in most specimens. The reniform spot is partially outlined in black with the outline most conspicuous on the margin adjacent to the orbicular spot. In most specimens the orbicular spot is oval and frequently tapered to a point that projects toward the reniform spot. Forewing length varies from 15 to 20 mm (17–19 mm in most specimens). Presence of a large subbasal cornutus in the vesica that is at least half as long as the apical diameter of the aedoeagus is distinctive as are the long, symmetrical saccular extensions that curl dorsally around the end of the cucullus. They can be observed, without dissection, by removing the scales from the end of the abdomen with a small brush. The shape of the corpus bursae is similar to that of *tessellata* but the posterior portion is slightly sclerotized. The long setae that form a subbasal row on the ovipositor lobe are stouter and spikelike near the ventral margin of the lobe than are those near the dorsal margin and the presence of a rounded, somewhat dog-eared sclerotized process at the apex of the lobe, characterize females.

The immature stages of *choris* are unknown.

Euxoa choris inhabits both dry coniferous forests and piñon-juniper woodland. It occurs from central Alberta and south-central Yukon southward to central New Mexico, northern Arizona, and east-central California. Its range extends into the Great Plains region in Alberta where it occurs in aspen groves in river bottoms. Adults of *choris* have been collected from early July until late September.

obeliscoides GROUP

This group includes two very similar North American species that can be distinguished with certainty only by characters of the male vesica.

The members of the *obeliscoides* group can readily be recognized both by wing markings and by genital characters. The forewing ground color varies from pale buffy brown through various shades of reddish brown and grayish brown to dark blackish brown. The basal and median areas are heavily dusted with black scales; the subterminal area appears paler and more evenly colored. The most distinctive wing marking feature is the large, somewhat rectangular, pale-buff or reddish-buff reniform spot. The forewing costa, and frequently the orbicular spot, are pale and contrasting as well. The forewing basal dash and claviform spot are black and prominent; in most specimens the median area distal to the claviform spot is similar in color to the subterminal area. The male genitalia can be recognized by the almost straight, non-pubescent harpe; the stout, slightly twisted saccular extension that is about 1⅓ times the length of the harpe; and shape of the vesica. In the female genitalia, the corpus bursae is bisaccate with the dorsal sac oval and positioned at an oblique angle to the ventral sac. This is one of three species groups with a bisaccate corpus bursae that also have a sclerotized process on each ovipositor lobe. They are elongate and separate in the *obeliscoides* group, elongate and fused together in the *punctigera* group, and form a rim on the ovipositor lobe in the *serricornis* group. The *obeliscoides* group is the only species-group with the combination of bisaccate corpus bursae and stout, spikelike subbasal setae on the ovipositor lobe.

KEY TO SPECIES OF THE *OBELISCOIDES* GROUP

1. Male vesica with large trilobed subbasal diverticulum and bilobed diverticulum on ventral curve; male hindwing white with smoky-brown shading on outer third of wing; widely distributed .... *obeliscoides* this page

— Male vesica with relatively small, bilobed subbasal diverticulum and without diverticulum on ventral curve; male hindwing white, some smoky-brown shading on wing margin of specimens from eastern Great Plains; known only from Great Plains region .... *oberfoelli* p. 116

Euxoa (*Euxoa*) *obeliscoides* (Guenée)

PL. 6, FIGS. 19–23; PL. K, FIG. 5; PL. Z, FIG. 7 (RWH 10817).

Agrotis obeliscoides Guenée, 1852, *Histoire Naturelle des Insectes. Species Général de Lépidoptères.* 5, *Noctuélites*, **1**: 293.
Type locality: North America. [BMNH]

Agrotis sexatilis Grote, 1873, *Bull. Buffalo Soc. Nat. Sci.*, **1**: 100.
Type locality: United States. [BMNH]

NOTE—It is uncertain how many specimens Grote had when he described *sexatilis*, but he stated that his material was from New York and Colorado (No. 66). Two of the original specimens are in the British Museum, but only one of them is labeled as type. This specimen, labeled "U.S. America, Grote Coll. 81–116/212/*Agrotis sexatilis* Grote [blue bordered label]/*Agrotis sexatilis* Grote, type [red bordered label]/Noctuidae, Brit. Mus. slide No. 6341," is here designated lectotype. The specimen is a male in good condition.

Carneades infusa Smith, 1902, *Can. Ent.*, **34**: 30.
Type locality: Cartwright, Manitoba. [USNM]

NOTE—The male lectotype of *infusa* was designated by Todd (1982: 108).

In most of its range, *obeliscoides* can easily be recognized by distinctive forewing markings and by genital characters discussed for the species-group. Forewing shade varies to a certain extent with habitat in that the palest specimens occur in open arid areas (plate 6, figure 22) and dark-colored specimens occur in forested areas (plate 6, figures 19, 20). Forewing length varies from 13 to 19 mm. In the Great Plains region, however, *obeliscoides* occurs sympatrically with *oberfoelli*, and in this area specimens of the two species can be difficult to distinguish without determining vesica shape. Most specimens of *obeliscoides* can be distinguished from those of *oberfoelli* by the presence of a dark marginal hindwing band in *obeliscoides* and by the paler forewing. Specimens of *oberfoelli* have forewings as dark as the forested form of *obeliscoides*. The darker forewing of *oberfoelli*, in combination with an almost pure white hindwing, can be used to identify most specimens.

The immature stages of *obeliscoides* are known only from laboratory reared material. The species overwinters in the egg stage (Hinks and Byers, 1976: 1349). The egg was illustrated by Salkeld (1975: 1152, figs. 58–60).

Euxoa obeliscoides is widely distributed in western North America where it occurs from Manitoba westward to British Columbia and southward to central Kansas, and southern New Mexico, Arizona, and California. Its range extends eastward in southern Canada and northern United States to southern Quebec and northern Pennsylvania (Lafontaine, 1982a: figure 27). It occurs in both forested and open habitats, but it is rarely collected in open aridlands. Adults have been collected from late July until late September.

Euxoa (*Euxoa*) *oberfoelli* Hardwick
PL. 6, FIGS. 24, 25; PL. K, FIG. 6; PL. FF, FIG. 3 (RWH 10818).

Euxoa oberfoelli Hardwick, 1973, *Can. Ent.*, **105**: 75.
Type locality: Fort Peck, Montana, 17 mi SE. [CNC]

Males of *oberfoelli* are very similar to those of the much more widespread species, *obeliscoides*, but can be distinguished from them by characters given in the key and under *obeliscoides*. Females of *oberfoelli* can be identified only by association with males. The two species are similar in size; forewing length in *oberfoelli* ranges from 13 to 18 mm. Attempts to hybridize the two species in the laboratory were unsuccessful; only unfertile eggs were obtained (J. R. Byers, personal communication).

The immature stages of *oberfoelli* are known only from laboratory reared material. Like *obeliscoides*, it overwinters in the egg stage. The larva has a longer summer aestivation period than does that of *obeliscoides* (Hinks and Byers, 1976: 1352); this is consistent with occurrence in drier, hotter habitats than *obeliscoides*. The egg of *oberfoelli* was illustrated by Salkeld (1975: 1152, figs. 55–57).

Euxoa oberfoelli has a relatively restricted range, both in terms of geographical distribution and in habitat preference. It occurs in the Great Plains region from the eastern Dakotas westward to west-central Montana and from southern Saskatchewan southward to central Kansas and northeastern New Mexico (Lafontaine, 1982a: figure 27). It has been collected only in areas of natural prairie. I have collected it at a number of locations in eastern South Dakota and Nebraska by locating areas of unbroken prairie; an excellent indicator of this is the presence of Grama Grass (*Bouteloua* spp.) and Buffalo Grass (*Buchloe dactyloides* (Nuttall) Engelmann). Adults of *oberfoelli* have been collected from late August until late September.

lillooet GROUP

The *lillooet* group includes only *lillooet*; the characters are given under the species.

Euxoa (*Euxoa*) *lillooet* McDunnough
PL. 6, FIGS. 26, 27; PL. K, FIG. 7; PL. Z, FIG. 8 (RWH 10808).

Euxoa lillooet McDunnough, 1937, *Can. Ent.*, **59**: 195.
Type locality: Seton Lake, British Columbia. [CNC]

Specimens of *lillooet* are similar to those of *declarata* (p. 79) and *albipennis* (p. 90) in wing markings but usually can be distinguished from them by the presence of pale-mauve shading on the forewing, particularly in the basal and subterminal areas and in the orbicular spot. The three species are similar in size; in *lillooet* forewing length varies from 14 to 17 mm. Males of *lillooet* may be distinguished from those of *albipennis* and *declarata* by the glabrous rather than pubescent harpe and by the shorter and straighter saccular extension. In *lillooet* the saccular extension is shorter than the sacculus and straight; in the other two species it is longer than the sacculus and curved. Also, males of *lillooet* have dark hindwings, similar to those of the females, whereas those of *albipennis* males are white. Females of *lillooet* differ from those of *albipennis* in having a bisaccate rather than unisaccate corpus bursae and in lacking the sclerotized process at the apex of each ovipositor lobe that is present in *albipennis* females. In *lillooet* the appendix bursae is dorsal and shorter than the elongate corpus bursae and at an oblique angle to it, whereas in *declarata* the appendix bursae is to the left of, and longer than, the corpus bursae.

The immature stages of *lillooet* are unknown.

Euxoa lillooet has a limited range in western North America. It occurs from southern British Columbia and eastern Washington southward in the Rocky Mountain region to northern New Mexico and in the Wasatch Range to southern Utah. An apparently disjunct population occurs in western South Dakota. The species inhabits coniferous forests, particularly pine forests. Adults of *lillooet* have been collected from late June until late August; in most areas the flight season of *lillooet* is several weeks before that of *albipennis* or *declarata.*

basalis GROUP

This group includes only *basalis*; the characters are given under the species.

Euxoa (*Euxoa*) *basalis* (Grote)

PL. 6, FIGS. 28–30; PL. K, FIG. 8; PL. AA, FIG. 1 (RWH 10798).

Agrotis basalis Grote, 1879, *North Amer. Ent.*, **1**: 38.
Type locality: Idaho Springs, Colorado. [BMNH]

NOTE—Grote described *basalis* from three specimens collected at Idaho Springs, Colorado by Snow. Of these, a male and a female are in the British Museum (Natural History). The male specimen, labeled "Type/Colorado, Snow, Grote Coll. 81–116/*Agrotis basalis* Grote, type Male/Noctuidae, Brit. Mus. slide No. 6336" is here designated lectotype. The specimen is in good condition except that the right antenna is broken.

Specimens of *basalis* can be recognized by the combination of reddish-brown forewing ground color and pale, contrasted, yellow-buff or pale-gray shading in the basal area and the reniform and orbicular spots. This two-toned forewing ground color persists through a variety of shades. In the most common and widespread form (plate 6, figure 28) the forewing ground color is bright reddish brown with the pale areas yellow buff. A less common, dark reddish-brown form (plate 6, figure 29) has pale-silver shading in the basal area and the reniform and orbicular spots. In very dry areas the forewing pattern appears to be pale and washed out (plate 6, figure 30). In very pale specimens, the hindwing in both sexes is pale buffy brown rather than smoky brown. Forewing length varies from 15 to 19 mm. Specimens of *basalis* are most likely to be confused with reddish-brown forms of *mimallonis* (p. 33) and *ochrogaster* (p. 112), but in these species the basal area of the forewing is not pale and contrasted with the median and subterminal areas. The valves are most similar to those of *messoria,* and to a lesser extent, those of *ochrogaster*; the shape of the vesica, however, is very distinctive. The female genitalia differ from those of other species in the shape of the corpus bursae: it is bisaccate, with the larger corpus bursae on the right curved to the left anteriorly around the end of the appendix bursae. The appendix bursae is triangular and fits into the concave wall of the corpus bursae so snugly that at first glance the bursa appears unisaccate rather than bisaccate.

The immature stages of *basalis* have been described in detail by Byers et al. (1975). The most unusual feature of them is in the coloration of the larva in the penultimate and ultimate instar. Although the mature larva is morphologically very similar to those of other *Euxoa* species, it differs from them in having orange rather than drably colored sclerotized areas on the body, particularly the head capsule and prothoracic and anal shields. The larvae of *basalis* have not been studied under natural conditions so it is not known if these brightly colored areas on the body reflect larval habits that differ from those of other *Euxoa* species.

Euxoa basalis occurs from central Saskatchewan and Alberta southward to the western Dakotas, central New Mexico, central Arizona, east-central Nevada, and northeastern Oregon. An apparently disjunct population occurs in relict prairie habitat near Fort Smith in the Northwest Territories. This species is most frequently collected in open grassland habitats; it inhabits open coniferous forests in the southern portion of its range. Adults of *basalis* have been collected from mid-July until early October.

detersa GROUP

The *detersa* group is the largest species-group in the genus; it contains 32 North American species. All species occur in western North America, although the ranges of four extend into the East as well. Members of the group can most easily be recognized by a combination of genital characters. They are: male genitalia with sacculus small and crescentic, saccular extensions slightly longer than the harpes in most species and about as stout as the harpes or slightly stouter; female genitalia with corpus bursae unisaccate and oval with the ductus bursae on the right side; ovipositor lobes not fused together on the dorsal margins. Males in the group can best be recognized by vesica shape but the use of this in the key to species-groups and key to species has been avoided when possible because it cannot be observed in genital preparations in which the vesica has not been everted.

Many species in the group have distinctive wing markings and can be identified by comparing specimens with the illustrations in the color plates.

The treatment is based on a recent revision of the group (Lafontaine, 1981).

KEY TO SPECIES OF THE *DETERSA* GROUP

1. Males 2
— Females 36

2. Right harpe at least 1.25× length of right sacculus (plate N, figure 1) .... *melura* p. 136
— Right harpe less than 1.25× length of right sacculus 3

3. Apical half of harpe very finely pubescent, and with much larger scattered setae 4
— Apical half of harpe without pubescence, only larger scattered setae present, these about half as long as diameter of harpe 9

4. Obicular spot elongate, barlike, two or three times as long as wide; right harpe less than 1.2 mm long 5
— Orbicular spot rounded or oval; elongate in some specimens in which it is fused with pale costa, if so, then right harpe more than 1.3 mm long 7

5. Right saccular extension about 1.5× length of right harpe (plate L, figure 4) .... *olivalis* p. 127
— Right saccular extension about 1.25× length of right harpe (plate L, figure 5) 6

6. Forewing with costa pale, evenly colored, pale shading extended uninterrupted to cubital vein on basal quarter of wing (plate 7, figures 5, 6) .... *oblongistigma* p. 128
— Forewing with costa brown like remainder of wing, not evenly colored; costal shading not extended to cubital vein near base of wing but separated from it by several light and dark lines (plate 7, figures 3, 4) .... *agema* p. 128

7. Forewing with black basal dash (plate 7, figures 34, 35) .... *laetificans* p. 133
— Forewing without black basal dash 8

8. Right saccular extension more than 1.5× length of right harpe; forewing dark brown with copper-colored basal and subterminal areas (plate 7, figure 33) .... *infracta* p. 132
— Right saccular extension less than 1.5× length of right harpe; forewing light brown or gray (plate 7, figure 32) .... *dodi* p. 132

9. Clavus prominent, sclerotized, much longer than wide; dorsal margin of sacculus invaginated at base of clavus 10
— Clavus inconspicuous and at most only slightly sclerotized; dorsal margin of sacculus not invaginated at base of clavus 13

10. Forewing without veins CuA_1 and M_3 pale lined and projecting into terminal area; occurring from Great Plains eastward to Great Lakes region .... *detersa personata* p. 137
— Forewing with veins CuA_1 and M_3 pale lined and projecting into terminal area; occurring either in eastern North America, or from western Great Plains westward 11

11. Occurring in eastern North America .... *detersa detersa* p. 137
— Occurring from western Great Plains westward 12

12. Pale areas of forewing white, cream colored, or pale orange brown; occurring throughout intermontane region and western Great Plains .... *cicatricosa* p. 137
— Pale areas of forewing yellow; occurring in southern and western portions of intermontane region .... *recula* p. 138

13. Forewing without basal dash, not longitudinally streaked 14
— Forewing either with black basal dash, or longitudinally streaked, particularly on veins CuA, CuA_1 and M_3, or both 21

14. Right saccular extension about as long as, or shorter than, right harpe 15
— Right saccular extension longer than right harpe 16

15. Hindwing white; forewing orange or yellow with maculation obscure except for prominent, black median line; apical half of vesica projecting at about right angle to aedoeagus .... *teleboa* p. 130
— Hindwing pale buff, trace of median line in many specimens; forewing brown or gray in most specimens, median line not particularly contrasting; vesica strongly bent subbasally so that apical portion projects almost parallel to aedoeagus .... *difformis* (part) p. 130

16. Uncus slightly constricted at about ¼ from apex (not visible in lateral view); setae on ventral surface of uncus near apex long and saberlike, longer than depth of uncus, densely crowded; right harpe with 10–15 setae; hindwing white or very pale buff in most specimens 17
— Uncus dilated at middle, then tapered to apex; setae on ventral surface of uncus near apex short, almost conical, shorter than depth of uncus; right harpe with 25–50 setae; hindwing brown 19

17. Apex of right saccular extension straight or bent away from cucullus; vesica with small, nipplelike diverticulum subbasally, in addition to normal footlike subbasal diverticulum; hindwing pale yellow buff. (Southern Californian populations of *difformis*) .... *difformis* (part) p. 130
— Apex of right saccular extension bent toward cucullus; vesica with only normal footlike diverticulum subbasally; hindwing pure white in southern California 18

18. Forewing yellow .... *citricolor* p. 129
— Forewing white or cream colored with black dusting .... *tronella* p. 129

19. Basal and subterminal areas of forewing orange or yellow, median area gray .... *murdocki* p. 132
— Basal, subterminal, and median areas of forewing similar in color 20

20. Vesica with small nipplelike diverticulum subbasally, in addition to normal footlike subbasal diverticulum; right harpe with about 30 setae; hindwing gray brown, paler toward base in most specimens; occurring in Great Plains and Great Basin .... *moerens* p. 131
— Vesica with only normal footlike diverticulum present subbasally; right harpe with 50–60 setae; hindwing dark brown; occurring along Cascades-Sierra Nevada axis from Washington southward to southern California, and in La Sal Mountains, Utah .... *latro* p. 131

21. Veins CuA_1 and M_3 pale lined, contrasting, projecting into terminal area 22
— Veins CuA_1 and M_3 not pale lined or contrasting; cubital vein pale lined to reniform spot in some specimens but no pale-lined veins projected into terminal area 26

22. Right saccular extension at least 1.5× length of right harpe .... *niveilinea* p. 135
— Right saccular extension at most 1.3× length of right harpe 23

23. Vesica with small nipplelike subbasal diverticulum on left side in addition to normal footlike subbasal diverticulum; right harpe 0.9–1.0 mm long .... *dargo* p. 135
— Vesica with only normal footlike diverticulum subbasally; right harpe 1.1–1.3 mm long 24

24. Forewing with costal area pale, evenly colored, pale shade extended uninterrupted to cubital

vein on basal quarter of wing .... *quadridentata* p. 133

— Forewing with costal area not particularly pale, or if so, then streaked and unevenly colored; costal shade interrupted from cubital vein on basal quarter of wing by a black line 25

25. Ground color of forewing brown with fine black streaks on veins .... *inscripta* (part) p. 134

— Ground color of forewing gray streaked with pale silver gray .... *brevipennis* (part) p. 125

26. Hindwing white with thin black terminal line .... *brevipennis* (part) p. 125

— Hindwing shaded with brown, at least on marginal third of wing 27

27. Hindwing pale with sharply contrasted dark marginal band on outer third 28

— Hindwing smoky brown, paler near base in many specimens 30

28. Right saccular extension as stout as, or thinner than, right harpe .... *redimicula* p. 126

— Right saccular extension stouter than right harpe 29

29. Basal half of prothoracic collar and costa of forewing pale yellow brown; forewing with transverse lines conspicuous; known only from vicinity of Saskatoon, Saskatchewan ...... *unica* p. 135

— Prothoracic collar and costa of forewing essentially same color as remainder of forewing; forewing with transverse lines obscure, veins overlaid with black scales; occurring from southern Montana southward to Colorado and westward to Lake Tahoe, California .... *inscripta* (part) p. 134

30. Forewing with transverse lines conspicuous; costal, subterminal, and basal areas paler than median area 31

— Forewing with transverse lines inconspicuous; subterminal and basal areas not paler than median area; costal area paler than remainder of forewing in many specimens 32

31. Terminal area of forewing blurred and streaked into subterminal area; postmedial line not scalloped between veins .... *servita* p. 125

— Terminal area of forewing with smooth, regular inner margin; postmedial line scalloped between veins .... *auripennis* p. 126

32. Forewing shaded with maroon or red 33

— Forewing shaded with buff, brown, or gray 34

33. Forewing light red with contrasting silver-gray reniform and orbicular spots and costa; occurring from British Columbia southward to southern California .... *costata* p. 123

— Forewing maroon dusted with dark scales; reniform and orbicular spots either not silver gray, or if so, then with cubital vein pale as well; occurring from western Canada southward in the Rocky Mountain region to eastern Arizona and northern New Mexico ......*castanea* p. 123

34. Forewing largely black with contrasting silver-gray reniform and orbicular spots and costa; occurring from northern Utah southward to Arizona and New Mexico .... *foeminalis* p. 123

— Forewing brown or with extensive brown shading, almost black in some specimens from Rocky Mountain region of British Columbia and Montana 35

35. Forewing, including terminal area, yellow buff; series of black, sagittate spots on proximal side of subterminal line; orbicular spot round; hindwing unevenly shaded with light and dark brown, dark median line in most specimens; occurring in Great Plains in Alberta and Montana .... *clausa* p. 124

— Forewing dark brown in most specimens; if light buff, then terminal area dark brown with dark streaks on proximal side of subterminal line; orbicular spot oval in most specimens; hindwing evenly colored, slightly paler near base; widespread .... *idahoensis* p. 124

36. Ovipositor lobe with stout setae dorsally or apically 37

— Ovipositor lobe without stout setae 42

37. Stout setae on dorsal margin of ovipositor lobe; ovipositor lobe without definite row of long subbasal setae 38

— Stout setae primarily on apical third of ovipositor lobe; ovipositor lobe with definite row of long subbasal setae 40

38. Stout setae scattered on dorsal margin of ovipositor lobe, not in single row 39
— Stout setae in single row on dorsal margin of ovipositor lobe; occurring from western Great Plains westward .... *cicatricosa* p. 137

39. Occurring east of Rocky Mountains .... *detersa* p. 136
— Occurring in western and southern intermontane region .... *recula* p. 138

40. Long subbasal setae on ovipositor lobe stout and spikelike .... *dargo* p. 135
— Long subbasal setae on ovipositor lobe thin and hairlike 41

41. Long subbasal setae on ovipositor lobe stouter on ventral portion of lobe than those on dorsal portion; forewing with cubital vein pale ... *melura* p. 136
— Long subbasal setae on ovipositor lobe thin and hairlike on both ventral and dorsal portions of lobe; forewing with cubital vein not pale .... *servita* p. 125

42. Sclerotized plates in wall of ductus bursae short, not extended as far anteriorly as anterior apophyses .... *quadridentata* p. 133
— Sclerotized plates in wall of ductus bursae longer, extended farther anteriorly than anterior apophyses 43

43. Ovipositor lobe with prominent, smooth, flangelike projection at apex 44
— Ovipositor lobe without flange, or at most, trace of flange at apex, its surface covered with minute setae 56

44. Forewing either with black basal dash, or prominently streaked longitudinally, or both 45
— Forewing without black basal dash and without longitudinal streaks 51

45. Orbicular spot elongate, almost barlike; junction of ductus bursae and corpus bursae anterior to middle of corpus bursae in most specimens 46
— Orbicular spot round or oval; junction of ductus bursae and corpus bursae at middle, or posterior to middle 48

46. Orbicular spot buff filled, not outlined in white; basal dash prominent; costa evenly colored, buff or pale gray .... *oblongistigma* p. 128
— Orbicular spot with dark center in most specimens, dark center outlined in white; basal dash thin and inconspicuous; costa not evenly colored but marked with white or brown 47

47. Forewing predominantly various shades of brown; hindwing dark brown .... *agema* p. 128
— Forewing streaked with gray and white in most specimens; hindwing gray or brown, paler at base .... *olivalis* p. 127

48. Forewing streaked longitudinally; basal dash inconspicuous if present 49
— Forewing not streaked longitudinally; basal dash prominent 50

49. Cubital vein and costa pale and contrasting with darker ground color .... *niveilinea* p. 135
— Cubital vein and costa not particularly pale .... *inscripta* p. 134

NOTE—The female of *unica* is unknown; however, it probably would key to this couplet.

50. Flangelike projection at apex of ovipositor lobe short and narrow, or rounded, about as long as wide; forewing with median area heavily dusted with black; occurring from the Great Plains westward .... *auripennis* p. 126
— Flangelike projection at apex of ovipositor lobe fingerlike, long and narrow; forewing with median area light gray or brown; occurring from Montana and Wyoming eastward .... *redimicula* p. 126

51. Anterior end of ovipositor lobe markedly produced ventrally, depth of lobe greater than its length 52
— Anterior end of ovipositor lobe, at most, slightly produced ventrally, depth of lobe less than its length 53

52. Forewing yellow .... *citricolor* p. 129

– Forewing cream colored or pale buff, dusted with black .... *tronella* p. 129

53. Ovipositor lobe with raised, finlike, sclerotized plate on posterior half of dorsal margin of lobe; dorsal edge of plate straight or slightly concave, curved abruptly at anterior end to lobe; sclerotized plate in ventral wall of ductus bursae extending as far anteriorly as plate in dorsal wall .... *teleboa* p. 130

– Ovipositor lobe with rounded, or fingerlike sclerotized projection at apex, this curved around dorsal margin of lobe, its dorsal edge convex and tapered anteriorly; tapered abruptly in some species with sclerotized plate in dorsal wall of ductus bursae extending farther anteriorly than plate in ventral wall 54

54. Forewing with median area pale gray, basal and subterminal areas orange .... *murdocki* p. 132

– Forewing with basal, median, and subterminal areas similar in color, if not in shade 55

55. Forewing dark brown, heavily dusted with gray in most specimens; hindwing dark brown to base; occurring along Cascades-Sierra Nevada axis from Washington southward to southern California; disjunct population in La Sal Mts., Utah .... *latro* p. 131

– Forewing light gray, light brown, or orange; hindwing gray, paler toward base in most specimens; occurring throughout Great Plains and Great Basin .... *moerens* p. 131

56. Forewing with prominent, black, basal dash 57

– Forewing without basal dash 63

57. Ovipositor lobe with short, conical setae apically; forewing with veins CuA_1 and M_3 not pale lined 58

– Ovipositor lobe with fine-tipped setae apically, or with veins CuA_1 and M_3 pale lined, pale streaks projecting into terminal area 62

58. Forewing shaded with maroon or reddish brown 59

– Forewing shaded with buff, brown, or gray 60

59. Ovipositor lobe rounded or somewhat truncate apically; forewing light reddish brown with contrasting silver-gray costa, reniform spot, and orbicular spot; cubital vein not pale; occurring from British Columbia southward to southern California .... *costata* p. 123

– Ovipositor lobe acutely angled apically; forewing maroon or chestnut brown, dusted with black; costa, reniform spot, and orbicular spot either without silver-gray shading, or with cubital vein pale as well; occurring from western Canada southward in the Rocky Mountain region to eastern Arizona and northern New Mexico .... *castanea* p. 123

60. Forewing extensively black with contrasting silver-gray reniform and orbicular spots and costa; occurring from northern Utah southward to New Mexico and Arizona ..... *foeminalis* p. 123

– Forewing brown or streaked with brown, almost black in some specimens from Rocky Mountain region of British Columbia and Montana 61

61. Forewing, including terminal area, yellow buff; series of black sagittate spots on proximal side of subterminal line; orbicular spot round; hindwing unevenly shaded with light and dark brown, dark median line in most specimens; occurring in Great Plains of Alberta and Montana .... *clausa* p. 124

– Forewing brown in most specimens; if light buff, then terminal area of forewing dark brown with dark streaks on proximal side of subterminal line; orbicular spot oval in most specimens; hindwing evenly colored, slightly lighter toward base; widely distributed .... *idahoensis* p. 124

62. Hindwing mottled gray brown with median line on ventral surface; forewing with inconspicuous, irregular, basal dash; forewing with dark streaks extending from terminal area into subterminal area interrupted by pale subterminal line .... *brevipennis* (part) p. 125

– Hindwing evenly colored, pale gray brown, slightly lighter at base, without trace of median line on ventral surface; forewing with conspicuous basal dash extending to antemedial line; forewing with dark streaks extending uninterrupted from terminal area into subterminal area, subterminal line not evident ..... *laetificans* p. 133

63. Forewing marked with silver gray; inner edge

of terminal area with W-mark where pale-lined veins CuA_1 and M_3 project into it .... *brevipennis* (part) p. 125

– Forewing brown or gray, without silver-gray streaks; inner edge of terminal area without W-mark, veins CuA_1 and M_3 not pale lined ... 64

64. Forewing with basal and subterminal areas similar in color to median area; claviform spot obscure or absent in most specimens; hindwing with dark median line in most specimens ... . *difformis* p. 130

– Forewing with basal and subterminal areas paler than median and terminal areas; claviform spot outlined in black in most specimens; hindwing without dark median line 65

65. Forewing with median area pale gray brown, basal and subterminal areas pale gray ...... *dodi* p. 132

– Forewing with median and terminal area dark brown, basal and subterminal areas paler brown with a copper suffusion .... *infracta* p. 132

Euxoa (*Euxoa*) *costata* (Grote)

PL. 6, FIGS. 31, 32; PL. K, FIG. 9; PL. AA, FIG. 2 (RWH 10825).

Agrotis costata Grote, 1876, *Bull. Buffalo Soc. Nat. Sci.,* **3**: 80.
Type locality: Vancouver Island, British Columbia. [BMNH]

This species and the next four form a group of closely related and frequently difficult to identify species. Until recently, most of them were considered forms of *costata*; all are similar in having a contrastingly pale costa that is fused with the pale orbicular spot in many specimens. Specimens of *costata* can be distinguished from those of the other related species by the bright orange-red color of the forewing and by lack of pale scales on the forewing cubital vein. Forewing length varies from 14 to 18 mm. In the male genitalia of *costata* and the next four species, the harpes and saccular extensions are long and thin and appear frail. Those of *costata* differ in that the harpes and saccular extensions are longer (length of right saccular extension 1.71 ± 0.08 mm, N = 20; length of right harpe 1.61 ± 0.04 mm). In the female genitalia, the ovipositor lobes are rounded and slightly truncate apically.

The immature stages of *costata* are unknown.

Euxoa costata occurs from southern British Columbia southward in the Cascades and Sierra Nevada Mountains to the Transverse Ranges of southern California. Its range lies within that of *idahoensis* except in southwestern British Columbia and possibly in southern California. It occurs sympatrically with *castanea* in the northeast portion of its range. Adults have been collected from mid-July until early September. The species inhabits dry coniferous forests.

Euxoa (*Euxoa*) *castanea* Lafontaine

PL. 6, FIGS. 33, 34; PL. FF, FIG. 4.

Euxoa castanea Lafontaine, 1981, *Quaestiones Entomologicae,* **17**: 19.
Type locality: Golden, British Columbia. [CNC]

Euxoa castanea is a widespread species that is called *costata* in many collections. It differs from *costata* in having a chestnut-brown rather than orange-red forewing and the cubital vein is pale lined. The forewing ground color is dusted with black scales in most specimens. Forewing length varies from 14 to 17 mm. The male genitalia are similar to those of *costata* but the harpes and saccular extensions are shorter (length of right saccular extension 1.33 ± 0.10 mm, N = 25; length of right harpe 1.28 ± 0.06 mm). The female genitalia differ from those of *costata* in that the ovipositor lobes are more triangular and tapered apically. Females from the southern Rocky Mountain region have a very small flangelike process apically.

The immature stages are known only from laboratory reared material. The species overwinters in the egg and has a very short prepupal aestivation period (Hinks and Byers, 1976, (as *costata*)).

This species occurs across the aspen parkland and southern boreal forest region of western Canada as far east as Manitoba; it occurs southward in the Rocky Mountain region to northern New Mexico and east-central Arizona. It also occurs in forested areas in the northern Great Plains. *Euxoa castanea* occurs sympatrically with *costata* in southern British Columbia and northern Idaho. The species inhabits dry aspen and coniferous forests. Adults have been collected from early July until late August.

Euxoa (*Euxoa*) *foeminalis* (Smith)

PL. 6, FIGS. 35, 36 (RWH 10828).

Carneades foeminalis Smith, 1900, *Proc. U. S. Natl. Mus.,* **22**: 454.
Type locality: Garfield County, Colorado. [USNM]

NOTE—The female lectotype was designated by Todd (1982: 88).

Specimens of *foeminalis* are similar to those of *idahoensis* but can be distinguished from them by the color of the forewing. The forewing is dark silver gray, heavily dusted with black. The subterminal area, costa, and reniform and orbicular spots are paler silver gray than the remainder of the forewing. The cubital vein is not pale lined. Forewing length varies from 14 to 18 mm. Specimens of *idahoensis* from the range of *foeminalis* have pale-brown forewings, and the cubital vein is pale in most specimens. A dark form of *idahoensis* occurs north of the range of *foeminalis* (see *idahoensis,* p. 124). The male genitalia are indistinguishable from those of *castanea* and the next two species. In the female genitalia, the ovipositor lobes are more broadly rounded than are those of *idahoensis.*

The immature stages are unknown.

Euxoa foeminalis occurs in southwestern United States from west-central Colorado and north-central Utah southward to northern New Mexico and central Arizona. Adults have been collected from late May until mid-July. The species inhabits pine forests and mixed forests of oak and pine.

Euxoa (*Euxoa*) *idahoensis* (Grote)

PL. 6, FIGS. 37–43; PL. FF, FIG. 5 (RWH 10826).

Agrotis idahoensis Grote, 1878, *Bull. U. S. Geol. Surv.,* **4**: 171.
Type locality: Idaho. [BMNH]

Agrotis furtivus Smith, 1890, *Trans. Amer. Ent. Soc.,* **17**: 56.
Type locality: California. [USNM]

NOTE—The male lectotype of *furtivus* was designated by Todd (1982: 92).

This species is more common, widespread, and variable than the other four similar species. In the most ubiquitous form (plate 6, figures 37, 38) the forewing is pale brown with darker brown streaks; the cubital vein is pale, there are dark-brown sagittate spots proximal to the subterminal line, and the costa and reniform and orbicular spots are pale. Specimens from the Great Plains region are pale buff (plate 6, figure 39) and can be confused with *clausa.* Specimens from the mountains of British Columbia, Alberta, South Dakota, Montana and Colorado (plate 6, figures 40, 41) are dark brown and resemble specimens of *foeminalis* except that in *idahoensis* the forewing is shaded with dark brown, not black; they lack pale silver-gray shading in the subterminal area, and the cubital vein is pale in most specimens of *idahoensis.* Some specimens from the Black Hills of South Dakota and from the Rocky Mountain region of Colorado have heavily streaked forewings (plate 6, figure 42). Many specimens from the Great Basin lack the pale-colored costa (plate 6, figure 43), and some lack pale coloration in the reniform and orbicular spots as well. Forewing length varies from 14 to 19 mm. The male genitalia are indistinguishable from those of *castanea, foeminalis,* and *clausa*; at locations where *idahoensis* occurs with one or more of these species, harpe and saccular extension lengths often differ between species. The range of variation in these structures over the entire distribution of the species, however, obliterates differences that may be observed in specimens from a single location. The same situation occurs with female genital characters. At some locations in the southern Rocky Mountains and the eastern Great Basin, specimens intermediate between *castanea* and *idahoensis* occur in addition to typical specimens; the two species may hybridize at these locations. Preliminary studies of pheromones of *idahoensis* and *castanea* not only revealed differences between these two species but suggested that *idahoensis* may include several cryptic species (J. R. Byers, personal communication).

The immature stages of *idahoensis* are unknown.

Euxoa idahoensis occurs from southeastern Saskatchewan northwestward to central Alaska and southward to southwestern South Dakota, northern New Mexico, central Arizona, southern Nevada, and central California. Adults have been collected from mid-June until late August; early dates are from the southern portion of the range. This species is most commonly collected in dry coniferous forest habitats; it also occurs less commonly in piñon-juniper woodland and in deciduous thickets.

Euxoa (*Euxoa*) *clausa* McDunnough

PL. 6, FIGS. 44, 45 (RWH 10827).

Euxoa clausa McDunnough, 1923, *Can. Ent.,* **55**: 163.
Type locality: Lethbridge, Alberta. [CNC]

This species is similar to the pale form of *idahoensis* (plate 6, figure 39) but can be distinguished from it by the characters given in the key. The male antenna is more prominently biserrate than is that of *idahoensis*; this is most evident when material of both species can be compared together. Forewing length varies from 13 to 18 mm.

The immature stages of *clausa* are unknown.

Euxoa clausa is known only from a small area in the northern Great Plains. It has been collected in southern Alberta, southern Saskatchewan, and southwestern Montana. Adults have been captured from mid-July until mid-August.

Euxoa (*Euxoa*) *brevipennis* (Smith)

PL. 6, FIGS. 46–48; PL. K, FIG. 10; PL. AA, FIG. 3; PL. FF, FIG. 6 (RWH 10824).

Agrotis brevipennis Smith, [1888] *Proc. U. S. Natl. Mus.*, **10**: 455.
Type locality: Colorado. [USNM]

NOTE—The male lectotype of *brevipennis* was designated by Todd (1982: 35).

Euxoa brevistriga Smith, 1910, *Trans. Amer. Ent. Soc.*, **36**: 257.
Type locality: Colorado (probably Denver). [AMNH]

NOTE—The male lectotype of *brevistriga* was designated by Todd (1982: 36).

Euxoa angulirena Smith, 1910, *Trans. Amer. Ent. Soc.*, **36**: 257.
Type locality: Colorado. [AMNH]

Euxoa brevipennis can be distinguished from other *Euxoa* species by a combination of large size, pattern of silver-gray and dark-gray forewing streaking, lack of transverse lines, and pale hindwing. In most specimens, the posterior half of the forewing and the subterminal area are pale silver gray with the veins dark gray; the area around the reniform and orbicular spots and the terminal area are dark gray. The dark terminal area has a W-shaped indentation where pale-lined veins CuA_1 and M_3 project into it. The reniform and orbicular spots have a pale line between the black outline and the gray center. In most specimens the costa and cubital vein are pale silver gray. Specimens of *brevipennis* are larger than those of most other pale-streaked species; forewing length varies from 13 to 17 mm. Individual variation in this species is primarily the result of variation between the amount of pale and dark shading. In the palest specimens the forewing is almost entirely pale silver gray with dark shading confined to the veins and terminal area. In the darkest specimens, the forewing is dark gray with a mottling of pale scales and pale shading on the subterminal line and in the reniform and orbicular spots. In most localities males tend to be paler than females. The male hindwing is white with a thin, black, terminal line; the female hindwing has dirty white shading on the outer half of the wing, or over the entire hindwing in some specimens.

The immature stages have not been described. The larva has been reared on tumblemustard (*Sisymbrium* spp.), hare's ear mustard (*Conringia orientalis* (Linnaeus) Dumort), Russian thistle (*Salsola kali* Linnaeus), and sweet clover (*Melilotus* spp.) in Montana (Cook, 1930).

Euxoa brevipennis occurs in dry habitats where sagebrush is common; this includes both open aridlands and piñon-juniper woodland. Its range extends from southwestern Saskatchewan westward to south-central British Columbia and southward to north-central New Mexico, central Arizona, and southern California. Adults have been collected from early September until mid-October in the North and early November in the South.

Euxoa (*Euxoa*) *servita* (Smith)

PL. 6, FIGS. 49–53; PL. L, FIG. 1; PL. AA, FIG. 4; PL. FF, FIG. 7 (RWH 10854).

Carneades servitus Smith, 1895, *Ent. News*, **6**: 336.
Type locality: Colorado. [AMNH]

NOTE—The male lectotype of *servitus* was designated by Todd (1982: 194).

Euxoa servita novangliae McDunnough, 1950, *Bull. Amer. Mus. Nat. Hist.*, **95**: 391.
Type locality: Franconia, New Hampshire. [AMNH]

NOTE—The dark form of *servita*, described as subspecies *novangliae*, is treated here as a dark morph associated with occurrence in mesic habitats and not as a subspecies.

Euxoa servita, *redimicula*, and *auripennis* have long been confused because of similarity in wing markings and lack of diagnostic male genital characters. Of the three species, *servita* is the most widely distributed and most variable in wing color and maculation. Males of *servita* can be distinguished from those of *redimicula* and *auripennis* by male genital characters but only by using numerical techniques (Lafontaine, 1974). In eastern North America, where *servita* occurs sympatrically with *redimicula*, males of the two species can be distinguished by hindwing color: the hindwing is smoky brown (*servita*); or the hindwing is pale buff basally with smoky-brown shading confined to the outer third (*redimicula*). In central and western North America, where *servita* occurs with *auripennis*, males of *servita* can be recognized by subtle differences in forewing markings given in the key. The forewing ground color varies with habitat; specimens from forested habitats are darker than those from open habitats. In *servita,* forewing length varies from 12 to 15 mm. Females of *servita* can readily be distinguished from those of other similar species by the presence of stout, conical setae on the apical third of the ovipositor lobes. The ovipositor lobes in *redimicula* and *auripennis* are clothed with fine, hairlike setae and there is a sclerotized, apical, flangelike process on each lobe.

The immature states of *servita* are known only from laboratory reared material.

Euxoa servita occurs from Nova Scotia westward to southern Yukon and southward to Massachusetts, southern New York, central Michigan, South Dakota, and southern British Columbia and in the mountains to central New Mexico, east-central Arizona, and southern Utah. *Euxoa servita* generally occurs in more mesic habitats than those preferred by the next two species. It occurs in forested areas of spruce and aspen in the East; in the West it occurs in areas of fir, Douglas-fir, and pine as well. Adults have been collected from mid-July until late August. In most areas, *servita* has an earlier flight period than that of the next two species, although there is usually some overlap.

Euxoa (*Euxoa*) *redimicula* (Morrison)
PL. 6, FIGS. 54–56; PL. L, FIG. 2; PL. AA, FIG. 5; PL. FF, FIG. 8 (RWH 10851).

Agrotis redimacula Morrison, 1874, *Proc. Boston Soc. Nat. Hist.,* **17**: 165.
Type locality: Cambridge, Massachusetts. [MSU]

NOTE—The lectotype of *redimicula* was designated by Lafontaine (1974: 412).

NOTE—The misspelled species name was corrected to *redimicula* by Morrison (1875: 57) and Lafontaine (1982, *Bull. Zool. Nomenclature,* **39**: 54–56).

Euxoa redimicula was confused with *servita* for many years. Males of *redimicula* can most easily be distinguished from those of *servita* and *auripennis* by their dirty-white hindwing with smoky-brown shading confined to the outer third of the wing. In the other species, the male hindwing is like that of the female; it is smoky brown, slightly paler toward the wing base. Also, the forewing markings of *redimicula* are more sharply defined than those of most *servita* specimens. The male genitalia can be distinguished from those of *servita* and *auripennis* only by numerical methods (Lafontaine, 1974). Females of *redimicula* have a long, fingerlike, sclerotized, flangelike process at the apex of each ovipositor lobe; this is not present in *servita* females and is short in those of *auripennis.* In females of *redimicula,* the subterminal area of the forewing is similar in color to the median area; in females of *auripennis* the median area is dark and contrasts with the pale subterminal area. Forewing length in *redimicula* varies from 13 to 16 mm.

The immature stages of *redimicula* have not been described.

Euxoa redimicula occurs from Nova Scotia southward to Virginia and westward to southwestern Saskatchewan, central Montana, western Wyoming, and central Missouri. It inhabits dry deciduous forests. In the Great Plains region, it occurs in groves of aspen, cottonwood, and ponderosa pine. Adults have been collected from late July until mid-September.

Euxoa (*Euxoa*) *auripennis* Lafontaine
PL. 6, FIGS. 57–60; PL. L, FIG. 3; PL. AA, FIG. 6; PL. FF, FIG. 9 (RWH 10852, 10853).

Euxoa auripennis Lafontaine, 1974, *Can. Ent.,* **106**: 412.
Type locality: Cranbrook, British Columbia. [CNC]

Euxoa arizonensis Lafontaine, 1974, *Can. Ent.,* **106**: 416. NEW SYNONYMY, NEW STATUS, SUBSPECIES.
Type locality: Greer, White Mountains, Apache County, Arizona. [CNC]

This species is similar to *redimicula* in wing markings and structural characters. Males can be distinguished by wing color and marking differences given in the key and by numerical methods using genital characters. In females, the flangelike process on the ovipositor lobe is short and rounded in northern populations (subspecies *auripennis*) or elongate and narrow in southern populations (subspecies *arizonensis*). The two taxa are treated as subspecies because they apparently do not occur together. Differences in ovipositor shape suggest that females of the two subspecies may oviposit in different soil types. In Colorado where their ranges approach each other, subspecies *auripennis* occurs in more mesic habitats, such as pine forests, than does subspecies *arizonensis,* which occurs in xeric piñon-juniper woodland. The range of *auripennis* overlaps that of *redimicula* in the northern Great Plains. In most of its range, specimens of *auripennis* are most likely to be confused with those of *servita* but can be distinguished from them by the even rather than streaked inner margin of the dark terminal area and by the toothed rather than wavy outer margin of the postmedial line. Forewing length in *auripennis* varies from 11 to 16 mm.

The immature stages are known only from laboratory reared material. The species overwinters in the egg stage, and the larva is reported to have only a short prepupal diapause (Hinks and Byers, 1976).

Euxoa auripennis occurs from south-central Manitoba, westward to central British Columbia, and southward to North Dakota, central New Mexico, central Arizona, and southern California. Adults have been collected from late July until late September.

Euxoa (*Euxoa*) *auripennis auripennis* Lafontaine

PL. 6, FIGS. 57–59; PL. L, FIG. 3; PL. AA, FIG. 6; PL. FF, FIG. 9.

Euxoa auripennis Lafontaine, 1974.

Subspecies *auripennis* occupies the northern and eastern portion of the range of the species, occurring in the West as far south as central Colorado, northern Utah, west-central Nevada, and east-central California. Adults of subspecies *auripennis* tend to have paler colored forewings than do those of subspecies *arizonensis* but the main difference between the two subspecies lies in the shape of the flangelike process on the ovipositor lobe; this is rounded in *auripennis,* about as wide as long, but is narrow in subspecies *arizonensis,* about twice as long as wide. In many other *Euxoa* species, differences in ovipositor shape are correlated with habitat differences, particularly in moisture and soil texture at oviposition sites. A similar situation apparently exists in subspecies *auripennis* and *arizonensis* in Colorado and Utah where their ranges approach each other. I have previously treated these two taxa as distinct species but prefer here to treat them as geographic and ecological races because of the lack of sympatry and of other structural differences.

Euxoa (*Euxoa*) *auripennis arizonensis* Lafontaine

PL. 6, FIG. 60.

Euxoa arizonensis Lafontaine, 1974.

This subspecies replaces subspecies *auripennis* in the South. Characters for distinguishing specimens of the two taxa are given under *auripennis*; numerical methods for distinguishing them were discussed in the original description (Lafontaine, 1974).

Subspecies *arizonensis* occurs from central Colorado westward to central Nevada and southward to central New Mexico, central Arizona, and southern California. In most of its range, *arizonensis* occurs in piñon-juniper woodland; in Colorado, where its range approaches that of subspecies *auripennis,* the two taxa do not occur together but *auripennis* occurs in pine forests at higher elevations than the habitats where *arizonensis* occurs. In the southern portion of its range, *arizonensis* occurs in open grassy clearings in aspen forests and ponderosa pine parkland. Adults of *arizonensis* have been collected from early August until mid-September.

Euxoa (*Euxoa*) *olivalis* (Grote)

PL. 7, FIGS. 1, 2; PL. L, FIG. 4; PL. AA, FIG. 7 (RWH 10833).

Agrotis olivalis Grote, 1879, *North Amer. Ent.,* **1**: 43.
Type locality: Colorado. [BMNH]

Euxoa mcdunnoughi Cook, 1930, *Can. Ent.,* **62**: 147.
Type locality: Jefferson County, Montana. [Montana Experiment Station Collection, Bozeman]

Euxoa olivalis, agema, and *oblongistigma* form a complex of closely related and morphologically similar species. Males can be recognized by the combination of longitudinally streaked forewing, barlike orbicular spot, and pubescent harpe in the male genitalia. Females can be recognized by the wing markings and by the junction of the ductus and corpus bursae being at the middle, or anterior to the middle, of the oval, or pear-shaped corpus bursae. Specimens of *olivalis* can be distinguished from those of *agema* and *oblongistigma* by characters given in the key. The long saccular extensions of males of *olivalis* curl dorsally around the "heel" of the cucullus. These can be observed in undissected specimens by removing scales from the end of the abdomen with a fine brush. In wing markings, specimens of *olivalis* are most likely to be confused with those of *agema.* Both species lack the pale, unmarked costa and orbicular spot that is characteristic of *oblongistigma,* but have fine cross lines on the costa, and most specimens have a dark central spot in the orbicular spot. Females of *olivalis* are indistinguishable from those of *agema* by genital characters. Most females of *olivalis* can be recognized by the gray and white streaking on the forewing and the pale color of the basal half of the hindwing (e.g., plate 7, figure 2). Specimens of *olivalis* from more mesic habitats, such as high elevation coniferous forests are darker and have more brown coloration (e.g., plate 7, figure 1) than do specimens from xeric, lowland habitats and closely resemble specimens of *agema.* Although males from these mesic habitats can be distinguished by genital characters, females from these areas sometimes can be identified only by association with males. Forewing length varies from 12 to 16 mm.

The immature stages of *olivalis* are unknown.

Euxoa olivalis occurs from southern Saskatchewan, westward to southern British Columbia, and southward to northern New Mexico, northern Arizona, and southern California. A disjunct population of *olivalis* has recently been found in relict aridland habitat in Yukon in northwestern Canada. Specimens of *olivalis* have been collected in a wide variety of habitats ranging from open sagebrush and prairie habitats to montane forests of ponderosa and lodgepole pine. In general *olivalis* occurs most commonly in open pine forests and piñon-juniper woodland, habitats intermediate between the open aridlands where *oblongistigma* occurs and the mesic coniferous forests inhabited by *agema.* Adults of *olivalis* have been collected from late June until early September.

Euxoa (*Euxoa*) *agema* (Strecker)
PL. 7, FIGS. 3, 4 (RWH 10832).

Agrotis agema Strecker, 1899, *Lepidoptera, Rhopaloceres and Heteroceres, Indigenous and Exotic,* Suppl., **2**: 5.
Type locality: Colorado. [FMNH]

This species is similar in forewing pattern to *olivalis,* but the ground color is dark brown with pale-brown longitudinal streaking; in most *olivalis* the forewing is streaked with dark gray, dark brown and pale gray. The hindwing of *agema* is dark smoky brown, slightly paler toward the base; in most *olivalis* specimens the hindwing is pale buff basally with smoky-brown shading on the outer third or quarter of the wing. In most areas where *agema* and *olivalis* occur together they can be distinguished by the color characters given above. At some localities in mesic coniferous forests, a dark form of *olivalis* occurs that is very similar in wing color to *agema* specimens, and the specimens must be identified by the male genital characters discussed under *olivalis.* In *agema,* the forewing length varies from 13 to 16 mm.

The immature stages of *agema* are unknown.

Euxoa agema occurs in mountain ranges that surround the Great Basin and Colorado Plateau from northern Oregon southward to southern California, and from central Idaho and southern Montana southward to southern Colorado and southern Utah. Within the Great Basin, it has been collected in the Ruby Mountains of Nevada. The species inhabits forested areas of spruce, aspen, fir, and pine. In general, *agema* occurs in more mesic habitats than those inhabited by *olivalis* and *oblongistigma*; however, the three species occasionally occur together. Adults of *agema* have been collected from mid-July until early September.

Euxoa (*Euxoa*) *oblongistigma* (Smith)
PL. 7, FIGS. 5, 6; PL. L, FIG. 5 (RWH 10834).

Agrotis oblongistigma Smith, [1888], *Proc. U. S. Natl. Mus.,* **10**: 454.
Type locality: Montana. [USNM]

NOTE—The female lectotype of *oblongistigma* was designated by Todd (1982: 155).

Specimens of *oblongistigma* are similar to those of *olivalis* and *agema* but may be distinguished from them by characters given in the key. In general, the pale, more evenly colored costa, reniform spot, and orbicular spot can be used to recognize specimens of *oblongistigma.* The forewing ground color is buff or pale gray. Forewing length varies from 12 to 15 mm. Males of *oblongistigma* can be distinguished from those of *olivalis* by longer harpe and shorter saccular extension length. Length of the right harpe is 1.15 ± 0.06 mm, N = 25 (1.03 ± 0.04 mm, N = 25, in *olivalis*); length of the right saccular extension is 1.32 ± 0.07 mm (1.57 ± 0.07 mm in *olivalis*). The male genitalia of *oblongistigma* are indistinguishable from those of *agema.* The female genitalia of all three species are indistinguishable.

The immature stages of *oblongistigma* are unknown.

Euxoa oblongistigma is widespread in dry areas in western North America. It occurs from southern Saskatchewan, westward to southern British Columbia, and southward to southern South Dakota, northern New Mexico, southern Utah, southern Nevada, and west-central California (Lafontaine, 1982a: figure 32). It is found most commonly in habitats where sagebrush is abundant, both in open habitats and in open coniferous forests. Adults have been collected from mid-July until late September.

Euxoa (*Euxoa*) *citricolor* (Grote)
PL. 7, FIGS. 7, 8; PL. L, FIG. 7; PL. BB, FIG. 1; PL. FF, FIG. 11 (RWH 10841).

Agrotis citricolor Grote, 1880, *Can. Ent.*, **12**: 154.
Type locality: Colorado. [BMNH]
NOTE—The male lectotype of *citricolor* was designated by Lafontaine (1981: 30).

Euxoa citricolor ab. *postmedialis* Strand, [1916], *Archiv für Naturgeschichte,* **81A12**: 144.
Type locality: Prescott, Arizona. [BMNH]
NOTE—The name *postmedialis* is an infrasubspecific name because it was described as an aberration.

Euxoa citricolor form *postmedialis* Draudt, 1924, *Die Gross-Schmetterlinge der Erde,* **7**: 39.
Type locality: Prescott, Arizona. [BMNH]
NOTE—Draudt validated the name when he used it as a geographical form of *citricolor.*

This species is closely related to *tronella,* and the two were considered to be color forms of a single species for many years. *Euxoa citricolor* is more restricted in habitat requirements than is *tronella*; the two species also differ in wing color, and in male genital details. In specimens of *citricolor,* forewing ground color varies from pale yellow to bright lemon yellow with a dusting of brown scales; the terminal area, reniform spot, and in most specimens the orbicular spot are dark grayish brown. In specimens of *tronella,* forewing ground color is pale buff or cream colored with a dusting of dark scales in most specimens; the terminal area, reniform spot, and orbicular spot are partially shaded with dark gray or concolorous with the remainder of the forewing. In both species the male hindwing is white in most specimens; in some males, and in most females, some dark shading is on the outer third of the hindwing. Forewing length is similar in the two species, 13 to 17 mm in *citricolor*; 12 to 16 mm in *tronella.* The harpe and saccular extension of *citricolor* are not as markedly curved basally as they are in *tronella* so that they form a V-shape or a more narrow U-shape than that formed by the harpe and saccular extension in *tronella.* This difference in shape is difficult to describe but can readily be recognized once learned (compare figures 7 and 8 on plate L). Also, the subbasal diverticulum in the vesica of *citricolor* is much smaller than is that of *tronella.* In *citricolor,* the subbasal diverticulum is slightly shorter than the aedoeagus width; it is longer in *tronella.* Females of the two species are indistinguishable by genital characters.

The immature stages of *citricolor* are unknown.

Euxoa citricolor inhabits open aridlands, particularly those where big sagebrush, *Artemisia tridentata* Nuttall, is common. It is widespread in the intermontane region where it occurs from southern Washington, southward in southern California to the Mexican border, and eastward to central Wyoming, central Colorado, and central New Mexico. In the Great Plains region it has been collected in heavily eroded areas in the Alberta badlands, the North Dakota badlands and near Mitchell, South Dakota (Lafontaine, 1982a: figure 32). Adults have been collected between late August and early October.

Euxoa (*Euxoa*) *tronella* (Smith)
PL. 7, FIGS. 9–12; PL. L, FIG. 8 (RWH 10842).

Carneades tronellus Smith, 1903, *Can. Ent.*, **35**: 11.
Type locality: Stockton, Utah. [AMNH]
NOTE—The male lectotype of *tronellus* was designated by Todd (1982: 215).

This species was confused with *citricolor* for many years, but specimens can be recognized by characters given in the key and discussed in more detail under *citricolor.* In most of the Great Basin region, specimens of *tronella* have pale-buff forewings with a dusting of dark scales (plate 7, figure 9); in specimens from southern California and southern Nevada the forewing is cream colored without darker scales (plate 7, figure 10); in those from the Great Plains region, the forewing is heavily dusted with dark scales, and the outer half of the hindwing is shaded with smoky brown in many specimens (plate 7, figure 11). A similar east-west cline is not evident in populations of *citricolor.* A rare form of *tronella* (plate 7, figure 12) with a dirty white forewing and dark-filled reniform and orbicular spots could easily be confused with specimens of *siccata* (p. 114) and *catenula* (p. 95). Specimens of these three species can readily be distinguished by both male and female genital characters.

The immature stages of *tronella* are unknown.

In the intermontane region, *tronella* has a range that is similar to that of *citricolor*; however, it has been collected at fewer localities in the northwestern and southeastern portions of the region. In the northern Great Plains region *tronella* is more widespread than is *citricolor* and occurs in shortgrass prairie in addition to heavily eroded sites where *citricolor* occurs. In the Great Plains region, *tronella* occurs from southwestern Saskatchewan and southeastern Alberta southward to southwestern South Dakota and southern Wyoming. In the intermontane region *tronella* occurs in habitats similar to those of *citricolor,* and the two species are frequently collected together. Adults of *tronella* have the same flight season as that of *citricolor,* late August until early October.

Euxoa (*Euxoa*) *teleboa* (Smith)
PL. 7, FIGS. 13–15; PL. L, FIG. 6; PL. AA, FIG. 8; PL. FF, FIG. 10 (RWH 10843).

Carneades teleboa Smith, 1890, *Bull. U. S. Natl. Mus.,* **38**: 219.
Type locality: Las Vegas, New Mexico. [USNM]

Carneades pedalis Smith, 1890, *Bull. U. S. Natl. Mus.,* **38**: 220.
Type locality: Colorado. [USNM]

Carneades recticincta Smith, 1895, *Ent. News,* **6**: 334.
Type locality: Calgary, Alberta. [USNM]

Specimens of *teleboa* can be distinguished from those of most other *Euxoa* species by the distinctive wing color and pattern. Forewing ground color may be cream colored, pale yellow, or reddish orange with the maculation generally obscure except for a prominent, black median line that is fused with the dark-filled reniform spot in many specimens. Forewing length varies from 12 to 16 mm. The male hindwing is white; that of most females is white with a dusting of dark-colored scales. Specimens of *teleboa* are most likely to be confused with those of *siccata* (p. 114), *citricolor* (p. 129), and *tronella* (p. 129) but can be distinguished from specimens of these species by the much more prominent median line and by genital characters. Males of *teleboa* can be distinguished from those of *siccata* by the disproportionately smaller sacculus in *teleboa* and from those of *tronella* and *citricolor* by the much shorter saccular extensions; these are slightly shorter than the harpe in *teleboa* but about 1.25× the length of the harpe in *tronella* and *citricolor.* Males of these four species can also be distinguished by details of the shape of the vesica. Females of *teleboa* can be distinguished from those of *siccata* by the presence of a sclerotized apical process rather than a sclerotized rim and by the ventrally expanded ovipositor lobe in *teleboa.* They can be distinguished from females of *tronella* and *citricolor* by the shorter sclerotized plate in the ductus bursae and shorter anterior apophysis in *teleboa.*

The immature stages of *teleboa* are unknown.

Euxoa teleboa occurs from southern Saskatchewan and southern Alberta southward and westward to northern Texas, central New Mexico, central eastern Nevada, and northern Arizona (Lafontaine, 1981: figure 171). Adults have been collected from mid-August until late September. In the Great Plains region, *teleboa* inhabits open arid areas where sagebrush (*Artemisia* spp.) is abundant. In the intermontane region, *teleboa* occurs in piñon-juniper woodland.

Euxoa (*Euxoa*) *difformis* (Smith)
PL. 7, FIGS. 16–25; PL. L, FIG. 9; PL. BB, FIG. 2 (RWH 10844, 10845).

Carneades difformis Smith, 1900, *Proc. U. S. Natl. Mus.,* **22**: 441.
Type locality: Oregon. [USNM]
NOTE—The female lectotype of *difformis* was designated by Todd (1982: 64).

Euxoa mercedes Barnes and McDunnough, 1912. *Contrib. Nat. Hist. Lep. N. Am.,* **1** (5): 5. NEW SYNONYMY.
Type locality: San Diego, California. [USNM]

This is a very appropriately named species; no other North American species of *Euxoa* is as variable in size and wing markings as *difformis.* Forewing ground color varies from pale gray or pale buff through various shades of brown and orange to dark gray or dark brown. The maculation may be obscure or prominent and contrasting. Forewing length is extremely variable, even among specimens from a single locality; it varies from 10 to 18 mm. The most distinctive wing marking character of *difformis* is the presence of a median line on the hindwing of most specimens. In males, the hindwing is dirty white or very pale buff with some pale smoky-brown shading on the wing margin in many specimens. The median line may be prominent, or may be reduced to a series of dark dots on the veins. In females, the hindwing is smoky brown with the median line indicated by slightly darker brown shading. Males of *difformis* can be distinguished from those of other species by a combination of genital characters, particularly relatively short saccular extensions, presence of a small diverticulum subbasally in the vesica in addition to the normal subbasal diverticulum, and the markedly reflexed vesica. Throughout most of the range of *difformis* the saccular extensions are slightly shorter than the harpes (plate L, figure 9); males from localities in southwestern California differ from those from other parts of the range in having saccular extensions that are longer than the harpes. Until recently, these southwestern populations were treated as a distinct species, *mercedes*. This character, however, is not constant; a cline exists between Los Angeles County and central California in which the length of the saccular extension is highly variable, but the shorter form of the saccular extension appears with increasing frequency northwards. The southwestern form may have been isolated from other populations of *difformis* with a broad zone of intergradation formed after secondary contact. The female genitalia differ from those of most other species with which *difformis* could be confused by the combination of unisaccate, mesially elbowed corpus bursae, and lack of a sclerotized process on the ovipositor lobe. Females of *difformis* are indistinguishable by genital characters from those of *dodi* and *infracta*; females of these three species must be distinguished by wing marking characters given in the key, or by association with males.

The immature stages of *difformis* are unknown.

Euxoa difformis occurs from central Saskatchewan and southern British Columbia southward to western South Dakota, southern Colorado, southern Arizona, and northern Baja California, Mexico (Lafontaine, 1982a, figure 39). It is predominantly a species of open aridland habitats but also occurs in piñon-juniper woodland and dry pine forests. Adults have been collected from mid-August until mid-October in most areas but from early October until late December in southern California.

Euxoa (*Euxoa*) *moerens* (Grote)
PL. 7, FIGS. 26–28; PL. L, FIG. 10; PL. BB, FIG. 3 (RWH 10847).

Carneades moerens Grote 1883, *Can. Ent.*, **15**: 4.
Type locality: Arizona. [BMNH]

Agrotis luteola Smith [1888], *Proc. U. S. Natl. Mus.*, **10**: 457.
Type locality: Arizona. [USNM]

NOTE—The female lectotype of *luteola* was designated by Todd (1982: 126).

Euxoa moerens is a small species (forewing length 11–14 mm) in which the forewing color is mottled, this tending to obscure the transverse lines. In most of its range, the forewing ground color is pale brown or pale gray, mottled with dark shading; specimens from Colorado and Arizona are reddish brown. The small, round orbicular spot, and in most specimens the reniform spot as well, has a white line inside the black outline that makes these spots prominent. The appearance of these spots in combination with small size and wing color allows most specimens of *moerens* to be recognized by wing markings. Specimens of *moerens* are most likely to be confused with those of *difformis* or *latro*. The three species can be distinguished by characters given in the key.

The immature stages of *moerens* are unknown although a single larva was collected on sweet clover (*Melilotus* sp.) and reared to adult (Cook, 1930: 263).

Euxoa moerens occurs in the western Great Plains from southern Saskatchewan and Alberta southward to central Colorado and in the intermontane region from northern Oregon southward and eastward to central Colorado, northern Arizona, and central Nevada (Lafontaine, 1981, figure 172). In the Great Plains region it occurs in open aridlands; in the intermontane region it inhabits piñon-juniper woodland and dry, open, pine forests. Adults of *moerens* have been collected from mid-August until late September.

Euxoa (*Euxoa*) *latro* (Barnes and Benjamin)
PL. 7, FIGS. 29, 30; PL. M, FIG. 1; PL. BB, FIG. 4 (RWH 10848).

Agrotis latro Barnes and Benjamin, (1927), *Can. Ent.*, **58**: 305.
Type locality: Truckee, California. [USNM]

Specimens of *latro* are similar to those of *moerens* but can be distinguished by the darker coloration of both forewing and hindwing and by the larger, paler orbicular spot of *latro*. In *latro*, the forewing is dark brown or gray with extensive dark-brown shading. The hindwing is dark smoky brown rather than pale gray as in that of *moerens*. Specimens of *latro* from a disjunct population in Utah resemble those of *moerens* in that the forewing is orange brown, and the orbicular spot is small. They can be distinguished from specimens of *moerens* by the dark hindwing and by the genital characters given in the key. Specimens of *latro* are similar in size to those of *moerens*; forewing length in *latro* varies from 12 to 15 mm.

The immature stages of *latro* are unknown.

The main range of *latro* is in the Cascades and Sierra Nevada Mountains; it occurs from central Washington southward to the Transverse Ranges of southern California and the mountains of northern Baja California (Lafontaine, 1981, figure 173). A disjunct population occurs in the La Sal Mountains of Utah. Generally, *latro* occurs at higher elevations than does *moerens* and inhabits pine and fir forests rather than piñon-juniper woodland; although the two species do occur together in the eastern Sierra Nevadas where the ponderosa pine zone meets the piñon-juniper zone. Adults of *latro* have been collected from mid-August until mid-September in the North and until mid-October in the South.

Euxoa (*Euxoa*) *murdocki* (Smith)
PL. 7, FIG. 31; PL. M, FIG. 2; PL. BB, FIG. 5 (RWH 10846).

Agrotis murdocki Smith, 1890, *Trans. Amer. Ent. Soc.*, **17**: 49.
Type locality: British Columbia. [USNM]

NOTE—The male lectotype of *murdocki* was designated by Todd (1982: 144).

This species is unmistakable: the combination of orange basal and subterminal areas and gray median and terminal areas allow specimens of *murdocki* to be recognized immediately. Forewing length varies from 12 to 15 mm. Males of *murdocki* are similar to those of *moerens* and *latro* in genital characters but differ from those of *moerens* in lacking the extra diverticulum in the vesica subbasally and from those of *latro* in having a longer, less setose harpe. Females of the three species are indistinguishable by genital characters.

The immature stages of *murdocki* are unknown.

Euxoa murdocki occurs from southern British Columbia and southwestern Alberta southward to west-central Colorado, central Utah, central Nevada, and southern California (Lafontaine, 1981: figure 174). In most of its range *murdocki* inhabits coniferous forests of pine and Douglas-fir; in the Great Basin, it occurs in piñon-juniper woodland as well. Adults of *murdocki* have been collected from late August until mid-October.

Euxoa (*Euxoa*) *dodi* (McDunnough)
PL. 7, FIG. 32; PL. M, FIG. 3; PL. BB, FIG. 6 (RWH 10849).

Euxoa dodi McDunnough, 1923, *Can. Ent.*, **55**: 163.
Type locality: Lethbridge, Alberta. [CNC]

Specimens of *dodi* could be confused with those of *declarata* (p. 79) but can be distinguished from them by: smaller size (forewing length 12–16 mm); more evenly colored, pale smoky-brown hindwing; apically straight, rather than incurved, saccular extensions and harpes in males; and unisaccate rather than bisaccate corpus bursae in females. In most of its range, specimens of *dodi* can be distinguished from those of *infracta* by the buffy-gray rather than dark-brown forewing color and by lack of copper suffusion on the forewing. In some areas of Colorado the two species cannot be distinguished by wing color, and it is necessary to rely on male genital differences. In males of *dodi*, the length of the right saccular extension (1.59 ± 0.10 mm, N = 20) is less than 1.5× the length of the harpe (1.17 ± 0.05 mm); in those of *infracta*, the length of the right saccular extension (1.62 ± 0.10 mm, N = 20) is more than 1.5× the length of the harpe (0.99 ± 0.06 mm). Differences in male genital characters are consistent, even in areas where wing color is intermediate between that typical for each species in other parts of their ranges. The two species have been hybridized in the laboratory; hybrid adults are intermediate both in wing color and male genital characters. The hybrids are apparently sterile since attempts to inbreed them, and backcross them with parental lines, were unsuccessful (J. R. Byers, personal communication).

The immature stages are known only from laboratory reared material and have not been described.

Euxoa dodi ranges from southern Alberta and Saskatchewan southward to southern South Dakota, southern New Mexico, and northern Arizona (Lafontaine, 1981, figure 175). In most of its range, *dodi* occurs in dry, shortgrass prairie; in the foothills of the Rocky mountains, it inhabits dry, open, pine forests. The flight season of *dodi* extends from mid-August until early September.

Euxoa (*Euxoa*) *infracta* (Morrison)
PL. 7, FIG. 33; PL. M, FIG. 4; PL. FF, FIG. 12 (RWH 10850).

Agrotis infracta Morrison, 1875, *Proc. Boston Soc. Nat. Hist.*, **18**: 115.
Type locality: Colorado. [MSU]

NOTE—The female lectotype of *infracta* was designated by Lafontaine (1981: 35). The lectotype of *infracta* is intermediate in color between typical specimens of *infracta* and those of *dodi*. The name *infracta* has been used for many years for the darker of the two species, and I retain this usage by considering the type to be an unusually pale specimen of *infracta*. The lectotype is a female; females of the two species are indistinguishable by genital characters.

Specimens of *infracta* are most likely to be confused with those of *campestris* (p. 79) and *dodi*. They can be distinguished from those of *campestris* by: the more evenly colored smoky-brown hindwing; the very differently proportioned male genitalia; and the unisaccate rather than bisaccate corpus bursae in females. They can be distinguished from specimens of *dodi* by characters given in the key and discussed in more detail under *dodi*. As mentioned under *dodi*, sterile hybrids have been produced in the laboratory by crossing *dodi* and *infracta*. The two species are similar in size; forewing length in *infracta* varies from 13 to 17 mm.

The immature stages of *infracta* have not been described; they are known only from laboratory reared material.

Euxoa infracta occurs in more mesic habitats than does *dodi*; it inhabits aspen forests and coniferous forests of pine and fir. The range of *infracta* overlaps that of *dodi* in the Great Plains; in this area *infracta* occurs in aspen and coniferous groves, whereas *dodi* inhabits the surrounding open prairie. The range of *infracta* extends from southern Manitoba, westward to southern British Columbia, and southward to western South Dakota, northern New Mexico, east-central Arizona, central Nevada, and southern California (Lafontaine, 1981, figure 176). Adults of *infracta* have been collected from mid-July until late September.

Euxoa (*Euxoa*) *laetificans* (Smith)

PL. 7, FIGS. 34, 35; PL. M, FIG. 5; PL. BB, FIG. 7 (RWH 10829).

Carneades laetificans, 1894, *Trans. Amer. Ent. Soc.*, **21**: 48.
Type locality: Glenwood Springs, Colorado. [USNM]

NOTE—The male lectotype of *laetificans* was designated by Todd (1982: 117).

Carneades masculinus Smith, 1903, *Jour. New York Ent. Soc.*, **11**: 6.
Type locality: Silverbow County, Montana. [AMNH]

NOTE—The male lectotype of *masculinus* was designated by Todd (1982: 132).

This species and *quadridentata* are sexually dimorphic. In males of *laetificans* the prothoracic collar and forewing ground color are light reddish brown; in females, these areas are gray. The forewing costa and cubital vein are pale yellow brown in males, silver gray in females. Specimens of *laetificans* are similar to those of *quadridentata* but can be distinguished from them by a number of characters. In *laetificans* the forewing cubital vein is pale from the wing base to the reniform spot; in *quadridentata* the pale shade extends beyond this on veins CuA_1 and M_3 through the subterminal area to the terminal area. In males of *laetificans* the harpe is pubescent (smooth in those of *quadridentata*), the right saccular extension is longer than that of *quadridentata* (1.71 ± 0.10 mm, $N = 20$, versus 1.22 ± 0.13 mm, $N = 20$) as is the right harpe (1.45 ± 0.05 mm versus 1.18 ± 0.05 mm). Females of *laetificans* differ from those of *quadridentata* in having the sclerotized plates in the ductus bursae extended farther anteriorly than the anterior apophyses rather than shorter as in *quadridentata*, and the ovipositor lobes are narrow and pointed apically rather than being broad and rounded. Forewing length in *laetificans* varies from 12 to 15 mm.

The immature stages have not been described; however, larvae have been reared on a variety of broad-leaved plants including: alfalfa, sweet clover, beet, lettuce, and Russian thistle (Cook, 1930: 261).

Euxoa laetificans inhabits pine and aspen forests in most of its range, and piñon-juniper woodland in the intermontane region. It occurs from southern Saskatchewan, westward to southern British Columbia, and southward to southern Colorado, southern Utah, central Nevada, and east-central California (Lafontaine, 1981, figure 177). Adults of *laetificans* have been collected from late July until mid-September.

Euxoa (*Euxoa*) *quadridentata* (Grote and Robinson)

PL. 7, FIGS. 36–38; PL. M, FIG. 6; PL. BB, FIG. 8 (RWH 10830).

Agrotis quadridentata Grote and Robinson, 1865, *Proc. Ent. Soc. Philadelphia*, **4**: 491.
Type locality: Colorado. [ANSP]

NOTE—The male lectotype of *quadridentata* was designated by Lafontaine (1981: 37).

Carneades pugionis Smith, 1900, *Proc. U. S. Natl. Mus.*, **22**: 419.
Type locality: Denver, Colorado. [USNM]

NOTE—The male lectotype of *pugionis* was designated by Todd (1982: 175).

Euxoa flutea Smith, 1910, *Trans. Amer. Ent. Soc.*, **36**: 255. SUBSPECIES.
Type locality: Sierra Nevada, California. [AMNH]

NOTE—The female lectotype of *flutea* was designated by Todd (1982: 87).

Specimens of *quadridentata* are similar to those of *laetificans* but can be distinguished from them by the characters given in the key and discussed under *laetificans.* As for that species, *quadridentata* is sexually dimorphic; the pale areas on the forewing and prothoracic collar are pale yellow or yellow buff in males and pale gray or silver gray in females. Forewing length varies from 11 to 15 mm. Females of *quadridentata* have a short sclerotized process at the apex of each ovipositor lobe in populations in the Great Plains and eastern portion of the intermontane region. This process is absent in females from the western portion of the intermontane region.

The immature stages have not been described; however, larvae have been collected and reared on wheat in Montana (Cook, 1930: 261).

Euxoa quadridentata is widely distributed in western North America; it occurs from southern Manitoba, westward through central Saskatchewan and central Alberta to southern British Columbia, and southward to central Nebraska, southern New Mexico, central Arizona, and southern California (Lafontaine, 1981, figure 178). Adults have been collected from mid-August until early October. In most of its range *quadridentata* occurs in open aridlands and piñon-juniper woodland; in the western portion of its range it occurs in pine forests as well. Differences in shape of the ovipositor lobe suggest that females may oviposit in different soil types in the eastern and western portions of the range. Populations of *quadridentata* are arranged in two subspecies.

Euxoa (*Euxoa*) *quadridentata quadridentata* (Grote and Robinson)

PL. 7, FIG. 36.

Agrotis quadridentata Grote and Robinson, 1865.
Carneades pugionis Smith, 1900.

Specimens of the nominate subspecies of *quadridentata* can be distinguished from those of the western subspecies only by the presence of a short, sclerotized apical process on the female ovipositor lobe; this process is absent in females of *flutea.* Subspecies *quadridentata* occurs in the Great Plains, the Rocky Mountain region, and the eastern intermontane region as far west as eastern British Columbia, western Montana and Wyoming, eastern Nevada, and northwestern Arizona (Lafontaine, 1982a, figure 35).

Euxoa (*Euxoa*) *quadridentata flutea* Smith

PL. 7, FIGS. 37, 38; PL. M, FIG. 6; PL. BB, FIG. 8.

Euxoa flutea Smith, 1910.

This subspecies occurs in the Cascade and Sierra Nevada Mountains and in the western portion of the intermontane region as far east as south-central British Columbia, western Idaho, and eastern Nevada (Lafontaine, 1982a, figure 35). Females of subspecies *flutea* lack the sclerotized process present at the apex of each ovipositor lobe of females of subspecies *quadridentata.* Specimens of subspecies *flutea,* unlike those of the nominate subspecies, have darker hindwings in more mesic habitats, such as pine forests.

Euxoa (*Euxoa*) *inscripta* Lafontaine

PL. 7, FIGS. 39–41; PL. M, FIG. 7; PL. CC, FIG. 1.

Euxoa inscripta Lafontaine, 1981, *Quaestiones Entomologicae,* **17**: 38.
Type locality: Craig, Colorado, 22 mi N. [CNC]

Specimens of *inscripta,* at first glance, resemble those of *Dicestra trifolii* (Hufnagel); this the result of brown forewing ground color, black veins lined with pale brown, pale W-mark in the subterminal area where pale-lined veins CuA_1 and M_3 project into it, and large prominent claviform spot. Within *Euxoa,* they are most likely to be confused with specimens of *olivalis* (p. 127) but can be recognized by the round rather than barlike orbicular spot and lack of longitudinal silver-gray streaking on the forewing. Forewing length varies from 12 to 15 mm. Specimens of *inscripta* from the Sierra Nevada Mountains differ from those from the eastern portion of its range in having darker, more gray forewings (plate 7, figure 41). Also, the pale shading between the dark central spot and black outline of the reniform and orbicular spots is white in most California specimens rather than pale brown as in those from more easterly populations. Specimens of *inscripta* can be distinguished from those of other similar species by characters given in the key and by male vesica shape.

The immature stages are unknown.

The range of *inscripta* is disjunct; the eastern portion of the range extends from southern Montana southward to west-central Colorado and central Utah and the western portion is in west-central Nevada and east-central California (Lafontaine, 1981, figure

179). The species inhabits arid areas where sagebrush (*Artemisia* spp.) is abundant, both in open areas and in open piñon-juniper woodland. Adults of *inscripta* have been collected only in the latter half of August although the flight season is undoubtedly much longer.

Euxoa (*Euxoa*) *unica* McDunnough
PL. 7, FIG. 42; PL. M, FIG. 8 (RWH 10837).

Euxoa unica McDunnough, 1940, *Can. Ent.*, **72**: 192.
Type locality: Saskatoon, Saskatchewan. [CNC]

This species is known from two male specimens. They were collected five years apart in a light trap that was run for many years in the vicinity of Saskatoon, Saskatchewan. The rarity of *unica* is puzzling, considering the extensive collecting that has been done in the northern Great Plains in general and in Saskatoon in particular. It is possible that *unica* may be of hybrid origin; the specimens differ from those of other members of the *detersa* group in that the male antenna is more prominently biserrate and the male saccular extensions are markedly asymmetrical. In both of these characters, specimens of *unica* are intermediate between those typical of species in the *detersa* group and those of *intrita* males (p. 69). Specimens of *unica* are most likely to be confused with those of *intrita* and *niveilinea* (p. 135). They can be distinguished from specimens of *intrita* by the narrower forewings with the pale yellow-buff costa, and from those of *niveilinea* by the brown rather than gray and black forewing ground color and by the pale yellow-buff rather than silver-gray costa. Males of *unica* are not likely to be confused with those of these two species on the basis of male genital characters. The male genitalia of *unica* are most similar to those of *inscripta,* but the saccular extensions in *unica* are shorter and markedly asymmetrical; the left saccular extension is about 4/5 as long as the right.

The two known males were collected in late August.

Euxoa (*Euxoa*) *niveilinea* (Grote)
PL. 7, FIGS. 43, 44; PL. M, FIG. 9; PL. CC, FIG. 2 (RWH 10831).

Agrotis niveilinea Grote, 1882, *Can. Ent.*, **14**: 216.
Type locality: Arizona. [USNM]
NOTE—The male lectotype of *niveilinea* was designated by Lafontaine (1981: 41).

Euxoa rabiata Smith, 1910, *Trans. Amer. Ent. Soc.*, **36**: 255.
Type locality: Volga, South Dakota. [AMNH]
NOTE—The male lectotype of *rabiata* was designated by Todd (1982: 180).

Specimens of *niveilinea* are similar to those of *dargo* and *cicatricosa*; each has longitudinally streaked forewings with yellowish-buff shading in the reniform and orbicular spots. In *niveilinea* forewing length varies from 13 to 18 mm. Specimens of *niveilinea* can be distinguished from those of *dargo* and *cicatricosa* by the lack of a prominent pale streak distal to the claviform spot. Further they can be distinguished from specimens of *dargo* by the pale rather than dark hindwing and from those of *cicatricosa* by the presence of a transverse band of white scales on the thorax posterior to the prothoracic collar that is absent in *cicatricosa.* Males of *niveilinea* differ from those of both *dargo* and *cicatricosa* in that the saccular extensions are much longer and bend dorsally around the "heel" of the cucullus; this can be observed in undissected males by brushing away scales from the end of the abdomen. Females of *niveilinea* differ from those of *dargo* and *cicatricosa* in that each ovipositor lobe has a sclerotized process apically rather than enlarged, stout setae apically.

The immature stages of *niveilinea* are unknown.

This species occurs in the Great Plains region from southern Saskatchewan and Alberta southward to northern Texas and southern New Mexico. Its range extends eastward from the Great Plains region into central Michigan and westward into the southeastern portion of the intermontane region as far as central Arizona. An apparently disjunct population occurs in southeastern Texas. *Euxoa niveilinea* inhabits dry, sandy areas where sagebrush (*Artemisia* spp.) is common. Adults have been collected from late August until late October.

Euxoa (*Euxoa*) *dargo* (Strecker)
PL. 7, FIG. 45; PL. M, FIG. 10; PL. CC, FIG. 3; PL. FF, FIG. 13 (RWH 10836).

Agrotis dargo Strecker, 1898, *Lepidoptera, Rhopaloceres and Heteroceres, Indigenous and Exotic,* Suppl. **1**: 6.
Type locality: Loveland, Colorado. [FMNH]
NOTE—The male lectotype of *dargo* was designated by Lafontaine (1981: 42).

Carneades rumatana Smith, 1903, *Trans. Amer. Ent. Soc.,* **29**: 203.
Type locality: Volga, South Dakota. [AMNH]

NOTE—The female lectotype of *rumatana* was designated by Todd (1982: 189).

This species is similar to *niveilinea* in forewing color and pattern and like that species, has a transverse white line on the thorax posterior to the prothoracic collar. Specimens of *dargo* differ from those of *niveilinea* in having a pale-buff streak on the forewing distal to the claviform spot, a darker hindwing, and in male and female genital characters. In most males the hindwing is buff basally with a sharply defined smoky-brown band on the outer third or half of the wing; in some males and in females the smoky-brown shading extends almost to the wing base. In both sexes of *niveilinea* dark smoky-brown shading is confined to the wing margin and veins. In males of *dargo* the right saccular extension is about 1.25× the length of the right harpe while it is more than 1.5× its length in *niveilinea* males. Females of the two species can readily be distinguished by the form of the ovipositor lobe; ovipositor lobe with stout setae apically (*dargo*), ovipositor lobe with fine setae apically and a sclerotized process at the apex (*niveilinea*). Most specimens of *dargo* are smaller than those of *niveilinea*; in *dargo* forewing length varies from 11 to 15 mm.

The immature stages have not been described although larvae have been found feeding on corn and Russian thistle in Montana (Cook, 1930: 261).

This species occurs from southern Manitoba, westward to southern British Columbia, and southward to central Nebraska, northern New Mexico, southern Idaho, and east-central Oregon (Lafontaine, 1981, figure 181). In most of its range *dargo* inhabits dry prairie and sagebrush areas; in the southern portion of its range it occurs in areas where prairie habitat occurs in open pine parkland. Adults have been collected from late August until early October.

Euxoa (*Euxoa*) *melura* McDunnough

PL. 7, FIG. 46; PL. N, FIG. 1; PL. CC, FIG. 4 (RWH 10835).

Euxoa melura McDunnough, 1932, *Can. Ent.,* **64**: 231.
Type locality: Eureka, Utah. [CNC]

Specimens of *melura* are similar to those of *olivalis* (p. 127) in size and wing markings but usually can be distinguished from those of *olivalis* by the rounded or oval rather than barlike orbicular spot. The female of *melura* illustrated is one of the few examined that has an elongate orbicular spot. Males of *melura* can be distinguished from those of *olivalis,* and any other superficially similar species, by the short saccular extensions and unusually long harpes. The right harpe is at least 1.25× longer than both the saccular extension and sacculus. Females of *melura* can be recognized by the presence of stout setae at the apex of the ovipositor lobe rather than a sclerotized process as in *olivalis* females. In this character females of *melura* are similar to those of *dargo,* but they lack the row of long, stout, spikelike setae subbasally on the ovipositor lobe. All these species are similar in size; forewing length of *melura* varies from 13 to 15 mm.

The immature stages of *melura* are unknown.

The range of *melura* is poorly known, possibly because of the early flight season. It has been collected from southern Washington and southwestern Montana southward to northwestern Colorado and south-central Utah (Lafontaine, 1981, figure 182). This species has an earlier flight period than that of similar species; adults have been collected from late May until early July.

Euxoa (*Euxoa*) *detersa* (Walker) (Sandhill Cutworm; Ver-gris Arénicole, m., Fr.)

PL. 7, FIGS. 47–50; PL. N, FIG. 2; PL. CC, FIG. 5; PL. FF, FIG. 14 (RWH 10838).

Charaeas detersa Walker, 1856, *List of the Specimens of Lepidopterous Insects in the Collection of the British Museum,* **9**: 212.
Type locality: Nova Scotia. [BMNH]

NOTE—The male lectotype of *detersa* was designated by Lafontaine (1981: 44).

Agrotis pitychrous Grote, 1873, *Bull. Buffalo Soc. Nat. Sci.,* **1**: 82.
Type locality: Long Island, New York. [lost]

Agrotis personata Morrison, 1876, *Proc. Boston Soc. Nat. Hist.,* **18**: 238. SUBSPECIES.
Type locality: Illinois. [MSU]

NOTE—The type locality of *personata* was stated by Morrison to be Illinois; however, the nominal holotype is labeled "Ohio."

Agrotis azif Strecker, 1898, *Lepidoptera, Rhopaloceres and Heteroceres, Indigenous and Exotic,* Suppl. **1**: 6.
Type locality: Clyde, New York. [FMNH]

Euxoa detersa is closely related to *cicatricosa* and *recula*. Males of the three species can be distinguished from those of other similar species by the apically incurved harpe and relatively large, heavily sclerotized clavus; females can be recognized by the presence of a row of long, stout setae on the dorsal margin of the ovipositor lobe. These setae are scattered on the dorsal margin of the lobe in *detersa* and *recula* females but are arranged in a single row in those of *cicatricosa*. Specimens of *detersa* are extremely variable in wing color and markings; fortunately, this seldom creates a problem in identification because the range of *detersa* overlaps that of *cicatricosa* only in a narrow band in the western Great Plains. In general, the forewing in specimens from localities on the Atlantic seaboard is longitudinally streaked (subspecies *detersa*) while that of specimens from the Great Plains and Great Lakes region is not streaked but is irregularly colored and blotchy in appearance (subspecies *personata*). In the Great Plains region, where the range of *detersa* overlaps that of *cicatricosa*, specimens of *detersa* differ from those of *cicatricosa* in the female genital character discussed above and in having darker hindwings and a markedly reduced, usually vestigial frontal tubercle. Most specimens of *cicatricosa* differ from Great Plains specimens of *detersa* in having a longitudinally streaked forewing. In some areas some specimens of *cicatricosa* lack the longitudinal streaking and look like pale specimens of *detersa*. These specimens may be of hybrid origin, but in the other characters listed above they are typical of *cicatricosa* (e.g., plate 7, figure 54).

The larva of *detersa*, commonly called the sandhill cutworm, is an important pest of crops grown in sandy soil. It has been reported to feed on a wide variety of plants including corn, tobacco, potato, strawberry, oats, wheat, and rye (Rings and Johnson, 1976). The species overwinters in the egg stage or as early instar larvae (Hinks and Byers, 1976: 1349). The egg was illustrated by Salkeld (1975: 1151, figures 49–51).

Euxoa detersa occurs from southwestern Newfoundland westward to Fort Smith in the Northwest Territories and to southwestern Alberta; it occurs southward to South Carolina, northern Missouri, southern Kansas, and northeastern Colorado (Lafontaine, 1981, figure 183; 1982a, figure 38). The species inhabits areas of loose, shifting sand such as beaches and dunes. The adult flight period extends from early August until early October.

Populations of *detersa* are arranged in two subspecies.

Euxoa (Euxoa) detersa detersa (Walker)

PL. 7, FIG. 47.

Charaeas detersa Walker, 1856.

Agrotis pitychrous Grote, 1873.

Specimens of subspecies *detersa* differ from those of subspecies *personata* in having the forewing longitudinally streaked with pale shading on the costa, veins CuA, CuA_1, and M_3, and distal to the claviform spot. The range of this subspecies extends from the Atlantic seaboard westward in the St. Lawrence Valley to the eastern Great Lakes region.

Euxoa (Euxoa) detersa personata (Morrison)

PL. 7, FIGS. 48–50; PL. N, FIG. 2; PL. CC, FIG. 5; PL. FF, FIG. 14.

Agrotis personata Morrison, 1876.

Agrotis azif Strecker, 1898.

Specimens of this subspecies lack the longitudinal streaking on the forewing that is present in those of subspecies *detersa*, and in most specimens the orbicular spot is much smaller. Subspecies *personata* ranges from the Great Plains region eastward in the Great Lakes region to the St. Lawrence River valley.

Euxoa (Euxoa) cicatricosa (Grote and Robinson)

PL. 7, FIGS. 51–54; PL. CC, FIG. 6; PL. FF, FIG. 15 (RWH 10839).

Agrotis cicatricosa Grote and Robinson, 1865, *Proc. Ent. Soc. Philadelphia*, **4**: 492.
Type locality: Colorado. [ANSP]

Carneades neomexicana Smith, 1890, *Bull. U. S. Natl. Mus.*, **38**: 218.
Type locality: New Mexico. [USNM]
NOTE—The female lectotype of *neomexicana* was designated by Todd (1982: 146).

Setagrotis ducalis Smith, 1907, *Trans. Amer. Ent. Soc.*, **33**: 128.
Type locality: Stockton, Utah. [USNM]

Euxoa teplia Smith, 1910, *Trans. Amer. Ent. Soc.*, **36**: 253.
Type locality: Stockton, Utah. [AMNH]
NOTE—The female lectotype of *teplia* was designated by Todd (1982: 207).

Specimens of *cicatricosa* are most likely to be confused with those of *niveilinea, dargo, detersa, recula,* and *cinereopallida.* They can readily be distinguished from specimens of *niveilinea, dargo,* and *cinereopallida* by the wing marking and genital characters given in the key and in the text under those species. Characters to distinguish specimens of *detersa* and *cicatricosa* are discussed under *detersa.* In the western portion of its range, *cicatricosa* occurs sympatrically with *recula,* and the two species are very similar in structural characters. Females of *cicatricosa* differ from those of *recula* in that the stout setae on the dorsal margin of the ovipositor lobe are in a single row, not scattered and irregular as in *recula* females. Males of the two species cannot safely be distinguished by genital characters, although the saccular extensions in *cicatricosa* are, on average, longer than those in *recula* males (length of right saccular extension 1.32 ± 0.12 mm, N = 25, in *cicatricosa* versus 1.02 ± 0.12 mm, N = 20, in *recula*). Specimens of *cicatricosa* can most easily be distinguished from those of *recula* by forewing color. In *cicatricosa* the forewing is dark gray brown with fine silver-gray streaking; the reniform and orbicular spots have dark central spots. In *recula,* the pale areas of the forewing are more extensive and are bright yellow; the reniform and orbicular spots and a broad streak extending from the claviform spot to the terminal line are yellow.

Euxoa cicatricosa shows a small amount of geographical variation. The most widespread form (plate 7, figure 51) occurs through the range of the species except for the Great Basin and southwestern United States. Specimens from the Great Basin (plate 7, figure 52) are paler and have more extensive white streaking than for the most widespread form. Specimens from New Mexico and Arizona (plate 7, figure 53) have the pale markings on the forewing pale orange brown rather than white. Specimens of *cicatricosa* are similar to those of *detersa* and *recula* in size; forewing length varies from 11 to 16 mm.

The immature stages of *cicatricosa* are unknown.

This species occurs from central Saskatchewan, westward to south-central British Columbia, and southward to northern Texas, southern New Mexico, central Arizona, and southern California (Lafontaine, 1981, figure 184). It occurs in very arid habitats where the vegetation is sparse, and the soil is loose sand or gravel. Adults of *cicatricosa* have been collected from late August until late October.

Euxoa (*Euxoa*) *recula* (Harvey)
PL. 7, FIGS. 55–57 (RWH 10840).

Agrotis recula Harvey, 1876, *Can. Ent.*, **8**: 37. Type locality: Oregon. [BMNH]

NOTE—The female lectotype of *recula* was designated by Lafontaine (1981: 47).

This distinctively marked species can be recognized by the brownish-gray and yellow pattern of the forewing and white hindwing. Specimens are likely to be confused only with those of *cicatricosa* but can be distinguished from them by the characters given in the key and discussed under *cicatricosa.*

The immature stages are unknown.

Euxoa recula occurs from central Oregon southward to southern California; in the South it occurs eastward through southern Arizona to central New Mexico (Lafontaine, 1981, figure 185). Like *cicatricosa,* it inhabits desert areas where vegetation is sparse. The two species are frequently collected together in Oregon and Nevada, but in the Mojave Desert and in southern Arizona it occurs in extremely xeric areas where *cicatricosa* does not occur. Adults are in flight from early September until late October.

aequalis GROUP

This group includes four North American species. The group can be characterized best by shape of the vesica; the vesica bends sharply ventrally above the apex of the aedoeagus, then curves in an arc to project dorsally; also, the apical third of the vesica is prominently enlarged. Members of the group can most easily be recognized by the combination of characters used in the key. The three brown-colored species are most likely to be confused with members of the *punctigera, comosa,* and *infausta* species-groups. Males can be distinguished from those of species in the first two groups by the relatively thin saccular extensions and from those in the *infausta* group by the longer sacculus. The species most likely to be confused with *cona* are discussed under that species. Females of all species in the group except *munis* can be distinguished from those of other similar species by the combination of fused dorsal margins of the ovipositor lobes and oval, unisaccate corpus bursae.

KEY TO SPECIES OF THE *AEQUALIS* GROUP

1. Males 2
— Females 5

2. Apical half of harpe very finely pubescent, in addition to much larger scattered setae 3
— Apical half of harpe without pubescence, only

larger scattered setae present, these about half as long as diameter of harpe 4

3. Saccular extension conspicuously stouter than harpe; hindwing with at most a small amount of brown shading on terminal line ..... *conjuncta* p. 141

– Saccular extension not stouter than harpe; hindwing buff with brown shading on marginal third of wing ..... *aequalis* this page

4. Apex of saccular extension curling dorsally toward cucullus, ground color of forewing creamy white with extensive black dusting ..... *cona* p. 142

– Apex of saccular extension straight or curling ventrally away from cucullus, ground color of forewing gray or brown ..... *munis* p. 142

5. Ovipositor lobe without sclerotized, apical process ..... *cona* p. 142

– Ovipositor lobe with sclerotized, apical, flange-like process 6

6. Sclerotized process at apex of each ovipositor lobe fused together along dorsal margins of lobes 7

– Sclerotized process at apex of each ovipositor lobe and dorsal margins of lobes not fused together but separated by non-sclerotized membrane ..... *munis* p. 142

7. Hindwing white or pale buff, veins brown in some specimens; occurring from Kansas and Texas westward to Arizona ..... *conjuncta* p. 141

– Hindwing dark brown, at least on outer two-thirds of wing; widespread in central and western North America ..... *aequalis* this page

Euxoa (*Euxoa*) *aequalis* (Harvey)

PL. 7, FIGS. 58–64; PL. N, FIG. 3; PL. CC, FIG. 7 (RWH 10776, 10777).

Agrotis aequalis Harvey, 1876, *Can. Ent.*, **8**: 36.
Type locality: California. [BMNH]

Carneades acornis Smith, 1895, *Ent. News*, **6**: 335. NEW SYNONYMY, NEW STATUS, SUBSPECIES.
Type locality: Calgary, Alberta. [USNM]

Agrotis alko Strecker, 1899, *Lepidoptera, Rhopaloceres and Heteroceres, Indigenous and Exotic*, Suppl. **2**: 5. NEW SYNONYMY, NEW STATUS, SUBSPECIES.
Type locality: Colorado. [FMNH]

Carneades naevulus Smith, 1900, *Proc. U. S. Natl. Mus.*, **22**: 424. NEW SYNONYMY.
Type locality: Placer County, California. [USNM]

NOTE—The male lectotype of *naevulus* was designated by Todd (1982: 145).

Carneades megastigma Smith, 1900, *Proc. U. S. Natl. Mus.*, **22**: 425. NEW SYNONYMY.
Type locality: Calgary, Alberta. [USNM]

NOTE—The female lectotype of *megastigma* was designated by Todd (1982: 136).

Carneades termessus Smith, 1900, *Proc. U. S. Natl. Mus.*, **22**: 426. NEW SYNONYMY.
Type locality: Yellowstone Park, Wyoming. [USNM]

NOTE—The male lectotype of *termessus* was designated by Todd (1982: 207).

Carneades sessile Smith, 1900, *Proc. U. S. Natl. Mus.*, **22**: 431. NEW SYNONYMY.
Type locality: Hall Valley, Colorado. [USNM]

NOTE—The female lectotype of *sessile* was designated by Todd (1982: 195).

Carneades testula Smith, 1900, *Proc. U. S. Natl. Mus.*, **22**: 440. NEW SYNONYMY.
Type locality: Calgary, Alberta. [USNM]

Euxoa (*Euxoa*) *aequalis yukonensis* Lafontaine, NEW SUBSPECIES.
Type locality: Yukon, 2 km N Carcross. [CNC]

This is one of the most drably colored species of *Euxoa*. The forewing color varies from pale sandy brown through various shades of brown and grayish brown to gray. The maculation may be obscure or outlined in black. In most specimens, the forewing has a dusting of pale-colored scales that gives it a mottled, powdery, and slightly hoary appearance. The hindwing is dirty white, or mottled buff, basally, with a mottled band of smoky-brown shading on the marginal third in males and on most of the wing in females. Specimens of *aequalis* are most likely to be confused with specimens of *comosa* (p. 97), *infausta* (p. 104), and species in the *punctigera* group (p. 83). The hoary appearance of the forewing and the uneven color of the hindwing can be used to distinguish specimens of *aequalis* from those of *comosa, infausta,* and most specimens of species in the *punctigera* group. In addition to the genital characters given for the species group, males of *aequalis* can be distinguished from those of *comosa, infausta,* and other similarly colored species other than those in the *punctigera* group, by the pubescent harpe. With the same exception, females can be recognized by the dorsally fused ovipositor lobes and the fused apical processes. Males of *aequalis* can be distinguished from those of species in the *punctigera* group, usually without dissection, by their relatively thin, apically straight, or downcurved, saccular extensions. Females can be distinguished from those in the *punctigera* group by the unisaccate rather than bisaccate corpus bursae. The species that creates most of the confusion is *punctigera* itself; however, it lacks the sclerotized apical processes on the ovipositor lobes that are present in *aequalis* females. Forewing length in *aequalis* varies from 14 to 18 mm.

The immature stages are known only from laboratory reared material (Hinks and Byers, 1976). The egg was illustrated by Salkeld (1975: 1145, figures 16–18).

This species is widespread in western North America. It occurs from southern Manitoba, westward to southern British Columbia, and southward to western Nebraska, northern New Mexico, northern Arizona, and southern California. A disjunct population occurs in southern Yukon. Adults have been collected from late July until mid-October.

Populations of *aequalis* are arranged in four subspecies.

KEY TO SUBSPECIES OF *AEQUALIS*

1. Forewing dark gray; vestiture of thorax mouse gray; occurring in Yukon .... *yukonensis* p. 141

– Forewing pale blue gray, grayish brown, or brown; vestiture of thorax a mixture of gray and brown scales; occurring from southern Canada southward 2

2. Forewing blue gray; known only from vicinity of San Francisco, California .... *aequalis* this page

– Forewing grayish brown or brown; widespread from southern and eastern California eastward to the central Great Plains 3

3. Forewing dark brown, maculation contrasting; occurring from Rocky Mountain region westward .... *alko* p. 141

– Forewing light, sandy brown; reniform and orbicular spots obscure in most specimens; occurring in Great Plains .... *acornis* this page

Euxoa (*Euxoa*) *aequalis aequalis* (Harvey)
PL. 7, FIGS. 58, 59.

Agrotis aequalis Harvey, 1876.

The nominate subspecies of *aequalis* is one of the rarest taxa of *Euxoa* and may be endangered in that it is known only from the vicinity of San Francisco. I have seen only five specimens of subspecies *aequalis*: the female holotype labeled "California," a pair from San Francisco in the collection of the California Academy of Sciences and a pair from Marin County in the collection J. G. Franclemont, Ithaca, New York.

Specimens of subspecies *aequalis* differ from those of other subspecies in being larger (forewing length: 17 to 19 mm), and in having blue-gray rather than gray-brown, or dark-brown, forewings. They may also differ in the relative length of harpe and saccular extension; in the only male genital preparation examined the saccular extension is about ¾ of the length of the harpe. It is rarely this short in other subspecies of *aequalis*; in most specimens it is similar in length or very slightly shorter. Additional collecting in coastal California is necessary in order to determine the consistency of these differences and to determine the complete range of this subspecies. The specimens were collected in September and October.

Euxoa (*Euxoa*) *aequalis acornis* (Smith)
PL. 7, FIG. 60; PL. N, FIG. 3.

Carneades acornis Smith, 1895.

Carneades megastigma Smith, 1900.

Carneades testula Smith, 1900.

Subspecies *acornis* differs from the other subspecies in having pale sandy-brown forewings and obscure maculation in most specimens.The forewing has an olive or pink flush in many specimens. This subspecies inhabits dry open prairie areas from southern Manitoba, eastward to south-central Alberta and southward to western Nebraska and southern Montana (Lafontaine, 1976b: 742, map 1). Where the range of *acornis* meets that of subspecies *alko* in the foothills of Alberta and Montana, many specimens are intermediate in color between those of the two subspecies. This is probably an area of intergradation because similarly colored forms have been reared in the laboratory by hybridizing specimens of *acornis* and *alko*.

Euxoa (*Euxoa*) *aequalis alko* (Strecker)
PL. 7, FIGS. 61, 62; PL. CC, FIG. 7.

Agrotis alko Strecker, 1899.

Carneades naevulus Smith, 1900.

Carneades termessus Smith, 1900.

Carneades sessile Smith, 1900.

This is the most widely distributed subspecies of *aequalis*. It occurs from southwestern Alberta and southern British Columbia southward to northern New Mexico, northern Arizona, and in eastern California to the Transverse Ranges of southern California (Lafontaine, 1976b, map 1). Specimens of subspecies *alko* differ from those of other subspecies in having dark-brown forewings with a dusting of gray scales. The thorax is similar in color to the forewings. Specimens of this subspecies are frequently confused with those of *punctigera* but can be recognized by the characters given for the species.

Euxoa (*Euxoa*) *aequalis yukonensis* Lafontaine, NEW SUBSPECIES
PL. 7, FIGS. 63, 64.

Specimens of subspecies *yukonensis* can be distinguished from those of other subspecies by the mouse-gray coloration of the thorax and forewings.

Antenna of male biserrate and bifasciculate; antenna of female filiform. Frontal tubercle large but obscured by surrounding vestiture. Eye slightly reduced from normal globular shape. Vestiture of head and thorax mouse gray with a sprinkling of black and white scales; scales of thorax very deeply forked, appearing hairlike. Ground color of forewing gray with dusting of black brown scales. Basal line indicated only by trace of black scales. Antemedial line black, double, almost straight, only slightly arced. Postmedial line black, double, scalloped between veins. Median line indicated by a diffuse dark line. Subterminal line made evident only by dark-gray shading in terminal area. Terminal line narrow, black. Fringe gray. Claviform spot outlined in black, only distal portion of spot present in most specimens. Reniform and orbicular spots partially or completely outlined in black. Forewing length 14 to 16 mm. Hindwing of male pale buff basally with smoky-brown shading on outer third of wing and on veins; discal lunule prominent. Hindwing of female smoky brown, slightly paler near base. Fringe in both sexes pale, buffy gray with smoky-brown subbasal line. Male and female genitalia indistinguishable from those of other subspecies.

TYPES. Holotype: ♂. Yukon, 2 km N Carcross, 650 meters; 7 Aug. 1980; Wood and Lafontaine. CNC (Type No. 19578). Paratypes: 38 ♂, 19 ♀. Same data as for holotype (10 ♂). Same data as for holotype but reared from egg, emerged May 1981 (10 ♂, 14 ♀). Yukon, Kluane Nat. Pk., base of Mt. Wallace, 1,050 meters; 6 Aug. 1980; Wood and Lafontaine (1 ♂). Yukon, Teslin; 13–14 Aug. 1948; M. T. Hughes and W. R. Mason (9 ♂, 1 ♀). Yukon, Whitehorse; 6–11 Aug. 1948; M. T. Hughes and W. R. Mason (8 ♂, 4 ♀). CNC, USNM.

Although specimens of *yukonensis* can most easily be recognized by color pattern discussed above, they also differ from those of other subspecies of *aequalis* in that the eyes are slightly reduced, and there is a tendency toward reduction of the wings in females. In reared specimens, about half of the females show a marked reduction in wing size. Specimens of subspecies *yukonensis* hybridize in the laboratory with those of subspecies *acornis* without apparent reduction in fertility.

Euxoa (*Euxoa*) *conjuncta* (Smith)
PL. 8, FIGS. 1, 2; PL. N, FIG. 4; PL. CC, FIG. 8 (RWH 10778).

Carneades conjuncta Smith, 1890, *Bull. U. S. Natl. Mus.*, **38**: 221.
Type locality: Las Vegas, New Mexico. [USNM]
NOTE—The female lectotype of *conjuncta* was designated by Todd (1982: 55).

Specimens of this species can readily be distinguished from those of other species in the *aequalis* group by the characters given in the key. They may, however, be confused with those of some species in the *punctigera* group because of similarity in both external appearance and some genital characters. Males of *conjuncta* could be confused with those of *aurantiaca* because both have pubescent harpes and a similar valve shape; in *conjuncta*, however, the saccular extension bends ventrally away from the valve apically while in *aurantiaca* it is straight apically or bends toward the valve. Females of *conjuncta* are similar to those of *aurantiaca* and *stigmatalis* in that the dorsal margins of the ovipositor lobes and the sclerotized apical processes are fused in all three species. They differ from females of these species in the *punctigera* group in that the corpus bursae is unisaccate rather than bisaccate. Many specimens of *conjuncta* can be recognized by the combination of powdery orange-brown forewing and a row of pale-colored spots along the subterminal line. The forewing length varies from 17 to 18 mm.

The immature stages of *conjuncta* are unknown.

This is a very uncommon species in collections. It occurs from west-central Kansas and central Colorado southward to northern Texas, south-central New Mexico, and central Arizona. Adults have been collected from mid-July until early October. On the basis of a limited amount of field data, it appears that this species inhabits dry, open brushy areas.

Euxoa (*Euxoa*) *cona* (Strecker)

PL. 8, FIGS. 3, 4; PL. N, FIG. 5; PL. DD, FIG. 1 (RWH 10779).

Agrotis cona Strecker, 1898, *Lepidoptera, Rhopaloceres and Heteroceres, Indigenous and Exotic,* Suppl., **1**: 6.
Type locality: Glenwood Springs, Colorado. [FMNH]

NOTE—Of the four specimens in the type series of *cona,* three are in the Field Museum of Natural History, Chicago, and one is in the United States National Museum of Natural History. Of these, a male in the Field Museum labeled "Glenwood Springs, Colo. 9 1892, W. Barnes /14/ *A. cona,* Orig. type/ *Euxoa* slide Strecker No. 3" is here designated lectotype. The specimen is in good condition.

Euxoa sotnia Smith, 1905, *Jour. New York Ent. Soc.,* **13**: 195.
Type locality: Stockton, Utah. [AMNH]

NOTE—The female lectotype of *sotnia* was designated by Todd (1982: 198).

Specimens of *cona* are frequently confused with those of the pale form of *misturata* and less frequently with those of *tronella* and *pallipennis.* All four species have dirty-white forewings and white hindwings in males and dirty-white hindwings in females. Most specimens of *cona* can be recognized by the combination: powdery forewing appearance, relatively prominent antemedial and postmedial lines, relatively pale-colored orbicular spot, and the more prominently biserrate male antennae. They can be positively identified, however, only by genital characters that can be observed in most specimens by brushing away the scales from the end of the abdomen. The saccular extensions are about as long as the harpes and about as stout (in males of *pallipennis* and *misturata* the saccular extensions are much shorter and much stouter than the harpes; in those of *tronella* the saccular extensions are longer and much stouter than the harpes). Females of *cona* lack the sclerotized apical process on each ovipositor lobe that is present in most similarly colored species; they differ from females of *pallipennis* in having the dorsal margins of the ovipositor lobes fused. Forewing length varies from 14 to 18 mm.

The immature stages of *cona* are unknown.

This species is primarily an inhabitant of the intermontane region where it occurs from northeastern Oregon and southern Idaho southward to west-central Colorado, central Utah, southernmost Nevada, and Inyo County, California. It has been collected in the Great Plains region only in eastern Montana and western North Dakota. *Euxoa cona* occurs in open, arid habitats. Adults have been collected from early September until mid-October.

Euxoa (*Euxoa*) *munis* (Grote)

PL. 8, FIGS. 5–9; PL. N, FIG. 6; PL. DD, FIG. 2 (RWH 10855).

Agrotis munis Grote, 1879, *North Amer. Ent.,* **1**: 38.
Type locality: Colorado. [BMNH]

NOTE—Grote described *munis* from an unknown number of specimens; three specimens, collected in Colorado by Snow, from the Grote collection are in the British Museum (Natural History). These specimens may represent the original type series although only one is labeled type. This latter specimen, a female labeled "Colorado, Snow, Grote Coll. 81–116/ *Agrotis munis* Grote/*Agrotis munis* Grote, type female/Noctuidae, Brit. Mus. slide No. 6326" is here designated lectotype. The specimen is in good condition.

Agrotis sublatis Grote, 1880, *North Amer. Ent.,* **1**: 91.
Type locality: Colorado. [USNM]

Agrotis rena Smith, 1890, *Trans. Amer. Ent. Soc.,* **17**: 53.
Type locality: Sierra Nevada, California. [USNM]

NOTE—The female lectotype of *rena* was designated by Todd (1982: 183).

Setagrotis dolens Smith, 1906, *Can. Ent.,* **38**: 226.
Type locality: Beulah, Manitoba. [AMNH]

NOTE—The male lectotype of *dolens* was designated by Todd (1968: 269).

Euxoa cervinea Smith, 1910, *Trans. Amer. Ent. Soc.,* **36**: 262.
Type locality: Bozeman, Montana. [AMNH]
NOTE—The male lectotype of *cervinea* was designated by Todd (1982: 44).

Euxoa munis is a difficult species to characterize because it has many forms. In the most common forms the forewing ground color is pale gray or pale reddish brown; the reniform and orbicular spots are outlined by a pale line with a partial black outline outside this in many specimens; the spots are disproportionately large, much wider than the space between them; and there is a conspicuous, dark-gray spot in the lower portion of the reniform spot. Other features of the forewing are extremely variable; the transverse lines may be prominent and deeply scalloped or obscure and barely traceable. Forewing length varies from 12 to 17 mm. Specimens of *munis* are most likely to be confused with those of *quebecensis* (p. 47) (subgenus *Pleonectopoda*), *intrita* (p. 69), and species in the *infausta* group (p. 103). Males of *munis* can be distinguished from those of other similar species by the relative proportions of the sacculus, saccular extension, and harpe and by the shape of the vesica. Females of *munis* differ from those of other similarly marked species in having a sclerotized flangelike process at the apex of each ovipositor lobe.

The immature stages of *munis* are essentially unknown although an adult was reared from a larva collected on sugarbeet in Montana (Cook, 1930: 266).

This species occurs from southern Northwest Territories, southern Yukon, and southeastern Alaska southward to southern Manitoba, central Colorado, southern Utah, and southern California. Most collections of *munis* were made in forested or brushy habitats, but it occasionally has been found in open areas. Adults of *munis* have been collected from mid-July until late September.

misturata GROUP

This group includes only two species, *melana,* a southern Great Plains species, and *misturata,* a northern Great Plains and intermontane region species. Although the two species are structurally indistinguishable, there is no evidence of intergradation where their ranges overlap in western Texas.

In external appearance, members of the *misturata* group are most likely to be confused with specimens of *pallipennis* (p. 86), *cona* (p. 142), and *mitis* (p. 146). Males of *misturata* and *melana* can be distinguished from those of other similar species by brushing away the scales from the end of the abdomen and examining the length and shape of the saccular extension: in the *misturata* group this is short and cylindrical, about two-thirds of the harpe length and bends inward apically toward the aedoeagus. In *pallipennis* it is short and apically flattened, about two-thirds of the length of the harpe, but it bends dorsally near the apex toward the cucullus; in *cona* it is longer, similar in length to the harpe; in *mitis* it is very short, about one-third the length of the harpe and like *pallipennis* it bends dorsally toward the cucullus. Males of *misturata* and *melana* can also be recognized by the shape of the vesica. The female genitalia of the two species in the group are characterized by the unisaccate corpus bursae, the short sclerotized plates in the wall of the ductus bursae, these about half of the length of the ductus, and by the presence of a rounded sclerotized process at the apex of each ovipositor lobe. Females of *pallipennis* and *cona* lack a sclerotized process on the ovipositor lobe. In those of *mitis,* the sclerotized processes on the ovipositor lobes are in the form of finlike plates on the dorsal margin of the lobe; these spread apart posteriorly and form a V-shaped gap between the dorsal margins that is half as long as the lobes.

KEY TO SPECIES OF THE *MISTURATA* GROUP

1. Widespread in western aridlands; forewing color and pattern extremely variable; populations in southern Great Plains with forewing orange or brownish orange and with orbicular spot similar in width to distance between it and reniform spot .... *misturata* this page
— Occurring in southern Great Plains region; forewing dark blackish brown with coppery suffusion; orbicular spot very small, about ½ as wide as distance between it and reniform spot .... *melana* p. 144

Euxoa (*Euxoa*) *misturata* (Smith)

PL. 8, FIGS. 10–17; PL. N, FIG. 7; PL. DD, FIG. 3 (RWH 10766).

Carneades misturata Smith, 1890, *Bull. U. S. Natl. Mus.,* **38**: 156.
Type locality: Colorado. [MSU]
NOTE—The male lectotype of *misturata* was designated by Todd (1982: 141). The name *misturata* was an unpublished manuscript name of Morrison.

Carneades perturbata Smith, 1890, *Bull. U. S. Natl. Mus.,* **38**: 222.
Type locality: Colorado. [USNM]

Carneades candida Smith, 1894, *Trans. Amer. Ent. Soc.*, **31**: 49.
Type locality: Boulder, Montana. [USNM]

Agrotis gian Strecker, 1898, *Lepidoptera, Rhopaloceres and Heteroceres, Indigenous and Exotic,* Suppl., **1**: 6.
Type locality: Arizona. [FMNH]

Carneades falerina Smith, 1900, *Proc. U. S. Natl. Mus.*, **22**: 429.
Type locality: Nevada. [USNM]

NOTE—The female lectotype of *falerina* was designated by Todd (1982: 80).

Euxoa vertesta Smith, 1910, *Trans. Amer. Ent. Soc.*, **36**: 254.
Type locality: Stockton, Utah. [USNM]

NOTE—The male lectotype of *vertesta* was designed by Todd (1982: 225).

This species is extremely variable in terms of forewing color; this may be white, gray, orange, or brown. To a certain extent, the forewing color varies with geographic location. In the northern Great Plains region, forms with a dirty-white forewing predominate; forms with gray forewings are most common in Washington and Oregon; a form with a mottled black and white forewing ground color occurs in southern California; and forms with a basically orange or brownish-orange forewing predominate in New Mexico, Colorado and Arizona. A mixture of all of these forms, together with all intermediate combinations, occurs in the Great Basin region. Adults vary greatly in size; as with forewing color, the variation is most marked in Nevada and Utah populations. Forewing length varies from 13 to 19 mm. The combination of mottled forewing appearance and white hindwing is a good rule of thumb way to recognize *misturata.*

The immature stages of *misturata* are unknown.

Euxoa misturata is a widespread and abundant species of western aridlands. It occurs from southwestern Saskatchewan and north-central Washington southward to western Texas, southern New Mexico, southern Arizona, and the Mojave Desert of southern California. Adults have been collected from late August until early October.

Euxoa (*Euxoa*) *melana* Lafontaine

PL. 8, FIG. 18 (RWH 10767).

Euxoa melana Lafontaine, 1975, *Can. Ent.*, **107**: 1329.
Type locality: Canadian, Texas, 6 mi E, 2,000′. [CNC]

This species is darker than any of the forms of *misturata.* The forewing is dark brown with a dusting of yellow scales that gives it a coppery suffusion. Also, the reniform spot is mesially constricted and very narrow, particularly near the costa. The orbicular spot is very small and round. Forewing length varies from 15 to 17 mm. In the Texas Panhandle, where the range of *misturata* overlaps that of *melana,* only the orange-brown form of *misturata* occurs, so specimens of the two species can be distinguished by forewing color as well as by the narrow reniform spot and small orbicular spot of *melana.*

The immature stages of *melana* are unknown.

Euxoa melana occurs in southwestern Kansas and northwestern Texas; an apparently disjunct population occurs in southeastern Texas. Adults have been collected from mid-September until early October.

atristrigata GROUP

This group includes only two species, *atristrigata* and *nevada.* Specimens of these species can usually be recognized by the peculiar, streaked forewing pattern; possible confusion between specimens of *nevada* and those of *brevipennis* and *misturata* is discussed under *nevada.*

Males of members of the group can be recognized by the relatively short, thin saccular extensions; these vary from being slightly shorter than half as long as the harpe to slightly longer than half and are thinner than the harpes. They differ from those of *mitis* in having a tapered rather than apically enlarged uncus, longer saccular extensions and a differently shaped vesica. Females of members of the group can be recognized by the combination of streaked forewing and large, heavily sclerotized processes on the ovipositor lobes. The processes are fused together in *atristrigata* so that the female genitalia resemble those of *aequalis.*

KEY TO SPECIES OF THE *ATRISTRIGATA* GROUP

1. Forewing pale brown with longitudinal black streaking; a series of prominent, black, sagittate marks in subterminal area adjacent to the terminal area; reniform and orbicular spots fused in most specimens; flangelike processes on ovipositor lobes fused together .... *atristrigata* p. 145

— Forewing grayish brown, heavily dusted with pale silver gray that gives it a hoary appearance; terminal area dark gray, streaked into subterminal area but without streaks ending in a series of black, sagittate spots; reniform and orbicular spots not fused in most specimens; flangelike processes on ovipositor lobes not fused together .... *nevada* this page

Euxoa (*Euxoa*) *atristrigata* (Smith)

PL. 8, FIGS. 19, 20; PL. N, FIG. 9; PL. DD, FIG. 5; PL. FF, FIG. 17 (RWH 10768).

Agrotis atristrigata Smith, 1890, *Ent. Amer.*, **6**: 76.
Type locality: British Columbia. [USNM]

Carneades collocata Smith, 1894, *Trans. Amer. Ent. Soc.*, **21**: 50.
Type locality: Colorado. [USNM]
NOTE—The male lectotype of *collocata* was designated by Todd (1982: 50).

Specimens of *atristrigata* are likely to be confused only with those of *nevada* but can be distinguished by the characters given in the key. Males of *atristrigata* cannot safely be distinguished from those of *nevada* by genital characters although the saccular extension tend to be longer in *atristrigata* than in *nevada* and not as prominently asymmetrical in length. The subbasal diverticulum in the vesica of *atristrigata* is foot shaped, or elbowed, and projects posteriorly; that of *nevada* is bilobed, or T-shaped, with pouches projecting both anteriorly and posteriorly. Forewing length varies from 13 to 17 mm.

The immature stages of *atristrigata* are unknown.

This species occurs from western North Dakota and northern Montana southward to northwestern New Mexico and from southern British Columbia southward to southern Nevada and northern Arizona (Lafontaine, 1982a, figure 42). It inhabits arid areas, both open aridlands and those with scattered conifers. Adults have been collected from late July until late September.

Euxoa (*Euxoa*) *nevada* (Smith)

PL. 8, FIGS. 21–23; PL. N, FIG. 10; PL. DD, FIG. 6; PL. FF, FIG. 18 (RWH 10769).

Carneades nevada Smith, 1900, *Proc. U. S. Natl. Mus.*, **22**: 420.
Type locality: Nevada. [USNM]
NOTE—The male lectotype of *nevada* was designated by Todd (1982: 148).

Euxoa floramina Smith, 1905, *Can. Ent.*, **37**: 202.
Type locality: Stockton, Utah. [AMNH]
NOTE—The female lectotype of *floramina* was designated by Todd (1982: 86).

Specimens of *nevada* are most likely to be confused with those of *atristrigata* but also may be confused with some specimens of *brevipennis* (p. 125) and *misturata* (p. 143). They may be distinguished from those of of *atristrigata* by characters given in the key and under *atristrigata.* Males of *nevada* differ from those of *misturata* in having thinner, more symmetrical saccular extensions that do not bend inward toward the aedoeagus apically; they differ from males of *brevipennis* in having saccular extensions that are much shorter rather than slightly longer than the harpes. Females of *nevada* can be distinguished from those of *misturata* by the more heavily sclerotized, and more finlike, processes on the ovipositor lobes and by the longer sclerotized plates in the wall of the ductus bursae that extend as far anteriorly as do the anterior apophyses; females of *brevipennis* lack sclerotized processes on the ovipositor lobes. Forewing length in *nevada* varies from 13 to 16 mm.

The immature stages of *nevada* are unknown.

This species has a very similar range to that of *atristrigata* except that it extends slightly farther both northward and southward. It occurs from southern Saskatchewan and Alberta and southern British Columbia southward to northern New Mexico, northern Arizona, and southern California (Lafontaine, 1982a, figure 42). As for *atristrigata, nevada* occurs in dry open areas, and the two species are frequently collected together. Specimens have been collected from mid-August until late September.

cinereopallida GROUP

This group includes only *cinereopallida*; the characters are given under the species.

Euxoa (*Euxoa*) *cinereopallida* (Smith)

PL. 8, FIGS. 24–26; PL. N, FIG. 8; PL. DD, FIG. 4; PL. FF, FIG. 16 (RWH 10770).

Carneades cinereopallidus Smith, 1903, *Can. Ent.*, **35**: 10.
Type locality: Stockton, Utah. [AMNH]
NOTE—The female lectotype of *cinereopallidus* was designated by Todd (1982: 47).

Although specimens of *cinereopallida* can be distinguished from those of all other *Euxoa* species by male and female genital characters, they are very similar to those of *cicatricosa* in external appearance and are frequently confused with them. Difficulties are further complicated by parallel geographical variation in the two species. In both species dark forms occur in the Northwest, pale forms occur in the Great Basin, forms intermediate between them occur in the Great Plains, and forms with orange-brown forewings occur in the South. Despite similarities in wing markings, specimens of the two species frequently can be distinguished by the narrower, more markedly bent, boomerang-shaped reniform spot in *cinereopallida*; also the pale streak distal to the claviform spot is less prominent than in *cicatricosa* and does not extend into the subterminal area. The two species are similar in size; in *cinereopallida* forewing length varies from 12 to 15 mm.

Males of *cinereopallida* can most easily be distinguished from those of other species by the small size of the harpes and saccular extensions; these are both much shorter than the sacculus, and the harpe is on the middle of the inner surface of the cucullus. Females of *cinereopallida* can be recognized by the shape of the sclerotized plates on the ovipositor lobes. In other species the dorsal margin of the plate is either flat and finlike, or convex and rounded; in *cinereopallida* the plate is enlarged at each end, giving it a two-humped profile. Females of *cicatricosa* have stout setae on the dorsal margin of the ovipositor lobe rather than a sclerotized, apical process. Also, the female of *cinereopallida* has a much more narrow eighth abdominal segment than that of females of other similar species.

The immature stages of *cinereopallida* are unknown.

Euxoa cinereopallida is an aridland species that occurs from southern Saskatchewan and Alberta southward to northern Texas and southern New Mexico and from south-central British Columbia southward to southern Nevada and central Arizona. Adults have been collected from mid-August until early October.

mitis GROUP

The *mitis* group includes only *mitis*; the characters are given under the species.

Euxoa (*Euxoa*) *mitis* (Smith)

PL. 8, FIGS. 27–29; PL. O, FIG. 1; PL. DD, FIG. 7; PL. FF, FIG. 19 (RWH 10774).

Carneades mitis Smith, 1894, *Trans Amer. Ent. Soc.*, **21**: 49.
Type locality: Glenwood Springs, Colorado. [USNM]

NOTE—The male lectotype of *mitis* was designated by Todd (1982: 142).

Agrotis colla Strecker, 1898, *Lepidoptera, Rhopaloceres and Heteroceres, Indigenous and Exotic*, Suppl., **1**: 6.
Type locality: Loveland, Colorado. [FMNH]

Euxoa ura Smith, 1905, *Can. Ent.*, **37**: 203.
Type locality: Stockton, Utah. [AMNH]

NOTE—The female lectotype of *ura* was designated by Todd (1982: 219).

Euxoa ura form *uramina* Smith, 1905, *Can. Ent.*, **37**: 204.
Type locality: Stockton, Utah. [AMNH]

NOTE—The female lectotype of *uramina* was designated by Todd (1982: 220).

Specimens of *mitis* are similar to those of some forms of *misturata* and, to a lesser extent, to specimens of *teleboa.* They can be distinguished from specimens of *teleboa* by the lack of a prominent, dark median line on the forewing. They can usually be distinguished from specimens of *misturata* by the more even forewing ground color, by the paler and more obscure reniform and orbicular spots, and by the more prominently biserrate male antenna. Males of *mitis* can be distinguished from those of *misturata* and *teleboa* by the very short, thin saccular extension, the shape of the harpe, and by shape of the vesica. In *mitis* the right saccular extension is less than one-half of the length of the harpe and thinner; in *misturata* the right sacular extension is about two-thirds the length of the harpe and stouter basally than the harpe; in *teleboa* the right saccular extension is similar in length to the harpe. In *mitis* the harpe is somewhat S-shaped, bending dorsally apically, whereas in *misturata* and *teleboa* the harpe is evenly incurved, or bow shaped. In females of *mitis* the dorsal margins of the ovipositor lobes spread apart posteriorly and form a V-shaped gap that extends to the middle of the dorsal margins; in those of *misturata* and *teleboa* the dorsal margins of the ovipositor lobes are parallel almost to the posterior end of the lobes.

There is some geographical variation in forewing color in *mitis.* In the northern portion of its range, northern Great Plains and northern Great Basin, the forewing is yellowish buff, grayish buff, or grayish orange in most specimens; in the southern half of its range, from central Nevada and Colorado southward, forms with bright pinkish-orange forewing predominate. Forewing length varies from 14 to 17 mm.

The immature stages of *mitis* are unknown.

This species inhabits open aridlands and piñon-juniper woodland from southern Saskatchewan and southern Alberta southward to central New Mexico and from southern British Columbia southward to southern Utah and southern Nevada. Adults have been collected from mid-August until early October.

luctuosa GROUP

This group includes only *luctuosa*; the characters are given under the species.

Euxoa (*Euxoa*) *luctuosa* Lafontaine

PL. 8, FIGS. 30–32; PL. O, FIG. 2; PL. DD, FIG. 8; PL. FF, FIG. 20 (RWH 10775).

Euxoa luctuosa Lafontaine, 1976, *Can. Ent.*, **108**: 746.

Type locality: Pagosa Springs, Colorado, 17 mi W, 6,600′. [CNC]

Specimens of *luctuosa* usually can be recognized by a combination of wing marking features. The forewing ground color is orange brown, heavily dusted with dark-brown scales. The dusting of dark scales is most prominent on the outer quarter of the wing and around the reniform and orbicular spots, tending to give these spots a sunken appearance. Forewing length varies from 13 to 17 mm. The male hindwing is white with smoky-gray shading on the veins and outer third of the wing; that of the female is smoky gray, with the basal two-thirds paler and with a slight pearly sheen. Specimens of *luctuosa* are most likely to be confused with those of *acornis* (p. 140), *conjuncta* (p. 141), or *comosa* (p. 97). They can be distinguished from those of *acornis* by the orange-brown rather than gray-brown forewing color, from those of *conjuncta* by the dark shading on the forewing and the outer third of the hindwing, and from those of *comosa* by the paler hindwing. They can readily be distinguished from them, and all other *Euxoa* species, by male and female genital characters. The male genitalia are somewhat similar to those of species in subgenus *Orosagrotis* in that the harpe lies on the inner surface of the cucullus parallel to the costal margin. In *luctuosa* the saccular extensions are longer than those of species of *Orosagrotis* and are markedly asymmetrical in length, and the uncus is enlarged apically. The shape of the vesica in *luctuosa* is unlike that of any other species. The female genitalia can be confused only with those of females in the subgenus *Orosagrotis* in that the corpus bursae is somewhat diamond shaped with the ductus seminalis arising from a pouch near the middle of one side. In *luctuosa* the pouch with the ductus seminalis is on the left side, whereas it is on the right side in *Orosagrotis*.

The immature stages of *luctuosa* are unknown.

This is a species of the American Southwest. It occurs from southern Colorado southward to southwestern Texas, central New Mexico, and southern Arizona. Adults have been collected from late July until late September.

SUBGENUS
Orosagrotis Hampson

Orosagrotis Hampson, 1903.

Menada Kozhanchikov, 1937.

The subgenus *Orosagrotis* includes 16 species: twelve nearctic, one holarctic, and three palearctic. The palearctic species are *filipjevi* Kozhanchikov, 1929, *subconspicua* Staudinger, 1888, and *varia* Alphéraky, 1889.

Members of the subgenus *Orosagrotis* can be recognized by both male and female genital characters. Males can most easily be distinguished from those of species in other subgenera by the vesica; this projects to the left in *Orosagrotis* rather than dorsally or to the right as in other subgenera. If the vesica has not been everted, males can still be associated with subgenus *Orosagrotis* by the combination of short saccular extension, orientation of the harpe on the inner side of the cucullus, and apically tapered rather than swollen uncus. Also, males of many species in the subgenus *Orosagrotis* have filliform or very slightly serrate antennae, a condition rare in other subgenera except *Chorizagrotis*.

Females of species in the subgenus *Orosagrotis* can be recognized by the presence of the ductus seminalis on the right side of the unisaccate corpus bursae; in other subgenera it is on the left side, or on the appendix bursae if the bursa is bisaccate. Females of most species in the subgenus have a sclerotized process at the apex of each ovipositor lobe and in all of these species except one these processes are fused together. In most species the sclerotized plates in the ductus bursae are short and triangular.

KEY TO SPECIES OF THE SUBGENUS *OROSAGROTIS*

1. Males with right saccular extension less than ⅔ length of harpe; females with sclerotized apical process on ovipositor lobe 2
— Males with right saccular extension more than ⅔ length of harpe; females without sclerotized apical process on ovipositor lobe (*wilsoni* group) 11

2. Males with right saccular extension very small, shorter than width of harpe; females with sclerotized processes on ovipositor lobes not fused together; ellipsoid-eyed species of alpine meadows in western Canada and Alaska (*nomas* group) .... *nomas* p. 149
— Males with right saccular extension longer than ⅓ length of harpe and longer than width of harpe; females with sclerotized processes on ovipositor lobes fused together; mostly round-eyed species but one ellipsoid-eyed species occurs in alpine meadows in Wyoming and Colorado (*ridingsiana* group) 3

3. Eyes greatly reduced in size, ellipsoid ... *montana* p. 151
— Eyes not greatly reduced, width similar to height 4

4. Forewing evenly colored, almost immaculate, either silky black or dark reddish brown; occurring from Newfoundland to British Columbia .... *perpolita* p. 151
— Forewing either with maculation prominent, or with ground color uneven and speckled; occurring from central Ontario and Michigan westward 5

5. Forewing with cubital vein white, contrasting with dark-brown forewing ground color and thorax 6
— Forewing with cubital vein not white and contrasted; may be slightly paler than ground color in species with pale-gray forewings and thorax 7

6. Basal half of prothoracic collar, and costa of forewing, yellow or white, contrasting with remainder of thorax and forewing; male antenna filiform, ciliate ventrally .... *flavicollis* p. 153
— Basal half of prothoracic collar, and costa of forewing, not pale and contrasting although costa may have some pale streaks; male antenna slightly to moderately serrate and bifasciculate .... *maimes, ridingsiana* pp. 153, 154

7. Forewing ground color pale, either yellowish buff or grayish buff 8
— Forewing ground color darker, chocolate brown, olive brown, or blackish brown 9

8. Male antenna essentially filiform, very slightly bifasciculate; known only from aridlands of southern Yukon .... *macrodentata* p. 150
— Male antenna slightly biserrate and prominently bifasciculate; occurring from southern Saskatchewan and Alberta southward ..... *taura* p. 152

9. Forewing with claviform spot obscure or absent and without a pale streak distal to claviform spot; forewing ground color gray brown heavily dusted with black and pale gray ..*aberrans* p. 150
— Forewing with claviform spot prominent and with a pale streak distal to claviform spot in most specimens; if pale streak not present, then forewing ground color chocolate brown 10

10. Forewing ground color olive brown; pale streak distal to claviform spot prominent, pale yellow .... *perolivalis* p. 151
— Forewing ground color chocolate brown; pale streak distal to claviform spot pale brown if present, inconspicuous or absent in most specimens .... *manitobana* p. 150

11. Hindwing white with buffy-brown shading on veins and marginal quarter of wing; discal spot inconspicuous or absent; male genitalia with right saccular extension 0.6–0.7× length of harpe; female genitalia with ductus seminalis near middle of right side of corpus bursae; occurring in coastal southern California from Channel Islands and Los Angeles County southward .... *riversii* p. 155
— Hindwing smoky brown, paler on basal ⅔ in some specimens; discal spot prominent, dark brown; male genitalia with right saccular extension 0.7× length of harpe to slightly longer than harpe, female genitalia with ductus seminalis at anterior end of corpus bursae on right side; occurring on Pacific coast from central

British Columbia southward to San Luis Obispo County in south-central California ..... *wilsoni* p. 154

nomas GROUP

This group includes only *nomas*; the characters are given under the species.

Euxoa (*Orosagrotis*) *nomas* (Erschov)
PL. 8, FIGS. 33, 34; PL. O, FIGS. 3, 7 (RWH 10866).

Agrotis nomas Erschov, 1874, *Cheshuekryla (Lepidoptera). In* A. P. Fedchenko, *Puteshestvie v. Turkestan.* pt. 2, vol. **3**, Zoogeographical Series, pt. 4, sec. 3: 38, pl. 3, fig. 36.
Type locality: Turkestan and Iran. [Zoological Institute, Leningrad]

Agrotiphila incognita Smith, 1893, *Ent. News,* **4**: 101. NEW SYNONYMY, NEW STATUS, SUBSPECIES.
Type locality: Laggan [Lake Louise], B.C. [Alberta]. [USNM]
NOTE—The male lectotype of *incognita* was designated by Todd (1982: 272).

Euxoa nomas is a small species with prominent forewing markings. Forewing length varies from 11 to 13 mm. It is one of only three *Euxoa* species that are diurnal and have reduced, ellipsoid eyes; the others are *montana* (p. 151) and *churchillensis* (p. 45). Males of *nomas* can be distinguished from those of *montana* and *churchillensis* by the much shorter saccular extensions and the differently shaped vesica. In *nomas* the right saccular extension is shorter than the basal width of the harpe and less than 0.2 × length of the harpe. In *montana* and *churchillensis* the right saccular extension is at least 2 × the basal width of the harpe and about one-third as long as the harpe (*montana*) or two-thirds as long (*churchillensis*). Females of *nomas* can be distinguished from those of *churchillensis* by the presence of sclerotized, flangelike projections on the ovipositor lobes and by the position of the ductus seminalis at the anterior rather than the posterior end of the corpus bursae. They can be distinguished from specimens of *montana* by the fact that the sclerotized processes on the ovipositor lobes in *nomas* are not fused together as they are in *montana*; also the ductus seminalis is at the middle of the right side of the corpus bursae in *montana,* not at the anterior end as in *nomas.*

The immature stages of *nomas* are unknown.

Euxoa nomas occurs in two widely separate areas; populations in these two areas are arranged in two subspecies. The nominate subspecies, *nomas* Erschov, occurs in the mountains of south-central Asia from Iran to Turkestan. The North American subspecies is discussed below.

Euxoa (*Orosagrotis*) *nomas incognita* (Smith)
PL. 8, FIGS. 33, 34; PL. O, FIGS. 3, 7.

Agrotiphila incognita Smith, 1893.

Specimens of *incognita* are likely to be confused only with those of the two other diurnal, ellipsoid-eyed species, *montana* (p. 151) and *churchillensis* (p. 45). The three species can be distinguished by characters given above under *nomas* and in the key.

Euxoa nomas incognita occurs from MacKenzie Delta in western Northwest Territories westward to northwestern Alaska and southward to southwestern Alberta and southern British Columbia. Its range does not overlap that of *montana* and overlaps that of *churchillensis* only in western Northwest Territories. Adults fly from late June until early August in the North and from late July until late August in the South. Adults are diurnal and usually fly only when it is sunny; they occur above treeline in dry, gravelly areas.

I can find no consistent differences between *nomas* and *incognita* in wing markings or genital characters. I tentatively retain the name *incognita* for North American populations because of the large gap in the distribution of the species; life history studies may reveal differences between *nomas* and *incognita.*

ridingsiana GROUP

This group includes ten species in the Nearctic Region and three in the Palearctic Region. All species are essentially indistinguishable by male and female genital characters although some differ slightly in shape and proportion of structures.

In the male genitalia, the saccular extensions are between ⅓ and ⅔ length of the harpe, longer than those of *nomas* but shorter than those of the two species in the *wilsoni* group. Males can be distinguished from those in other groups by details of the shape of the vesica. Females can be distinguished from those in other species-groups by the presence of sclerotized flangelike processes on the ovipositor lobes that are fused together. Within the species-group, species can be distinguished by wing color and markings and by details of the male antennae.

It is fortunate that most species in the *ridingsiana* group are sympatric in the northern Great Plains; otherwise, it would be difficult to determine whether the various forms represent distinct species or geographical forms.

I have made no attempt to arrange species in a phylogenetic sequence; they are placed in alphabetical order in two groups, those without a white cubital vein first and then those with a white cubital vein.

Euxoa (*Orosagrotis*) *aberrans* McDunnough
PL. 8, FIGS. 35–37 (RWH 10862).

Euxoa aberrans McDunnough, 1932, *Can. Ent.*, **64**: 231.
Type locality: Jefferson Co., Montana. [CNC]

This is the most nondescript species in the subgenus. The forewing is gray brown with a dusting of black and pale gray that gives the wing a hoary, or "pepper and salt," appearance. The maculation is generally obscure except for a pale line that surrounds the dark-gray centers of the reniform and orbicular spots. Unlike its two close relatives, *manitobana* and *taura*, *aberrans* apparently lacks a form with pale contrasting wing markings. Forewing length varies from 14 to 18 mm.

Specimens of *aberrans* are most likely to be confused with those of the evenly colored forms of *manitobana* and *taura* and with some specimens of *tessellata* (p. 87), *spumata* (p. 85), and *aurantiaca* (p. 84). They can be distinguished from those of the first two species by the characters given in the key. Males of *aberrans* can be distinguished from those of *tessellata, spumata,* and *aurantiaca* by the very short saccular extensions in *aberrans* and by the very slightly, rather than prominently, biserrate antennae. Females can be distinguished from similarly marked forms of *tessellata* by the presence of sclerotized flangelike processes on the ovipositor lobes in *aberrans*. Females of *aberrans*, like those of *spumata* and *aurantiaca*, have fused flangelike processes on the ovipositor lobes. They can usually be distinguished from females of these two species by forewing color and the pale outline reniform and orbicular spots. Positive identification must be made by examination of the corpus bursae; unisaccate in *aberrans*, rather than bisaccate in *spumata* and *aurantiaca*.

The immature stages of *aberrans* are unknown.

Euxoa aberrans occurs in dry coniferous forests and aspen groves from southern Alberta and south-central British Columbia southward to southwestern Montana and southern Washington. It has also been collected in two areas east of the Great Plains in southern Manitoba and northern Michigan. Adults have been collected from mid-July until mid-September.

Euxoa (*Orosagrotis*) *macrodentata* Hardwick
PL. 8, FIG. 38 (RWH 10859).

Euxoa macrodentata Hardwick, 1965, *Can. Ent.*, **97**: 1223.
Type locality: Whitehorse, Yukon. [CNC]

This species looks like diminutive form of *taura* and may be a well-defined population of that species. As for *taura*, it occurs in two forms: one with a pale gray-brown forewing with black maculation and a dusting of black scales (plate 8, figure 38) and a second form with pale-buff shading in the reniform and orbicular spots, on the costa, and distal to the claviform spot. Specimens of *macrodentata* tend to be smaller than are those of *taura*; the forewing length is 13 to 14 mm (14 to 17 in *taura*). I treat *macrodentata* as a species distinct from *taura* because of differences in the male antenna: filiform and slightly bifasciculate in *macrodentata*, prominently biserrate and bifasciculate in *taura*. Similar differences are found between species in the *ridingsiana* species complex where females are known to produce different pheromones and interspecific hybrids are infertile.

The immature stages of *macrodentata* are unknown.

Euxoa macrodentata is known only from southern Yukon in northwestern Canada where it occurs on dry grass and sagebrush (*Artemisia* spp.) covered hillsides. Adults have been collected from late July until late August.

Euxoa (*Orosagrotis*) *manitobana* McDunnough
PL. 8, FIGS. 39, 40 (RWH 10863).

Euxoa perolivalis var. *manitobana* McDunnough, 1925, *Can. Ent.*, **57**: 243.
Type locality: Miniota, Manitoba. [CNC]

This species is most similar to *perolivalis* and *aberrans*. All three species occur together in the western Great Plains. Specimens of *manitobana* differ from those of *perolivalis* in that: the forewing ground color is chocolate brown rather than paler olive brown or gray brown; the pale streak distal to the claviform spot is narrow and obscure, or absent, and pale-buff colored rather than broad and pale yellow as in *perolivalis*. Specimens of *manitobana* that lack a pale streak distal to the claviform spot could be confused with specimens of *aberrans* but differ from them in having chocolate-brown rather than paler gray-brown forewings and in lacking the extensive black and white speckling present on the forewing in *aberrans*. Forewing length in *manitobana* varies from 14 to 16 mm.

The immature stages of *manitobana* are unknown.

Euxoa manitobana was described as an eastern race of *perolivalis*. It was raised to full species status by Hardwick (1965: 1223) when it was discovered that it occurs sympatrically with *perolivalis* without evidence of intergradation. The range of *manitobana* extends from southwestern Ontario and northern Michigan westward to southern Alberta, western Wyoming, and central Utah. It has also been collected at Fort Smith in southwestern Northwest Territories. Adults have been collected from early July until late August; most records are between mid-July and early August.

Euxoa (*Orosagrotis*) *montana* (Morrison)
PL. 8, FIGS. 41, 42 (RWH 10856).

Agrotis montana Morrison, 1875, *Ann. Lyceum Nat. Hist. N.Y.*, **11**: 94.
Type locality: Colorado. [MCZ]
NOTE—The lectotype of *montana* was designated by Smith (1903: 204). The second syntype of *montana* is a specimen of *Agrotiphila colorado* Smith.

Agrotiphila rigida Smith, 1891, *Trans. Amer. Ent. Soc.*, **18**: 134.
Type locality: Colorado. [USNM]

This species looks like a miniature form of *perolivalis*; it has a forewing length of 10 to 12 mm. On the forewing of specimens of *montana* the black and white antemedial and postmedial lines are more prominent and less scalloped than are those of other species in the group. *Euxoa montana* can be distinguished from all other *Euxoa* species in western United States, except *E. churchillensis alpina* (p. 45), by the reduced, ellipsoid eyes. Specimens of *montana* can be distinguished from those of *churchillensis* by the subgeneric characters, and also by the shape and proportions of structures on the male valve and by the presence of sclerotized processes on the female ovipositor lobes in *montana*.

The immature stages of *montana* are unknown.

Euxoa montana is a day-flying species that occurs above treeline in the mountains of Colorado and in the Bighorn Mountains of Wyoming. It has been collected at elevations of 12,000 to 13,000 feet. Adults fly in late July and early August.

Euxoa (*Orosagrotis*) *perolivalis* (Smith)
PL. 8, FIGS. 43, 44 (RWH 10860).

Rhizagrotis perolivalis Smith, 1905, *Jour. New York Ent. Soc.*, **13**: 194.
Type locality: Calgary, Alberta. [AMNH]
NOTE—The male lectotype of *perolivalis* was designated by Todd (1968: 274).

This species can be recognized by the combination of pale olive-brown or gray-brown forewing ground color with a prominent pale-yellow streak distal to the claviform spot. Unlike *macrodentata*, *manitobana*, and *taura*, *perolivalis* occurs in only one form, apparently lacking a form without pale markings. Specimens of *perolivalis* are most likely to be confused with those of *manitobana* and *taura*. They can be distinguished from those of *manitobana* by the characters given in the key and discussed under *manitobana*. Specimens of *perolivalis* differ from the streaked form of *taura* in lacking the pale yellow-buff shading on the costa and subterminal area of the forewing. This gives the forewing ground color of *taura* a blotchy appearance. The two species also can be distinguished by the male antenna: filiform and very slightly bifasciculate in *perolivalis*, conspicuously biserrate and prominently bifasciculate in *taura*. Forewing length in *perolivalis* varies from 13 to 16 mm.

The immature stages of *perolivalis* are unknown.

Euxoa perolivalis inhabits dry coniferous forests and aspen groves in the western Great Plains and eastern Rocky Mountain region. Most collections were made in an area that extends from central Saskatchewan, central Alberta, and southeastern British Columbia southward to central Colorado and northern Utah. North and west of this region, it has been collected in southern Northwest Territories, southern Yukon, and central Washington. Adults have been collected from late June until early September; most records are between late July and late August.

Euxoa (*Orosagrotis*) *perpolita* (Morrison)
PL. 8, FIGS. 45, 46 (RWH 10865).

Agrotis perpolita Morrison, 1876, *Proc. Boston Soc. Nat. Hist.*, **18**: 237.
Type locality: Orono, Maine. [MSU]

Carneades exculta Smith, 1900, *Proc. U. S. Natl. Mus.*, **22**: 424.
Type locality: British Columbia. [USNM]

Euxoa criddlei Smith, 1908, *Ann. New York Acad. Sci.*, **18**: 97.
Type locality: Aweme, Manitoba. [AMNH]
NOTE—The female lectotype of *criddlei* was designated by Todd (1968: 268).

Euxoa perpolita is the most distinctive species in the subgenus *Orosagrotis* in terms of wing markings. Specimens of *perpolita* can be recognized by the even, silky-black, or reddish-brown color of the forewing and the smoky-brown color of the hindwing in both sexes. The forewing is devoid of maculation except for a black outline around the reniform and orbicular spots and a trace of the transverse lines. The form with black forewings occurs throughout the range of the species except in the Great Plains; the form with reddish-brown forewings is found in sandy areas of coastal Massachusetts and in the Great Plains; forms intermediate in color occur in coastal New Brunswick, in the northern Great Plains, and in the Black Hills of South Dakota. Specimens of *perpolita* are most likely to be confused with those of *velleripennis* because both species have black forewings. Males of *perpolita* can readily be distinguished from those of *velleripennis* by: brown rather than white hindwing; slightly rather than prominently biserrate male antenna; male genital characters; and by the maculation characters discussed below. Females of *perpolita* differ from those of *velleripennis* in having the claviform spot and transverse lines on the forewing absent, or barely traceable, not black and well defined as in *velleripennis*; and the reniform spot in *perpolita* lacks white speckling inside the black outline that is present in *velleripennis*. The sclerotized, flangelike processes on the ovipositor lobes are short, broad and fused together in *perpolita*, long, narrow and not fused together in *velleripennis*. Forewing length in *perpolita* varies from 13 to 19 mm.

The immature stages of *perpolita* are unknown.

Euxoa perpolita occurs in dry, usually sandy areas from Newfoundland, southward to Massachusetts, and westward to south-central British Columbia, western Montana, and western South Dakota. Forbes (1954: 39) also reports it from Colorado. Adults have been collected from early August until mid-September.

Euxoa (*Orosagrotis*) *taura* Smith
PL. 8, FIGS. 47–49 (RWH 10857, 10858).

Agrotis orbicularis Smith, [1888], *Proc. U. S. Natl. Mus.*, **10**: 460. NEW SYNONYMY.
Type locality: Nevada. [MSU]
NOTE—*Agrotis orbicularis* Smith [1888] is a junior primary homonym of *Agrotis orbicularis* Walker, 1865.

Euxoa taura Smith, 1905, *Can. Ent.*, **37**: 202.
Type locality: Regina, Saskatchewan. [AMNH]

Euxoa cooki McDunnough, 1925, *Can. Ent.*, **57**: 243. NEW SYNONYMY.
Type locality: Jefferson County, Montana. [CNC]

Euxoa taura is the palest species in the *ridingsiana* group, and specimens usually can be recognized by the pale buffy-brown or yellow-brown ground color of the forewing. The species occurs in two forms, one with a predominantly pale forewing (plate 8, figure 47) and one with dark shading in the median area with contrasting pale shading on the costa, the orbicular and reniform spots, and distal to the claviform spot (plate 8, figures 48, 49). These two forms appear to be equally common. In most of its range, specimens of *taura* are most likely to be confused with those of *aberrans* and *perolivalis* but can be distinguished from them by the characters in the key and as discussed under those species. *Euxoa taura* is most similar to *macrodentata* and is replaced by it in northwestern North America. Its relationship to this species is discussed under *macrodentata*. Males of *taura* can be distinguished from those of other similar species in the *ridingsiana* group by their more prominently biserrate and bifasciculate antennae. The antenna is about three times, rather than twice, as wide as the central shaft. Forewing length varies from 14 to 18 mm.

Hardwick (1970: 19) treated *taura* as a species distinct from *cooki* and known only from one specimen because the forewing ground color of the holotype of *taura* is more evenly colored and the dark hindwing band is wider and more sharply defined than in specimens of *cooki*. Although no other specimens match the holotype of *taura* in both of these details, a number of specimens from Saskatchewan and Montana approach it closely in one character or the other. As a result of this, and because of the variability of the species, I consider the holotype of *taura* to be a slightly atypical specimen of the species previously known as *cooki*.

The immature stages of *taura* are unknown.

Euxoa taura occurs in dry, open areas in the western Great Plains and eastern intermontane region.

In the Great Plains it occurs from southern Saskatchewan and Alberta southward to central Nebraska and central Colorado. Its range extends into the intermontane region to central Idaho and Nevada. Adults have been collected from mid-July until late September.

Euxoa (*Orosagrotis*) *flavicollis* (Smith)
PL. 8, FIGS. 50, 51 (RWH 10864).

Agrotis flavicollis Smith, [1888], *Proc. U. S. Natl. Mus.*, **10**: 456.
Type locality: Montana. [MSU]

Euxoa flavicollis, maimes, and *ridingsiana* form a complex of closely related and frequently confused species. Specimens can be associated with this complex by the presence of prominent white shading on the forewing cubital vein from the wing base at least to the reniform spot and frequently to the postmedial line. Of the three species in the complex *flavicollis* is the easiest to recognize because of the presence of extensive pale orange-brown (most) or silver-brown (some females) shading on the forewing costa and the basal half of the prothoracic collar. In *maimes* and *ridingsiana* these areas are similar in color to the forewing ground color with at most scattered yellow or silver scales on the posterior margin of the costa. Specimens of *flavicollis* also tend to have a narrower forewing and broader hindwing than do those of *ridingsiana* and *maimes.* Males of *flavicollis* differ from the others in that the antenna is essentially filiform and ciliate ventrally rather than slightly biserrate and bifasciculate ventrally. Females of *flavicollis* have between nine and twelve coils in the spermathecal duct rather than five or six as in *maimes* and *ridingsiana.* This last character can only be determined in freshly killed specimens. Forewing length varies from 13 to 17 mm.

The immature stages of *flavicollis* are known only from laboratory reared material. The larva has a short period of aestivation before pupation takes place (Hinks and Byers, 1976: 1352). The egg was illustrated by Salkeld (1976: 1380, figures 20–23).

Euxoa flavicollis occurs from southern Northwest Territories southward in the Great Plains and Rocky Mountain regions to southern North Dakota, southern Colorado, central Utah, and southern Washington. It is found most commonly in dry, forested areas but has also been collected in open areas in the Great Plains. Adults have been found from mid-June until mid-September; although most were collected in July and August.

Euxoa (*Orosagrotis*) *maimes* (Smith)
PL. 8, FIGS. 52, 53 (RWH 10861, part).

Carneades maimes Smith, 1903, *Can. Ent.*, **35**: 131.
Type locality: Calgary, Alberta. [AMNH]

NOTE—The female lectotype of *maimes* was designated by Todd (1968: 272).

This species and the next have been confused for many years, and until recently the two names were treated as synonyms. The first indication that two species might be involved came with the discovery that at some localities, large, broad-winged forms and smaller, narrower winged forms occurred together without intermediate forms. Subsequent experiments on pheromone specificity and experimental hybridization (Byers, Struble, and Lafontaine, 1985) have clearly demonstrated that two species are involved.

In areas where the two species occur together, or if comparative material is available, *maimes* usually can be recognized by its smaller size (forewing length 11 to14 mm versus 13 to 16 mm in *ridingsiana*). Although the two species overlap in size when the entire range of variation is considered, there is rarely any overlap among specimens from any particular area. Some subtle differences are useful for distinguishing the two species: 1) the male antenna in *maimes* is slightly biserrate and slightly bifasciculate as opposed to moderately biserrate and bifasciculate in *ridingsiana*; 2) the forewing of *maimes* appears to be slightly narrower than does that of *ridingsiana*; 3) the thoracic vestiture is dark gray rather than reddish brown in about 90 per cent of *maimes* specimens, but it is reddish brown in about half of the *ridingsiana* specimens. The two species cannot be distinguished by male or female genital characters, but the two species sometimes differ in shape of the harpe at individual localities.

The immature stages of *maimes* are known only from laboratory reared material. The larva of *maimes* differs from that of *ridingsiana* only in size, and *maimes* larvae take about eight days longer to reach maturity than do those of *ridingsiana.*

Euxoa maimes is usually found in drier, more open areas than is *ridingsiana.* The two species are frequently found together in the northern Great Plains where *maimes* inhabits the open prairie and *ridingsiana* occurs in aspen groves and river bottoms. It occurs from central Yukon and southern Northwest Territories southward to southern North Dakota and southern Montana in the Great Plains, and to southern British Columbia west of the Rockies. An apparently disjunct population of *maimes* occurs from central Colorado and southeastern Utah southward to central western New Mexico and central eastern Arizona. Adults of *maimes* have been collected from early August until early September. In areas where *maimes* occurs with *ridingsiana,* the flight season of *maimes* starts about a week later than does that of *ridingsiana.*

Euxoa (*Orosagrotis*) *ridingsiana* (Grote)
PL. 8, FIGS. 54, 55; PL. O, FIGS. 4, 8 (RWH 10861, part).

Agrotis ridingsiana Grote, 1875, *Bull. Buffalo Soc. Nat. Sci.,* **2**: 305.
Type locality: Colorado and Nevada. [? lost]

NOTE—Grote described *ridingsiana* from three specimens, two from Colorado and one from Nevada. These specimens apparently have been lost. The only specimen in the British Museum (Natural History) from the Grote collection is from the Sierra Nevada of California; this specimen is labeled as type but if so, then the locality label is incorrect. I use the name *ridingsiana* for the larger of two very similar species because: this matches the identity of the nominal type; also only this species occurs in Nevada and in most areas of Colorado where early collections were made.

This species and *maimes* have been confused for many years, and references to *ridingsiana* may refer to either species. The distinguishing characters for the two species are discussed in detail under *maimes.*

Euxoa ridingsiana is more widely distributed than is *maimes.* It occurs from central Alaska and southern Northwest Territories southward to northern North Dakota and central Montana in the Great Plains, to northern New Mexico and east-central Utah in the Rocky Mountain region, and to central California in the Sierra Nevada Mountains. In Canada the ranges of *ridingsiana* and *maimes* almost completely overlap but in western United States *maimes* is largely restricted to the Great Plains and Rocky Mountain foothills while *ridingsiana* is widely distributed in montane areas of western United States from the Rocky Mountain foothills westward. *Euxoa ridingsiana* is found most commonly in dry, forested areas. Adults have been collected from mid-July until early September; the flight season starts about a week before that of *maimes* in areas where the two occur together.

wilsoni GROUP

The *wilsoni* group includes two species, both occurring in the vicinity of the Pacific coast from central British Columbia southward to the Mexican border.

The two species in the group can be distinguished from those in other species groups of *Orosagrotis,* and other similarly marked species with which they could be confused, by the fact that the frontal tubercle in *wilsoni* and *riversii* is absent or vestigial. They can also be recognized by their relatively large size, forewing pattern, and by details of the male and female genitalia. In the male genitalia the saccular extensions are longer than are those of species in the *ridingsiana* group; they vary from 2/3 the length of the harpe to slightly longer than the harpe rather than from 1/3–2/3 as long as in the *ridingsiana* group. Females of species in the *wilsoni* group can be distinguished from those in the *ridingsiana* group by the lack of sclerotized flangelike processes on the ovipositor lobes and by the disproportionately larger corpus bursae.

Euxoa wilsoni and *riversii* are the only two species in the subgenus *Orosagrotis* that occur west of the Cascades and Sierra Nevada Mountains.

Euxoa (*Orosagrotis*) *wilsoni* (Grote)
PL. 8, FIGS. 56, 57; PL. O, FIGS. 5, 9 (RWH 10867).

Agrotis wilsoni Grote, 1873, *Bull. Buffalo Soc. Nat. Sci.* **1**: 135.
Type locality: California. [lost, see Hardwick, 1970: 40]

Agrotis specialis Grote, 1874, *Bull. Buffalo Soc. Nat. Sci.,* **2**: 62.
Type locality: California. [BMNH]

This species is closely related to *riversii* but can be distinguished from it by the characters given in the key and by the generally darker color of *wilsoni.* The forewing ground color varies from dark gray with pale silver-gray markings, to dark reddish brown with paler yellow-brown markings. Forewing length varies from 15 to 19 mm. The hindwing is smoky brown, slightly paler toward the base, with a prominent, dark-brown discal spot. The variability of this species was discussed by Buckett (1964: 104–107). In most of the range of *wilsoni,* the reddish-brown form can be confused with males of *perexcellens.* It can be distinguished from specimens of *perexcellens* by the much shorter saccular extensions of the male genitalia, by the lack of a frontal tubercle, and by the lack of a pale subterminal line on the forewing that in *perexcellens* interrupts the dark streaks from the terminal area that project into the subterminal area.

The immature stages have not been described or preserved. The larva has been reported to be a pest of lupine in the San Francisco area (Hardwick, 1970: 42).

Euxoa wilsoni occurs in the vicinity of the Pacific coast from the Queen Charlotte Islands in central British Columbia southward to San Luis Obispo County in south-central California. Its range is apparently separated from that of *riversii* by the lack of suitable beach habitat in Santa Barbara County. Adults of *wilsoni* have a very long flight period; they have been collected from early June until early October.

Euxoa (*Orosagrotis*) *riversii* (Dyar)
PL. 8, FIGS. 58–60; PL. O, FIGS. 6, 10 (RWH 10868).

Carneades perexcellens var. *riversii* Dyar, 1903, *Proc. Ent. Soc. Wash.*, **5**: 137.
Type locality: Santa Monica, California. [USNM]

NOTE—Although there is only one specimen labeled type in the United States National Museum of Natural History, Dyar stated that he was proposing the name on the basis of "... a series ... showing much variety, collected ... by Mr. J. J. Rivers." To avoid any possible confusion, the specimen in USNM labeled "Santa Monica, Cal., J. J. Rivers/Type No. 7339 USNM/ *Carneades perexcellens* var. *riversii*, Type Dyar/♀ genitalia on slide Jan. 1964 ELT 1805" is here designated as lectotype. The specimen is a female in good condition except that the right antenna is missing and the corpus bursae is missing from the female genitalia.

This species is the southern counterpart of *wilsoni* and can be distinguished from that species by its paler coloration and by the genital characters given in the key. *Euxoa riversii* has a similar range of forms as has *wilsoni* but the predominantly white hindwing, pale forewing ground color, and dusting of white scales on the forewing, combine to give *riversii* the appearance of being bleached specimens of *wilsoni*. In general, *riversii* is larger and has narrower forewings than does *wilsoni*; forewing length ranges from 17 to 21 mm.

The immature stages are unknown.

Euxoa riversii occurs in coastal southern California from Los Angeles County southward to San Diego; it also occurs on the Channel Islands as far north as San Miguel Island. Adults have been collected from early May until early August.

LITERATURE

ARNOLD, J. W. 1976. Biosystematics of the genus *Euxoa* (Lepidoptera: Noctuidae) VII. The hemocytological position of *E. rockburnei* in the "*declarata* group." *Can. Ent.*, **108**: 1387–1390.

ARNOLD, J. W., and C. F. HINKS. 1975. Biosystematics of the genus *Euxoa* (Lepidoptera: Noctuidae) III. Hemocytological distinctness between two closely related species *E. campestris* and *E. declarata*. *Can. Ent.*, **107**: 1095–1100.

BANG-HAAS, O. 1922. Die typen der gattung *Agrotis* der collection Staudinger und collection Bang-Haas in Dresden–Blasewitz. *Deutsche Ent. Zeits., Iris*, **35**: 31–39, pl. 3–17.

BARNES, W., and F. H. BENJAMIN. 1926. Notes and new species. *Can. Ent.*, **58**: 303–310.

BEIRNE, B. P. 1971. Pest insects of annual crop plants in Canada. I. Lepidoptera, II. Diptera, III. Coleoptera. *Mem. Ent. Soc. Can.*, **78**: 1–124.

BERRY, R. E. 1975. Redbacked cutworm: flight period and egg development under field conditions. *Environmental Ent.*, **4**: 603–605.

BOURSIN, C. 1964. Noctuidae Trifinae. Zweiter Beitrag zur Kenntnis der Fauna der Noctuidae von Nepal (Beiträge zur Kenntnis der "Noctuidae Trifinae" 146). Lepidoptera der deutschen Expedition 1955. Teil II. *Veröffentlichungen der Zoologischen Staatssamlung München*, **8**: 3–40.

BRETHERTON, R. F., B. GOATER, and R. I. LORIMER. 1979. Noctuidae: Noctuinae and Hadeninae. *In* HEATH, J. and A. M. EMMET, editors. *The Moths and Butterflies of Great Britain and Ireland*. **9**: 120–278, figs. 11a–19, maps 87–203, pl. 7–13. Curwen Books, London.

BUCKETT, J. S. 1964. Identity of the moth, *Euxoa wilsoni* (Lepidoptera: Noctuidae). *Pan-Pacific Ent.*, **40**: 104–107.

BYERS, J. R., and C. F. HINKS. 1978. Biosystematics of the genus *Euxoa* (Lepidoptera: Noctuidae) XI. Mating discrimination between three closely related species of the *declarata* group. *Canadian Jour. Zool.*, **56**: 1981–1987.

BYERS, J. R., C. F. HINKS, and J. D. LAFONTAINE. 1975. Biosystematics of the genus *Euxoa* (Lepidoptera: Noctuidae) II. A description of the immature stages of *Euxoa basalis* and a redescription of the adult. *Can. Ent.*, **107**: 1083–1094.

BYERS, J. R., E. W. UNDERHILL, W. F. STECK, M. D. CHISHOLM, and P. E. A. TEAL. 1981. Biosystematics of the genus *Euxoa* (Lepidoptera: Noctuidae) XV. Sex pheromone cross attractancy among the three closely related species in the *declarata* group. *Can. Ent.*, **113**: 235–243.

BYERS, J. R., and J. D. LAFONTAINE. 1982. Biosystematics of the genus *Euxoa* (Lepidoptera: Noctuidae) XVI. Comparative biology and experimental taxonomy of four subspecies of *Euxoa comosa*. *Can. Ent.*, **114**: 551–565.

BYERS, J. R., D. L. STRUBLE, and J. D. LAFONTAINE. 1985. Biosystematics of the genus *Euxoa* (Lepidoptera: Noctuidae) XVIII. Comparative biology and experimental taxonomy of the sibling species *Euxoa ridingsiana* (Grt.) and *Euxoa maimes* (Sm.). *Can. Ent.*, **117**: 481–493.

COOK, W. C. 1930. An ecologically annotated list of the Phalaenidae of Montana (Lepidoptera). *Can. Ent.*, **62**: 257–277.

CRUMB, S. E. 1929. Tobacco cutworms. *U. S. Dept. Agric. Tech. Bull.*, **88**: 1–179.

CRUMB, S. E. 1932. The more important climbing cutworms. *Bull. Brooklyn Ent. Soc.*, **27**: 73–100.

CRUMB, S. E. 1956. The larvae of the Phalaenidae. *U. S. Dept. Agric. Tech. Bull.*, **1135**: 1–356, pl. 1–11.

FONTANA, P. G. 1976. Improved resolution of the meiotic chromosomes in both sexes of *Euxoa* species and their hybrids (Lepidoptera: Noctuidae). *Canadian Jour. Genetics and Cytology*, **18**: 537–544.

FORBES, W. T. M. 1934. A grouping of Agrotine genera. *Ent. Americana*, **14** (n.s.): 1–39.

FORBES, W. T. M. 1954. Lepidoptera of New York and neighboring states. Part 3 Noctuidae. *Cornell Univ. Agric. Experiment Station, Mem.*, **329**: 1–433, figs. 1–290.

GROTE, A. R. 1874. Preliminary catalogue of the Noctuidae of California. Pt. 2. *Can. Ent.*, **6**: 154–157.

GUENÉE, A. 1852. *Histoire Naturelle des Insectes. Species général de Lépidoptères*. 5, *Noctuélites*, **1**: 1–407.

HAMPSON, G. F. 1903. *Catalogue of the Lepidoptera Phalaenae in the British Museum*, **4**: 1–689, pl. 55–77.

HARDWICK, D. F. 1950. Preparation of slide mounts of lepidopterous genitalia. *Can. Ent.*, **82**: 231–235.

HARDWICK, D. F. 1965. A synopsis of the *ridingsiana* group of the genus *Euxoa* (Lepidoptera: Noctuidae) with a description of a new species. *Can. Ent.*, **97**: 1221–1226.

HARDWICK, D. F. 1970. The genus *Euxoa* (Lepidoptera: Noctuidae) in North America 1. Subgenera *Orosagrotis, Longivesica, Chorizagrotis, Pleonectopoda* and *Crassivesica. Mem. Ent. Soc. Can.*, **67**: 1–177, figs. 1–326.

HARDWICK, D. F. 1971. The "phenological date" as an indicator of the flight period of noctuid moths. *Can. Ent.*, **103**: 1207–1216.

HARDWICK, D. F. 1973. A synopsis of the *albipennis-lillooet-declarata* series of the genus *Euxoa* (Lepidoptera: Noctuidae) with descriptions of a new species and a new subspecies. *Can. Ent.*, **105**: 493–500.

HARDWICK, D. F., and L. P. LEFKOVITCH. 1973. Species separation in the *declarata* group of the genus *Euxoa*, a computer analysis based on structural characters. *Can. Ent.*, **105**: 501–508.

HARVEY, L. F. 1876. New Californian and Texas moths. *Can. Ent.*, **8**: 35–38.

HINKS, C. F., and J. R. BYERS. 1976. Biosystematics of the genus *Euxoa* (Lepidoptera: Noctuidae) V. Rearing procedures and life cycles of 36 species. *Can. Ent.*, **108**: 1345–1357.

HUDSON, A., and P. Y. JUI. 1976. Biosystematics of the genus *Euxoa* (Lepidoptera: Noctuidae) VI. Esterase isozymes of natural populations of *Euxoa declarata* and *Euxoa campestris. Can. Ent.*, **108**: 1359–1370.

JACOBSON, L. A. 1969. Biology of *Euxoa tristicula* in the field and laboratory. *Jour. Econ. Ent.*, **62**: 1053–1054.

KOVÁCS, L., and Z. VARGA. 1973. Ergebnisse der zoologischen Forschungen von Dr. Z. Kaszab in der Mongolei. 316. Noctuidae: Noctuinae (Lepidoptera). *Folia Entomologica Hungarica* (new series), **26**: 287–343.

KOZHANCHIKOV, I. 1929. Zur Kenntnis der Agrotiden (Lepidoptera, Noctuidae). 1. Übersicht der Gattung *Euxoa* Hb. *Annuaire du Musée Zoologique de l' Academie des Sciences de l' U. R. S. S.*, **30**: 141–216, pl. 12–26.

KOZHANCHIKOV, I. 1937. Fam. Noctuidae (Subfam Agrotinae). *Faune de l' U. R. S. S. Insectes Lepidoptères,* **XIII**, No. 3 (New Series 15): 1–690, figs. 1–306, pl. 1–13. Edition de L' Academie des Sciences de l' U. R. S. S. Moscow-Leningrad.

LAFONTAINE, J. D. 1974. The *punctigera* group of the genus *Euxoa* (Lepidoptera: Noctuidae) with descriptions of two new species. *Can. Ent.*, **106**: 1233–1240.

LAFONTAINE, J. D. 1975. The *mimallonis* group of the genus *Euxoa* Hbn. (Lepidoptera: Noctuidae) with descriptions of three new species. *Can. Ent.*, **107**: 155–165.

LAFONTAINE, J. D. 1976a. Two new species of *Euxoa* (Lepidoptera: Noctuidae) from southern United States allied to *Euxoa camalpa. Can. Ent.*, **108**: 663–668.

LAFONTAINE, J. D. 1976b. A synopsis of the *aequalis* group of the genus *Euxoa* Hbn. (Lepidoptera: Noctuidae) with a description of a new species from southwestern United States. *Can. Ent.*, **108**: 741–750.

LAFONTAINE, J. D. 1976c. A new species of *Euxoa* (Lepidoptera: Noctuidae) allied to *Euxoa olivia* and *Euxoa septentrionalis. Can. Ent.*, **108**: 1275–1280.

LAFONTAINE, J. D. 1981. Classification and phylogeny of the *Euxoa detersa* group (Lepidoptera: Noctuidae). *Quaestiones Entomologicae*, **17**: 1–120.

LAFONTAINE, J. D. 1982a. Biogeography of the genus *Euxoa* (Lepidoptera: Noctuidae) in North America. *Can. Ent.*, **114**: 1–53.

LAFONTAINE, J. D. 1982b. *Agrotis redimicula* Morrison, 1875 (Insecta: Lepidoptera): proposed conservation from 1874. Z.N.(S.) 2305. *The Bull. of Zool. Nomenclature*, **39**: 54–56.

LAFONTAINE, J. D., and J. R. BYERS. 1982. A revision of the *comosa* group of the genus *Euxoa* Hbn. (Lepidoptera: Noctuidae), with descriptions of two new species. *Can. Ent.*, **114**: 575–589.

MCDUNNOUGH, J. 1923. New Canadian Lepidoptera. *Can. Ent.*, **55**: 163–168.

MCDUNNOUGH, J. 1950. Species of *Euxoa* of eastern North America, with particular reference to genitalic characters (Lepidoptera, Phalaenidae). *Bull. Amer. Mus. Nat. Hist.*, **95**: 359–408.

MORRISON, H. K. 1875. Notes on the Noctuidae with descriptions of certain new species. *Proc. Acad. Nat. Sci., Philadelphia*, **27**: 55–71.

NORDSTRÖM, F. 1946. *Euxoa norvegica* Stgr., dess stallning till närstående arter samt uppträdande i Fennoskandia. *Ent. Tidskr.*, **67**: 184–194.

PRUESS, K. P. 1967. Migration of the army cutworm, *Chorizagrotis auxiliaris* (Lepidoptera: Noctuidae). 1. Evidence for a migration. *Ann. Ent. Soc. Amer.*, **60**: 910–920.

RINGS, R. W., and B. A. JOHNSON. 1976. An annotated bibliography of the sandhill cutworm *Euxoa de-*

tersa (Walker). *Ohio Agric. Research and Development Center, Research Circular* **215**: 1–16.

SALKELD, E. H. 1975. Biosystematics of the genus *Euxoa* (Lepidoptera: Noctuidae) IV. Eggs of the genus *Euxoa* Hbn. *Can. Ent.*, **107**: 1137–1152.

SALKELD, E. H. 1976. Biosystematics of the genus *Euxoa* (Lepidoptera: Noctuidae) VII. Eggs of the subgenera *Chorizagrotis, Crassivesica, Longivesica, Orosagrotis,* and *Pleonectopoda. Can. Ent.*, **108**: 1371–1385.

SALKELD, E. H. 1977. Biosystematics of the genus *Euxoa* (Lepidoptera: Noctuidae) IX. Morphological and electrophoretic characteristics of the eggs of the "*declarata* group." *Can. Ent.*, **109**: 1137–1144.

SALKELD, E. H. 1984. A catalogue of the eggs of some Canadian Noctuidae (Lepidoptera). *Mem. Ent. Soc. Can.*, **127**: 1–167, pl. 1–124.

SMITH, J. B. 1890. Contribution toward a monograph of the insects of the lepidopterous family Noctuidae of temperate North America.—Revision of the species of the genus *Agrotis. Bull. U. S. Natl. Mus.*, **38**: 1–237.

SMITH, J. B. 1893. Catalogue of the lepidopterous superfamily Noctuidae found in boreal America. *Bull. U.S. Natl. Mus.*, **44**: 1–424.

SMITH, J. B. 1903. New noctuids for 1903, no. 4, with notes on certain described species. *Trans. Amer. Ent. Soc.*, **29**: 191–224.

TAMAKI, G., H. R. MOFFITT, and J. E. TURNER. 1975. The influence of perennial weeds on the abundance of the redbacked cutworm on Asparagus. *Environmental Ent.*, **4**: 274–276.

TEAL, P. E. A., and J. R. BYERS. 1980. Biosystematics of the genus *Euxoa* (Lepidoptera: Noctuidae) XIV. Effect of temperature on calling behavior and temporal partitioning in three sibling species of the *declarata* group. *Can. Ent.*, **112**: 113–117.

TEAL, P. E. A., J. R. BYERS, and B. J. R. PHILOGENE. 1978. Differences in female calling behavior of three interfertile sibling species of *Euxoa* (Lepidoptera: Noctuidae). *Ann. Ent. Soc. Amer.*, **71**: 630–634.

TODD, E. L. 1968. Selection of lectotypes for some species of *Euxoa* Hübner described by J. B. Smith (Lepidoptera: Noctuidae). *Proc. Ent. Soc. Washington,* **70**: 263–280.

TODD, E. L. 1982. The noctuid type material of John B. Smith (Lepidoptera). *U. S. Dept. Agric. Tech. Bull.*, **1645**: 1–228.

UNDERHILL, E. W., W. F. STECK, J. R. BYERS, and M. D. CHISHOLM. 1981. Biosystematics of the genus *Euxoa* (Lepidoptera: Noctuidae) XVI. (Z)-5-Decenyl acetate, a sex attractant for three closely related species, *Euxoa declarata, Euxoa campestris,* and *Euxoa rockburnei. Can. Ent.*, **113**: 245–249.

WALKDEN, H. H. 1950. Cutworms, armyworms and related species attacking cereal and forage crops in the central Great Plains. *U. S. Dept. Agric. Circular* **849**: 1–52.

WALKER, F. 1856 [1857]. *List of the Specimens of Lepidopterous Insects in the Collection of the British Museum,* **10**: 253–491. London.

WARREN, W. 1909–1914. Die palaearktischen eulenartigen nachtfalter. Noctuidae. *In* SEITZ, A. *Die Gross-Schmetterlinge der Erde,* **3**: 9–511, pl. 1–75.

WOLFF, N. L. 1964. The Lepidoptera of Greenland. *Meddelelser om Grønland,* **159** (11): 1–74, figs. 1–55, pl. 1–21.

MONOCHROME PLATES

NOTE—Male genitalia are shown with the genital capsule to the left and the aedoeagus, with vesica everted, to the right.

PLATE A: MALE GENITALIA OF *EUXOA* SPECIES

1. *Euxoa lidia thanatologia* (Dyar); Bicknell, Utah; CNC 6915. (p. 30).
2. *Euxoa auxiliaris* (Grote); Boulder, Colorado; CNC 3409. (p. 30).
3. *Euxoa inconcinna* (Harvey); Big Bend National Park, Texas; CNC 3323. (p. 31).
4. *Euxoa terrealis* (Grote); Apache Co., Arizona; CNC 3291. (p. 31).
5. *Euxoa shasta shasta* Lafontaine, paratype; Mt. Shasta, California; CNC 3817. (p. 32).
6. *Euxoa biformata* Smith; Mt. Shasta, California; CNC 10483. (p. 33).
7. *Euxoa intermontana* Lafontaine, paratype; Suttle Lake, Oregon; CNC 10486. (p. 33).
8. *Euxoa mimallonis mimallonis* (Grote); Scout Lake, Saskatchewan; CNC 6123. (p. 34).
9. *Euxoa vernalis* Lafontaine, paratype; Parks Co., Arizona; CNC 10547. (p. 35).
10. *Euxoa septentrionalis* (Walker); Toppenish, Washington; CNC 3671. (p. 36).

PLATE B: MALE GENITALIA OF *EUXOA* SPECIES

1. *Euxoa olivia* (Morrison); Frenchglen, Oregon; CNC 1468. (p. 36).
2. *Euxoa messoria* (Harris); Lac La Hache, British Columbia; CNC 2178. (p. 37).
3. *Euxoa divergens* (Walker); Buena Vista, Colorado; CNC 6693. (p. 39).
4. *Euxoa edictalis* (Smith); Toppenish, Washington; CNC 3374. (p. 40).
5. *Euxoa westermanni* (Staudinger); Banff, Alberta; CNC 610. (p. 44).
6. *Euxoa churchillensis churchillensis* (McDunnough); Eskimo Point, Northwest Territories; CNC 2146. (p. 45).
7. *Euxoa dissona* (Möschler); Mt. Washington, New Hampshire; CNC 8141. (p. 46).
8. *Euxoa hyperborea* Lafontaine, paratype; Colville River, Alaska; CNC 7796. (p. 46).

PLATE C: MALE GENITALIA OF *EUXOA* SPECIES

1. *Euxoa quebecensis* (Smith); Forestville, Quebec; CNC 1278. (p. 47).
2. *Euxoa scandens* (Riley); Port Colborne, Ontario; CNC 1261. (p. 47).
3. *Euxoa aurulenta* (Smith); Calgary, Alberta; CNC 841. (p. 48).
4. *Euxoa vallus bivittata* Lafontaine, paratype; Lee Vining, California; CNC 7979. (p. 49).
5. *Euxoa macleani macleani* McDunnough; Slate Peak, Okanogan Co., Washington; CNC 2137. (p. 50).
6. *Euxoa trifasciata* (Smith); Mt. Shasta, California; CNC 3856. (p. 50).
7. *Euxoa lewisi lewisi* (Grote); Banff, Alberta; CNC 573. (p. 51).
8. *Euxoa altens* McDunnough; Mt. Shasta, California; CNC 3625. (p. 51).
9. *Euxoa cryptica* Hardwick; Bishop, Inyo Co., California; CNC 3924. (p. 52).
10. *Euxoa leuschneri* Lafontaine, paratype; Barton Flats, San Bernardino Co., California; CNC 9300. (p. 52).

PLATE D: MALE GENITALIA OF *EUXOA* SPECIES

1. *Euxoa extranea* (Smith); Mt. Shasta, California; CNC 2545. (p. 53).
2. *Euxoa tristicula* (Morrison); Debert, Nova Scotia; CNC 3381. (p. 54).
3. *Euxoa vetusta* (Walker); Cathlamet, Washington; CNC 3016. (p. 54).
4. *Euxoa fuscigera* (Harvey); Mojave, California; CNC 3488. (p. 55).
5. *Euxoa atomaris* (Smith); Keremeos, British Columbia; CNC 3167. (p. 55).
6. *Euxoa pleuritica* (Grote); Attons Lake, Saskatchewan; CNC 3175. (p. 56).
7. *Euxoa pestula* Smith; Milo, Alberta; CNC 3260. (p. 57).
8. *Euxoa simona* McDunnough; Mt. Evans, Colorado; CNC 3201. (p. 58).
9. *Euxoa hardwicki* Lafontaine, paratype; Walla Walla, Washington; CNC 6084. (p. 63).
10. *Euxoa camalpa* (Dyar); White Mts., Arizona; CNC 10602. (p. 64).

PLATE E: MALE GENITALIA OF *EUXOA* SPECIES

1. *Euxoa serotina* Lafontaine, paratype; Welder Wildlife Refuge, Sinton, Texas; CNC 10561. (p. 65).
2. *Euxoa bifasciata* (Smith); Mt. Lowe, California; CNC 3113. (p. 67).
3. *Euxoa bostoniensis* (Grote); Chatham, Ontario; CNC 6515. (p. 66).
4. *Euxoa medialis* (Smith); Scott City, Kansas; CNC 6305. (p. 66).
5. *Euxoa perexcellens* (Grote); Hoodsport, Washington; CNC 6201. (p. 67).
6. *Euxoa sculptilis* (Harvey); Portal, Arizona; CNC 4986. (p. 67).
7. *Euxoa ustulata* Lafontaine, holotype; Cedarville, California; CNC 10548. (p. 68).
8. *Euxoa rufula basiflava* (Smith); Mt. Shasta, California; CNC 10632. (p. 69).
9. *Euxoa intrita* (Morrison); Craig, Colorado; CNC 6263. (p. 69).
10. *Euxoa annulipes annulipes* (Smith); Eureka, Utah; CNC 3080. (p. 70).

PLATE F: MALE GENITALIA OF *EUXOA* SPECIES

1. *Euxoa scholastica* McDunnough; Niagara Falls, Ontario; CNC 890. (p. 71).
2. *Euxoa terrena* (Smith); Golden, British Columbia; CNC 655. (p. 71).
3. *Euxoa antica* Lafontaine, paratype; Eureka, Utah; CNC 10403. (p. 72).
4. *Euxoa franclemonti* Lafontaine, paratype; Prescott, Arizona; CNC 7999. (p. 72).
5. *Euxoa absona* Lafontaine, holotype; Angel Creek, SSW of Wells, Elko Co., Nevada; USNM 10655. (p. 73).
6. *Euxoa scotogrammoides* McDunnough; Oliver, British Columbia; CNC 3727. (p. 74).
7. *Euxoa serricornis* (Smith); Oracle Junction, Arizona; CNC 6958. (p. 74).
8. *Euxoa tocoyae* (Smith); Sonoma Co., California; CNC 6956. (p. 74).
9. *Euxoa pluralis* (Grote); Lund, Nevada; CNC 3781. (p. 75).
10. *Euxoa cinnabarina* Barnes and McDunnough; Upper Santa Ana River, San Bernardino Co., California; CNC 563. (p. 75).

PLATE G: MALE GENITALIA OF *EUXOA* SPECIES

1. *Euxoa permixta* McDunnough; Mayfield, Utah; CNC 6914. (p. 76).
2. *Euxoa setonia* McDunnough; State Bridge, Colorado; CNC 6939. (p. 76).
3. *Euxoa pallidimacula* Lafontaine, paratype; Stockton, Utah; CNC 6938. (p. 77).
4. *Euxoa mojave* Lafontaine, paratype; Apple Valley, California; CNC 6942. (p. 77).
5. *Euxoa declarata* (Walker); Cedarville, California; CNC 6583. (p. 79).
6. *Euxoa campestris* (Grote); Lac La Hache, British Columbia; CNC 6644. (p. 79).
7. *Euxoa rockburnei* Hardwick, paratype; Hamilton, Montana; CNC 6592. (p. 80).
8. *Euxoa flavidens* (Smith); Taos, New Mexico; CNC 6210. (p. 80).
9. *Euxoa silens* (Grote); Lee Vining, California; CNC 6799. (p. 81).
10. *Euxoa pimensis* Barnes and McDunnough; Prescott, Arizona; CNC 6971. (p. 81).

PLATE H: MALE GENITALIA OF *EUXOA* SPECIES

1. *Euxoa immixta* (Grote); St. Paul, Minnesota; CNC 6972. (p. 81).
2. *Euxoa simulata* McDunnough; Frijoles Canyon, New Mexico; CNC 10477. (p. 82).
3. *Euxoa punctigera* (Walker); Lyle, Washington; CNC 10456. (p. 83).
4. *Euxoa cana* Lafontaine, paratype; Grasmere, Idaho; CNC 10436. (p. 84).
5. *Euxoa aurantiaca* Lafontaine, paratype; Alcova, Wyoming; CNC 10509. (p. 84)
6. *Euxoa spumata* McDunnough; Wolf Creek, Montana; CNC 10445. (p. 85).
7. *Euxoa pallipennis* (Smith); Modoc Co., California; CNC 4478. (p. 86).
8. *Euxoa baja* Lafontaine, holotype; San Diego Co., California; USNM 10648. (p. 86).
9. *Euxoa tessellata* (Harris); Kenworth, California; CNC 10553. (p. 87).
10. *Euxoa plagigera* (Morrison); Milk River, Alberta; CNC 6114. (p. 89).

PLATE I: MALE GENITALIA OF *EUXOA* SPECIES

1. *Euxoa albipennis* (Grote); Miles City, Montana; CNC 923. (p. 90).
2. *Euxoa henrietta* (Smith); Ripon, California; CNC 6226. (p. 91).
3. *Euxoa cincta* Barnes and Benjamin; Paradise, Arizona; CNC 1954. (p. 92).
4. *Euxoa coconino* Lafontaine, holotype; Kaibab National Forest, Coconino Co., Arizona; CNC 10315. (p. 92).
5. *Euxoa hollemani* (Grote); Grantsville, Utah; CNC 6486. (p. 93).
6. *Euxoa subandera* Lafontaine, paratype; Preston, Nevada; CNC 6872. (p. 94).
7. *Euxoa xasta* Barnes and McDunnough; Lincoln Co., Nevada; California Dept. Agric. slide. (p. 94).
8. *Euxoa catenula* (Grote); Dixon, New Mexico; CNC 4915. (p. 95).
9. *Euxoa comosa comosa* (Morrison); Apache Co., Arizona; CNC 3731. (p. 99).
10. *Euxoa lucida* Barnes and McDunnough; Logan, Utah; CNC 1811. (p. 101).

PLATE J: MALE GENITALIA OF *EUXOA* SPECIES

1. *Euxoa fumalis* (Grote); Ithaca, New York; CNC 1789. (p. 101).
2. *Euxoa occidentalis* Lafontaine, paratype; Wrightwood, California; CNC 6913. (p. 102).
3. *Euxoa velleripennis* (Grote); Lac Mondor, Quebec; CNC 1748. (p. 102).
4. *Euxoa infausta* (Walker); Manyberries, Alberta; CNC 855. (p. 104).
5. *Euxoa satis* (Harvey); Manning Provincial Park, British Columbia; CNC 1298. (p. 105).
6. *Euxoa selenis* (Smith); Riverside Co., California; CNC 3128. (p. 108).
7. *Euxoa piniae* Buckett and Bauer; Almanor, California; CNC 3754. (p. 108).
8. *Euxoa satiens* (Smith); Oliver, British Columbia; CNC 830. (p. 108).
9. *Euxoa violaris* (Grote and Robinson); Lakehurst, New Jersey; CNC 3013. (p. 109).
10. *Euxoa cursoria wirima* Hardwick; Fort Smith, Northwest Territories; CNC 2099. (p. 111).

PLATE K: MALE GENITALIA OF *EUXOA* SPECIES

1. *Euxoa ochrogaster ochrogaster* (Guenée); Manyberries, Alberta; CNC 2092. (p. 113).
2. *Euxoa nostra* (Smith); Logan, Utah; CNC 1932. (p. 113).
3. *Euxoa siccata* (Smith); Scott City, Kansas; CNC 6304. (p. 114).
4. *Euxoa choris* (Harvey); Lee Vining, California; CNC 6542. (p. 114).
5. *Euxoa obeliscoides* (Guenée); Mount Laguna, California; CNC 4940. (p. 115).
6. *Euxoa oberfoelli* Hardwick, paratype; Sand Springs, Montana; CNC 4948. (p. 116).

7. *Euxoa lillooet* McDunnough; Walla Walla, Washington; CNC 4828. (p. 116).
8. *Euxoa basalis* (Grote); Wolf Creek, Montana; CNC 6499. (p. 117).
9. *Euxoa costata* (Grote); Greenville, California; CNC 10651. (p. 123).
10. *Euxoa brevipennis* (Smith); Tempiute, Nevada; CNC 6533. (p. 125).

PLATE L: MALE GENITALIA OF *EUXOA* SPECIES

1. *Euxoa servita* (Smith); Debert, Nova Scotia; CNC 10215. (p. 125).
2. *Euxoa redimicula* (Morrison); Montreal, Quebec; CNC 10229. (p. 126).
3. *Euxoa auripennis auripennis* Lafontaine; Princeton, British Columbia; CNC 10222. (p. 127).
4. *Euxoa olivalis* (Grote); Malta, Montana; CNC 6041. (p. 127).
5. *Euxoa oblongistigma* (Smith); Seneca, Oregon; CNC 6044. (p. 128).
6. *Euxoa teleboa* (Smith); Lehi, Utah; CNC 1828. (p. 130).
7. *Euxoa citricolor* (Grote); Lyman, Wyoming; CNC 6352. (p. 129).
8. *Euxoa tronella* (Smith); Theodore Roosevelt National Park, North Dakota; CNC 10642. (p. 129).
9. *Euxoa difformis* (Smith); Oliver, British Columbia; CNC 6405. (p. 130).
10. *Euxoa moerens* (Grote); Gibbs, Nevada; CNC 6154. (p. 131).

PLATE M: MALE GENITALIA OF *EUXOA* SPECIES

1. *Euxoa latro* (Barnes and Benjamin); Chiloquin, Oregon; CNC 6290. (p. 131).
2. *Euxoa murdocki* (Smith); Burns, Oregon; CNC 6144. (p. 132).
3. *Euxoa dodi* McDunnough; Scout Lake, Saskatchewan; CNC 6422. (p. 132).
4. *Euxoa infracta* (Morrison); Ruth, Nevada; CNC 6244. (p. 132).
5. *Euxoa laetificans* (Smith); Lund, Nevada; CNC 6271. (p. 133).
6. *Euxoa quadridentata flutea* Smith; Bend, Oregon; CNC 6995. (p. 134).
7. *Euxoa inscripta* Lafontaine, paratype; Kemmerer, Wyoming; CNC 6886. (p. 134).
8. *Euxoa unica* McDunnough; Saskatoon, Saskatchewan; CNC 6815. (p. 135).
9. *Euxoa niveilinea* (Grote); Prescott, Arizona; CNC 10649. (p. 135).
10. *Euxoa dargo* (Strecker); Kamloops, British Columbia; CNC 6381. (p. 135).

PLATE N: MALE GENITALIA OF *EUXOA* SPECIES

1. *Euxoa melura* McDunnough; State Bridge, Colorado; CNC 6091. (p. 136).
2. *Euxoa detersa personata* (Morrison); Valentine, Nebraska; CNC 6396. (p. 137).
3. *Euxoa aequalis acornis* (Smith); Calgary, Alberta; CNC 1753. (p. 140).
4. *Euxoa conjuncta* (Smith); Capulin Canyon National Monument, New Mexico; CNC 10568. (p. 141).
5. *Euxoa cona* (Strecker); McDermitt, Nevada; CNC 4542. (p. 142).
6. *Euxoa munis* (Grote); McGill, Nevada; CNC 3737. (p. 142).
7. *Euxoa misturata* (Smith); Jemez Springs, New Mexico; CNC 4493. (p. 143).
8. *Euxoa cinereopallida* (Smith); Follyfarm, Oregon; CNC 6335. (p. 145).
9. *Euxoa atristrigata* (Smith); Helena, Montana; CNC 4545. (p. 145).
10. *Euxoa nevada* (Smith); Eureka, Utah; CNC 4560. (p. 145).

PLATE O: MALE AND FEMALE GENITALIA OF *EUXOA* SPECIES

1. *Euxoa mitis* (Smith), ♂; Burns, Oregon; CNC 4508. (p. 146).
2. *Euxoa luctuosa* Lafontaine, paratype, ♂; Kayenta, Arizona; CNC 10589. (p. 147).
3. *Euxoa nomas incognita* (Smith), ♂; Firth River, British Mountains, Yukon; CNC 3355. (p. 149).
4. *Euxoa ridingsiana* (Grote), ♂; Lethbridge, Alberta; CNC 8065. (p. 154).
5. *Euxoa wilsoni* (Grote), ♂; Neah Bay, Washington; CNC 3979. (p. 154).
6. *Euxoa riversii* (Dyar), ♂; San Miguel Island, California; CNC 3980. (p. 155).
7. *Euxoa nomas incognita* (Smith), ♀; Mt. McLean, British Columbia; CNC 3365. (p. 149).
8. *Euxoa ridingsiana* (Grote), ♀; Seneca, Oregon; CNC 9324. (p. 154).
9. *Euxoa wilsoni* (Grote), ♀; Neah Bay, Washington; CNC 3972. (p. 154).
10. *Euxoa riversii* (Dyar), ♀; San Diego, California; CNC 3965. (p. 155).

PLATE P: FEMALE GENITALIA OF *EUXOA* SPECIES

1. *Euxoa lidia thanatologia* (Dyar); Lethbridge, Alberta; CNC 10533. (p. 30).
2. *Euxoa auxiliaris* (Grote); Mt. Evans, Colorado; CNC 3297. (p. 30).
3. *Euxoa inconcinna* (Harvey); Apache Co., Arizona; CNC 3462. (p. 31).
4. *Euxoa terrealis* (Grote); McKinley Co., New Mexico; CNC 3438. (p. 31).
5. *Euxoa biformata* Smith; Mt. Shasta, California; CNC 10516. (p. 33).
6. *Euxoa vernalis* Lafontaine, paratype; Durango, Mexico; CNC 10542. (p. 35).
7. *Euxoa septentrionalis* (Walker); Eureka, Utah; CNC 10550. (p. 36).
8. *Euxoa olivia* (Morrison); Frenchglen, Oregon; CNC 1448. (p. 36).

PLATE Q: FEMALE GENITALIA OF *EUXOA* SPECIES

1. *Euxoa messoria* (Harris); Keremeos, British Columbia; CNC 2150. (p. 37).
2. *Euxoa divergens* (Walker); Brandon, Manitoba; CNC 1252. (p. 39).

3. *Euxoa edictalis* (Smith); Glenwood Springs, Colorado; CNC 3454. (p. 40).
4. *Euxoa westermanni* (Staudinger); Banff, Alberta; CNC 607. (p. 44).
5. *Euxoa churchillensis churchillensis* (McDunnough); Chesterfield, Northwest Territories; CNC 1633. (p. 45).
6. *Euxoa dissona* (Möschler); Hopedale, Labrador; CNC 1640. (p. 46).
7. *Euxoa hyperborea* Lafontaine, paratype; Kulu, Magadanskaya Oblast', U.S.S.R.; slide in Far East Science Center, Vladivostok. (p. 46).
8. *Euxoa quebecensis* (Smith); Thunder Bay, Ontario; CNC 1636. (p. 47).

PLATE R: FEMALE GENITALIA OF *EUXOA* SPECIES

1. *Euxoa scandens* (Riley); Brandon, Manitoba; CNC 1217. (p. 47).
2. *Euxoa aurulenta* (Smith); Aweme, Manitoba; CNC 3026. (p. 48).
3. *Euxoa vallus bivittata* Lafontaine, paratype; Lee Vining, California; CNC 7977. (p. 49).
4. *Euxoa macleani macleani* McDunnough, holotype; Mt. McLean, British Columbia; CNC 138. (p. 50).
5. *Euxoa trifasciata* (Smith); Mt. Shasta, California; CNC 3553. (p. 50).
6. *Euxoa lewisi lewisi* (Grote); Nordegg, Alberta; CNC 1234. (p. 51).
7. *Euxoa altens* McDunnough; Hamilton, Montana; CNC 1772. (p. 51).
8. *Euxoa cryptica* Hardwick; Mt. Shasta, California; CNC 6935. (p. 52).

PLATE S: FEMALE GENITALIA OF *EUXOA* SPECIES

1. *Euxoa leuschneri* Lafontaine, paratype; Barton Flats, San Bernardino Co., California; CNC 9301. (p. 52).
2. *Euxoa extranea* (Smith); Mt. Shasta, California; CNC 3592. (p. 53).
3. *Euxoa tristicula* (Morrison); Brandon, Manitoba; CNC 3391. (p. 54).
4. *Euxoa vetusta* (Walker); Glacier, British Columbia; CNC 3441. (p. 54).
5. *Euxoa fuscigera* (Harvey); Soledad, California; CNC 3498. (p. 55).
6. *Euxoa atomaris* (Smith); Garfield, Washington; CNC 3216. (p. 55).
7. *Euxoa pleuritica* (Grote); Richland, Washington; CNC 3233. (p. 56).
8. *Euxoa simona* McDunnough; Lee Vining, California; CNC 9325. (p. 58).

PLATE T: FEMALE GENITALIA OF *EUXOA* SPECIES

1. *Euxoa hardwicki* Lafontaine, paratype; Prosser, Washington; CNC 6095. (p. 63).
2. *Euxoa camalpa* (Dyar); Alpine, Texas; CNC 4563. (p. 64).
3. *Euxoa serotina* Lafontaine, paratype; Welder Wildlife Refuge, Sinton, Texas; CNC 10599. (p. 65).
4. *Euxoa bifasciata* (Smith); Mt. Lowe, California; CNC 10624. (p. 67).
5. *Euxoa bostoniensis* (Grote); Wareham, Massachusetts; CNC 10681. (p. 66).
6. *Euxoa perexcellens* (Grote); Leavenworth, Washington; CNC 10594. (p. 67).
7. *Euxoa sculptilis* (Harvey); Prescott, Arizona; CNC 10683. (p. 67).
8. *Euxoa ustulata* Lafontaine, paratype; Plumas Co., California; CNC 1319. (p. 68).

PLATE U: FEMALE GENITALIA OF *EUXOA* SPECIES

1. *Euxoa rufula basiflava* (Smith); Mt. Shasta, California; CNC 6904. (p. 69).
2. *Euxoa intrita* (Morrison); Kalispell, Montana; CNC 1901. (p. 69).
3. *Euxoa annulipes annulipes* (Smith); Eureka, Utah; CNC 1820. (p. 70).
4. *Euxoa scholastica* McDunnough; Mt. Pocono, Pennsylvania; CNC 10685. (p. 71).
5. *Euxoa terrena* (Smith); Cedar Pass, California; CNC 10415. (p. 71).
6. *Euxoa antica* Lafontaine, paratype; Boulder, Colorado; CNC 10409. (p. 72).
7. *Euxoa scotogrammoides* McDunnough, paratype; Hamilton, Montana; CNC 10622. (p. 74).
8. *Euxoa serricornis* (Smith); Apple Valley, California; CNC 47. (p. 74).

PLATE V: FEMALE GENITALIA OF *EUXOA* SPECIES

1. *Euxoa pluralis* (Grote); Eureka, Utah; CNC 1317. (p. 75).
2. *Euxoa cinnabarina* Barnes and McDunnough; Buckhorn Flats, Los Angeles Co., California; UCB. (p. 75).
3. *Euxoa permixta* McDunnough, allotype; Eureka, Utah; CNC 10623. (p. 76).
4. *Euxoa setonia* McDunnough; Drumheller, Alberta; CNC 485. (p. 76).
5. *Euxoa pallidimacula* Lafontaine, paratype; Stockton, Utah; CNC 6840. (p. 77).
6. *Euxoa mojave* Lafontaine, paratype; Apple Valley, California; CNC 6941. (p. 77).
7. *Euxoa declarata* (Walker); Cartwright, Manitoba; CNC 6746. (p. 79).
8. *Euxoa rockburnei* Hardwick, paratype; Oliver, British Columbia; CNC 6684. (p. 80).

PLATE W: FEMALE GENITALIA OF *EUXOA* SPECIES

1. *Euxoa flavidens* (Smith); White Mts., Arizona; CNC 10677. (p. 80).
2. *Euxoa silens* (Grote); Lee Vining, California; CNC 6803. (p. 81).
3. *Euxoa simulata* McDunnough; Lee Vining, California; CNC 10473. (p. 82).
4. *Euxoa punctigera* (Walker); Mount Laguna, California; CNC 10488. (p. 83).
5. *Euxoa aurantiaca* Lafontaine, paratype; Alcova, Wyoming; CNC 10510. (p. 84).
6. *Euxoa spumata* McDunnough; Three Forks, Montana; CNC 10431. (p. 85).
7. *Euxoa pallipennis* (Smith); Swift Current, Saskatchewan; CNC 4482. (p. 86).
8. *Euxoa tessellata* (Harris); Radium Hot Springs, British Columbia; CNC 727. (p. 87).

PLATE X: FEMALE GENITALIA OF *EUXOA* SPECIES

1. *Euxoa plagigera* (Morrison); Joliet, Montana; CNC 6124. (p. 89).
2. *Euxoa albipennis* (Grote); Lloydminster, Alberta; CNC 1836. (p. 90).
3. *Euxoa henrietta* (Smith); Bass Lake, California; CNC 10708. (p. 91).
4. *Euxoa cincta* Barnes and Benjamin; Paradise, Arizona; CNC 1955. (p. 92).
5. *Euxoa hollemani* (Grote); Follyfarm, Oregon; CNC 10696. (p. 93).
6. *Euxoa subandera* Lafontaine, paratype; Atascadero, California; CNC 10691. (p. 94).
7. *Euxoa xasta* Barnes and McDunnough; Callao, Utah; CNC 1655. (p. 94).
8. *Euxoa catenula* (Grote); Keremeos, British Columbia; CNC 1697. (p. 95).

PLATE Y: FEMALE GENITALIA OF *EUXOA* SPECIES

1. *Euxoa comosa annir* (Strecker); Helena, Montana; CNC 10674. (p. 99).
2. *Euxoa lucida* Barnes and McDunnough; Baker, Nevada; CNC 10679. (p. 101).
3. *Euxoa fumalis* (Grote); Meech Lake, Quebec; CNC 1784. (p. 101).
4. *Euxoa occidentalis* Lafontaine, paratype; Seneca, Oregon; CNC 10627. (p. 102).
5. *Euxoa velleripennis* (Grote); Wareham, Massachusetts; CNC 10707. (p. 102).
6. *Euxoa brunneigera excogita* (Smith); Logan, Utah; CNC 10162. (p. 107).
7. *Euxoa selenis* (Smith); Riverside Co., California; CNC 10703. (p. 108).
8. *Euxoa satiens* (Smith); Kamloops, British Columbia; CNC 1076. (p. 108).

PLATE Z: FEMALE GENITALIA OF *EUXOA* SPECIES

1. *Euxoa violaris* (Grote and Robinson); Wareham, Massachusetts; CNC 10698. (p. 109).
2. *Euxoa cursoria wirima* Hardwick, paratype; Whitehorse, Yukon; CNC 1165. (p. 111).
3. *Euxoa ochrogaster ochrogaster* (Guenée); Forestville, Quebec; CNC 1179. (p. 113).
4. *Euxoa nostra* (Smith); Truckee, California; CNC 10695. (p. 113).
5. *Euxoa siccata* (Smith); Big Spring, Texas; CNC 1041. (p. 114).
6. *Euxoa choris* (Harvey); Lee Vining, California; CNC 10705. (p. 114).
7. *Euxoa obeliscoides* (Guenée); Lytton, British Columbia; CNC 4905. (p. 115).
8. *Euxoa lillooet* McDunnough; Walla Walla, Washington; CNC 6679. (p. 116).

PLATE AA: FEMALE GENITALIA OF *EUXOA* SPECIES

1. *Euxoa basalis* (Grote); Last Mountain Lake, Saskatchewan; CNC 6505. (p. 117).
2. *Euxoa costata* (Grote); Golden, British Columbia; CNC 633. (p. 123).
3. *Euxoa brevipennis* (Smith); Fallon, Nevada; CNC 6537. (p. 125).
4. *Euxoa servita* (Smith); Radium Hot Springs, British Columbia; CNC 742. (p. 125).
5. *Euxoa redimicula* (Morrison); Mont St. Hilaire, Quebec; CNC 10196. (p. 126).
6. *Euxoa auripennis auripennis* Lafontaine; Lower Post, British Columbia; CNC 1382. (p. 127).
7. *Euxoa olivalis* (Grote); Lee Vining, California; CNC 6026. (p. 127).
8. *Euxoa teleboa* (Smith); Ackman, Colorado; CNC 1823. (p. 130).

PLATE BB: FEMALE GENITALIA OF *EUXOA* SPECIES

1. *Euxoa citricolor* (Grote); Silver Springs, Nevada; CNC 1685. (p. 129).
2. *Euxoa difformis* (Smith); Follyfarm, Oregon; CNC 10704. (p. 130).
3. *Euxoa moerens* (Grote); Burns, Oregon; CNC 10715. (p. 131).
4. *Euxoa latro* (Barnes and Benjamin); Upper Santa Ana River, San Bernardino Co., California; CNC 6299. (p. 131).
5. *Euxoa murdocki* (Smith); Lillooet, British Columbia; CNC 1356. (p. 132).
6. *Euxoa dodi* McDunnough; Missoula, Montana; CNC 1906. (p. 132).
7. *Euxoa laetificans* (Smith); Riverton, Wyoming; CNC 6283. (p. 133).
8. *Euxoa quadridentata flutea* Smith; Follyfarm, Oregon; CNC 1654. (p. 134).

PLATE CC: FEMALE GENITALIA OF *EUXOA* SPECIES

1. *Euxoa inscripta* Lafontaine, paratype; Kemmerer, Wyoming; CNC 10666. (p. 134).
2. *Euxoa niveilinea* (Grote); Corona, New Mexico; CNC 10712. (p. 135).
3. *Euxoa dargo* (Strecker); Valentine, Nebraska; CNC 6388. (p. 135).
4. *Euxoa melura* McDunnough, paratype; Eureka, Utah; CNC 6094. (p. 136).
5. *Euxoa detersa personata* (Morrison); Atkinson, Nebraska; CNC 10716. (p. 137).
6. *Euxoa cicatricosa* (Grote and Robinson); Twentynine Palms, California; CNC 7348. (p. 137).
7. *Euxoa aequalis alko* (Strecker); Eureka, Utah; CNC 6461. (p. 141).
8. *Euxoa conjuncta* (Smith); Hot Springs, New Mexico; CNC 1739. (p. 141).

PLATE DD: FEMALE GENITALIA OF *EUXOA* SPECIES

1. *Euxoa cona* (Strecker); Stockton, Utah; CNC 4601. (p. 142).
2. *Euxoa munis* (Grote); Truckee, California; CNC 1376. (p. 142).
3. *Euxoa misturata* (Smith); Lyman, Wyoming; CNC 4533. (p. 143).
4. *Euxoa cinereopallida* (Smith); Follyfarm, Oregon; CNC 6344. (p. 145).
5. *Euxoa atristrigata* (Smith); Helena, Montana; CNC 1657. (p. 145).

6. *Euxoa nevada* (Smith); Lillooet, British Columbia; CNC 1663. (p. 145).
7. *Euxoa mitis* (Smith); Swift Current, Saskatchewan; CNC 4517. (p. 146).
8. *Euxoa luctuosa* Lafontaine, paratype; Pagosa Springs, Colorado; CNC 10578. (p. 147).

PLATE EE: OVIPOSITOR LOBES OF *EUXOA* SPECIES

1. *Euxoa intrita* (Morrison); Kamloops, British Columbia; CNC 7191. (p. 69).
2. *Euxoa annulipes oncocnemoides* (Barnes and Benjamin); La Tuna Canyon; Los Angeles Co., California; CNC 7190. (p. 70).
3. *Euxoa scholastica* McDunnough; Hectanoosa, Nova Scotia; CNC 7221. (p. 71).
4. *Euxoa antica* Lafontaine, paratype; McGaffey, New Mexico; CNC 7186. (p. 72).
5. *Euxoa serricornis* (Smith); Apple Valley, California; CNC 7187. (p. 74).
6. *Euxoa pluralis* (Grote); Lee Vining, California; CNC 7223. (p. 75).
7. *Euxoa setonia* McDunnough; Colfax Co., New Mexico; CNC 7222. (p. 76).
8. *Euxoa declarata* (Walker); Lloydminster, Alberta; CNC 7078. (p. 79).
9. *Euxoa silens* (Grote); Apple Valley, California; CNC 7193. (p. 81).
10. *Euxoa simulata* McDunnough; Lee Vining, California; CNC 10657. (p. 82).
11. *Euxoa cana* Lafontaine; Hill City, Idaho; CNC 7084. (p. 84).
12. *Euxoa stigmatalis* (Smith); Eagar, Arizona; CNC 7086. (p. 85).
13. *Euxoa tessellata* (Harris); Lake Louise, Alberta; CNC 561. (p. 87).
14. *Euxoa albipennis* (Grote); Beaver, Utah; CNC 7181. (p. 90).
15. *Euxoa cincta* Barnes and Benjamin; Sierra Vista, Arizona; CNC 7244. (p. 92).
16. *Euxoa catenula* (Grote); Ft. Peck, Montana; CNC 7232. (p. 95).
17. *Euxoa comosa annir* (Strecker); Wolf Creek, Montana; CNC 10670. (p. 99).
18. *Euxoa piniae* Buckett and Bauer; Almanor, California; CNC 10690. (p. 108).
19. *Euxoa satiens* (Smith); Lyman, Wyoming; CNC 7220. (p. 108).
20. *Euxoa ochrogaster ochrogaster* (Guenée); Meech Lake, Quebec; CNC 7227. (p. 113).

PLATE FF: OVIPOSITOR LOBES OF *EUXOA* SPECIES

1. *Euxoa siccata* (Smith); Big Spring, Texas; CNC 1041. (p. 114).
2. *Euxoa choris* (Harvey); Sapinero, Colorado; CNC 7240. (p. 114).
3. *Euxoa oberfoelli* Hardwick; Webster, South Dakota; CNC 7238. (p. 116).
4. *Euxoa castanea* Lafontaine; Golden, British Columbia; CNC 693. (p. 123).
5. *Euxoa idahoensis* (Grote); Keremeos, British Columbia; CNC 10665. (p. 124).
6. *Euxoa brevipennis* (Smith); Follyfarm, Oregon; CNC 7168. (p. 125).
7. *Euxoa servita* (Smith); Debert, Nova Scotia; CNC 7182. (p. 125).
8. *Euxoa redimicula* (Morrison); Chatham, Ontario; CNC 10693. (p. 126).
9. *Euxoa auripennis auripennis* Lafontaine; Tonasket, Washington; CNC 1857. (p. 127).
10. *Euxoa teleboa* (Smith); Ft. Defiance, Arizona; CNC 7153. (p. 130).
11. *Euxoa citricolor* (Grote); Mojave, California; CNC 10709. (p. 129).
12. *Euxoa infracta* (Morrison); Princeton, British Columbia; CNC 7154. (p. 132).
13. *Euxoa dargo* (Strecker); Biggs, Oregon; CNC 10718. (p. 135).
14. *Euxoa detersa personata* (Morrison); Simcoe, Ontario; CNC 10687. (p. 137).
15. *Euxoa cicatricosa* (Grote and Robinson); Corona, New Mexico; CNC 6326. (p. 137).
16. *Euxoa cinereopallida* (Smith); Ft. Wingate, New Mexico; CNC 7202. (p. 145).
17. *Euxoa atristrigata* (Smith); Jefferson Co., Montana; CNC 7208. (p. 145).
18. *Euxoa nevada* (Smith); Stockton, Utah; CNC 7209. (p. 145).
19. *Euxoa mitis* (Smith); Stockton, Utah; CNC 7211. (p. 146).
20. *Euxoa luctuosa* Lafontaine, paratype; Kayenta, Arizona; CNC 7205. (p. 147).

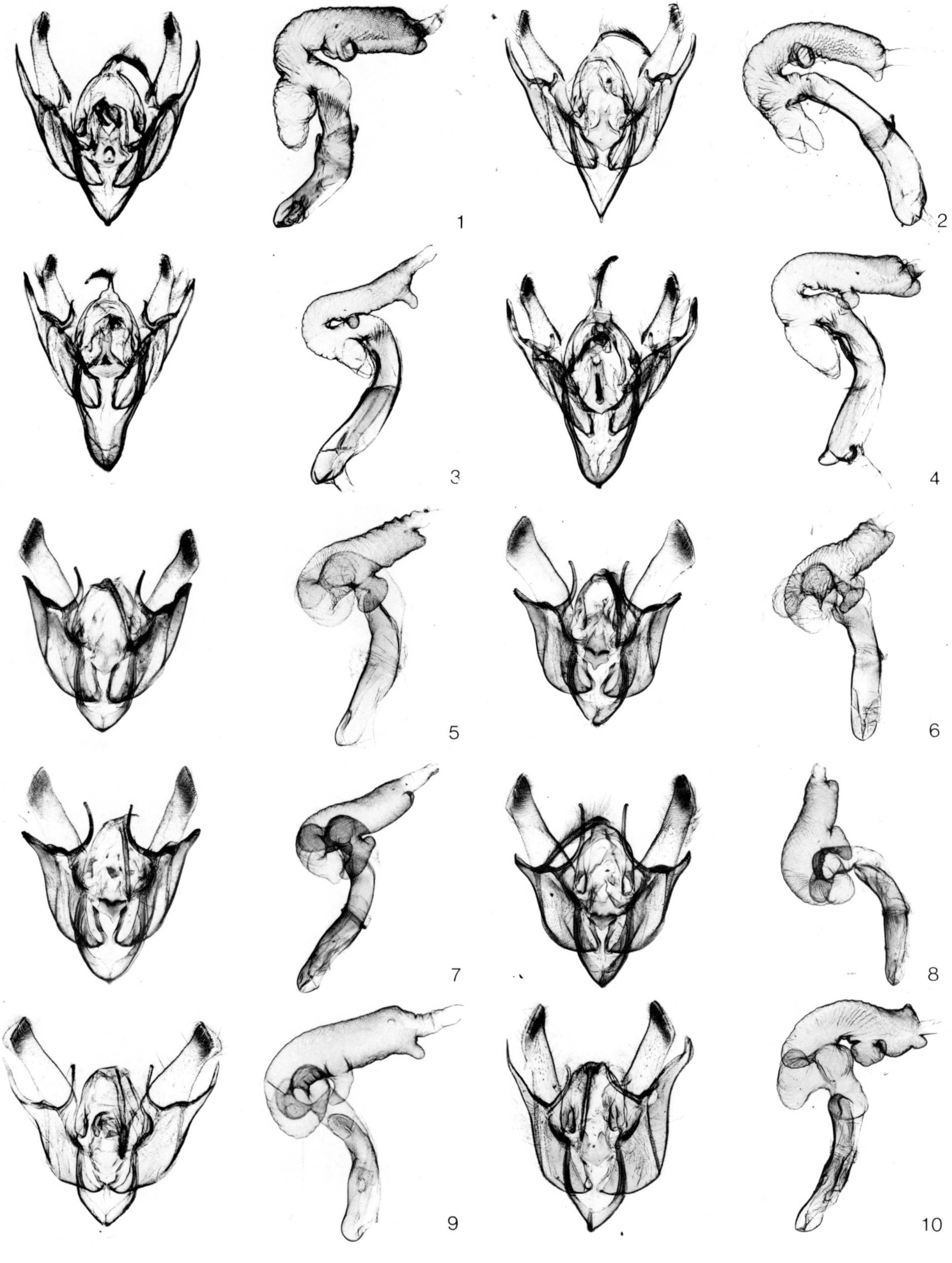
1
2
3
4
5
6
7
8
9
10

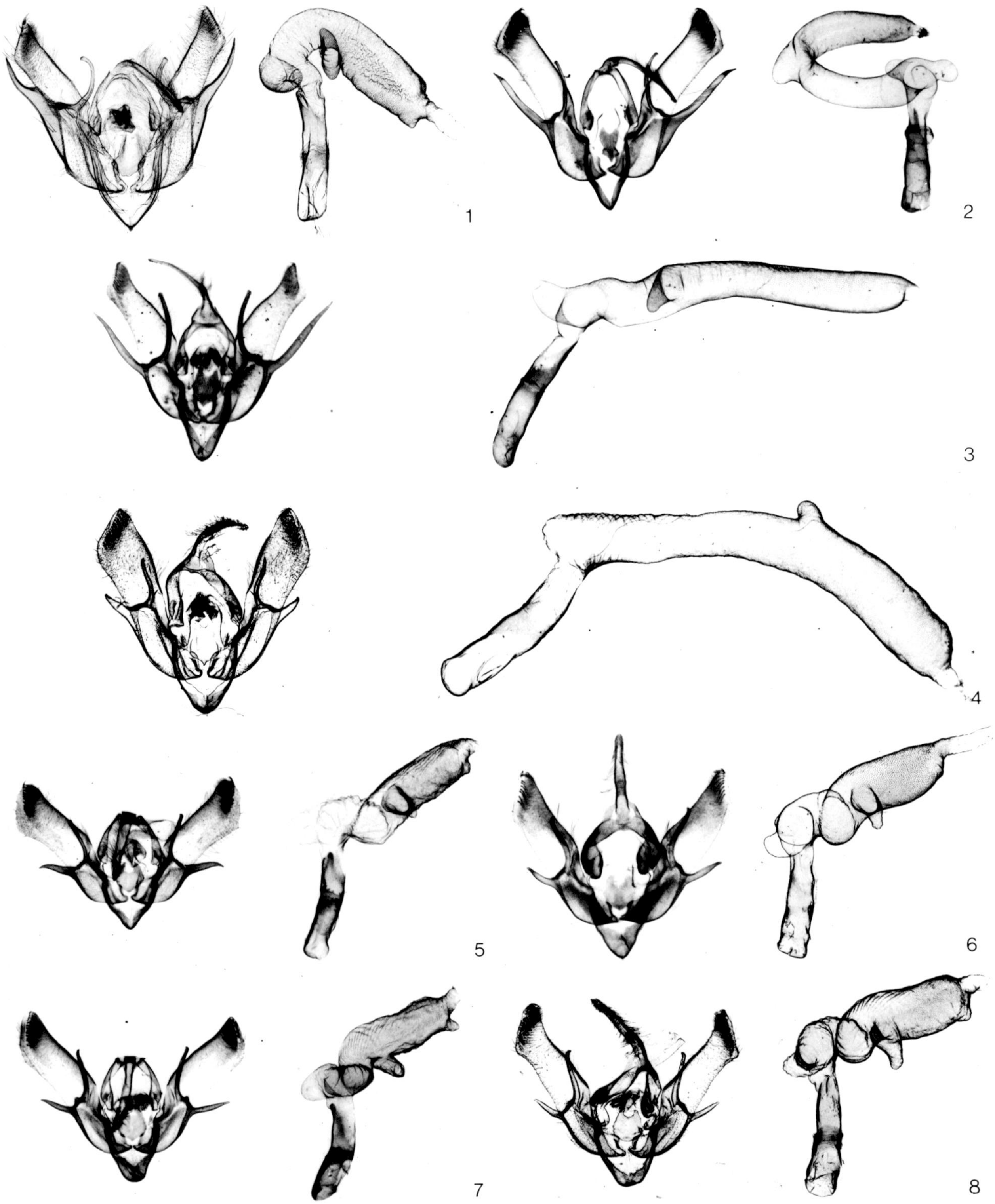
1
2
3
4
5
6
7
8

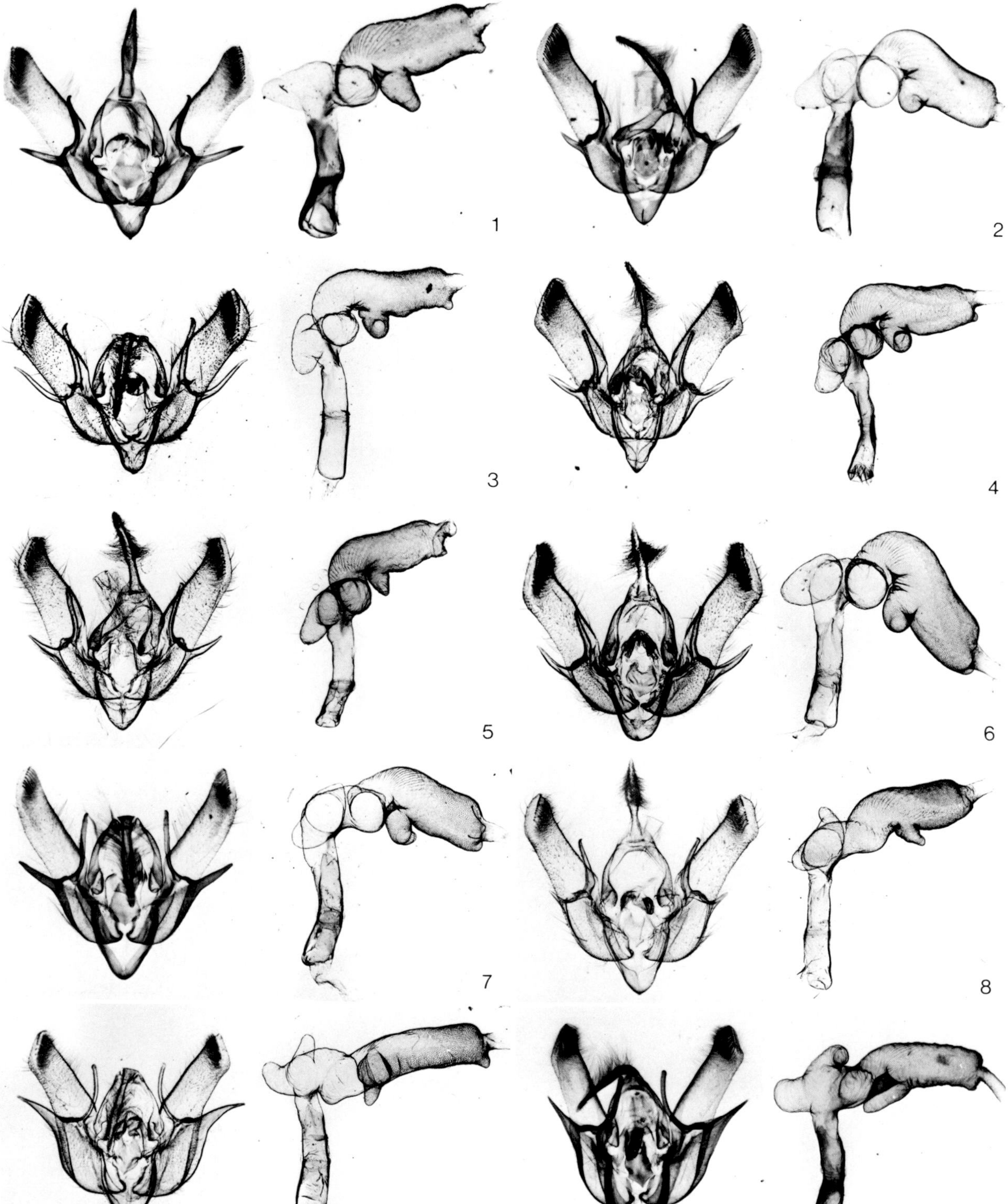
1
2
3
4
5
6
7
8
9
10

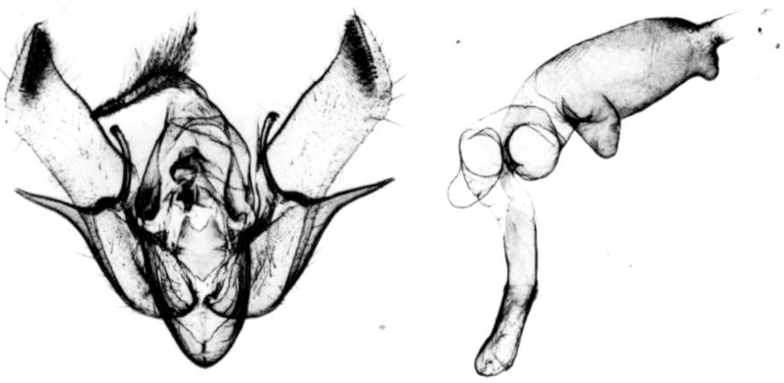
1

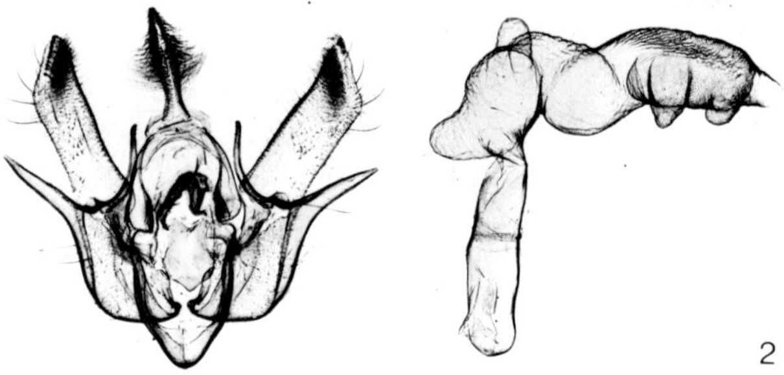
2

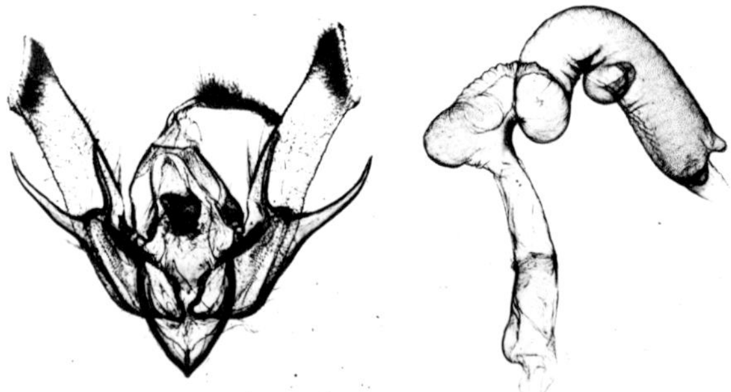
3

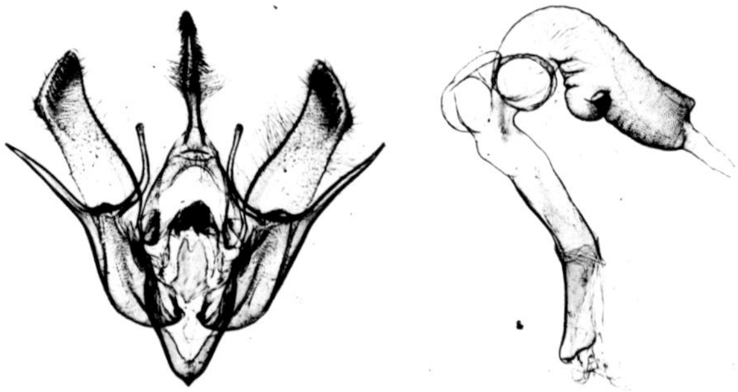
4

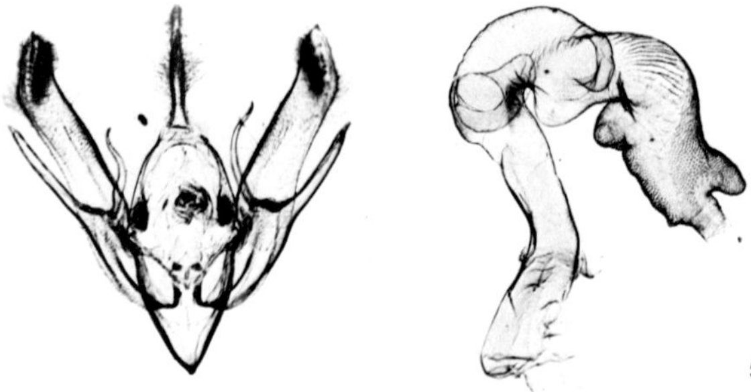
5

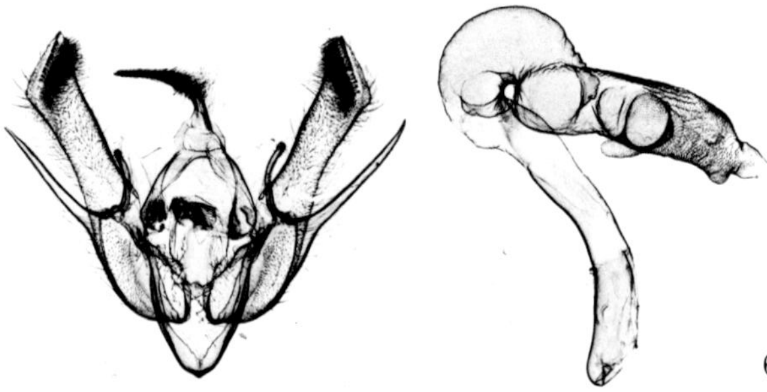
6

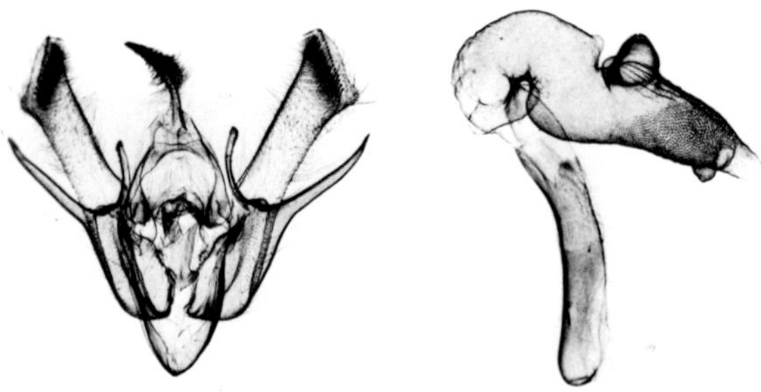
7

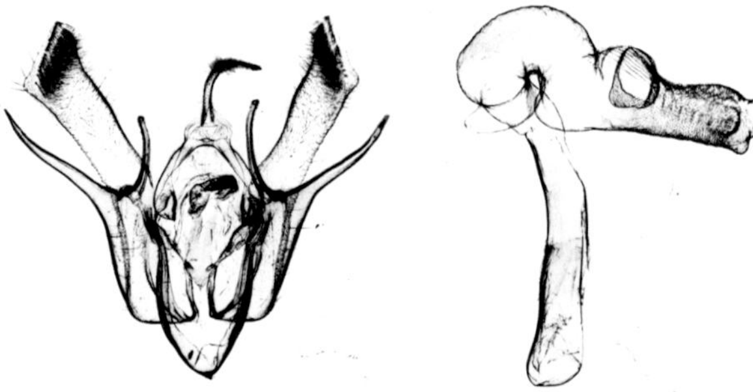
8

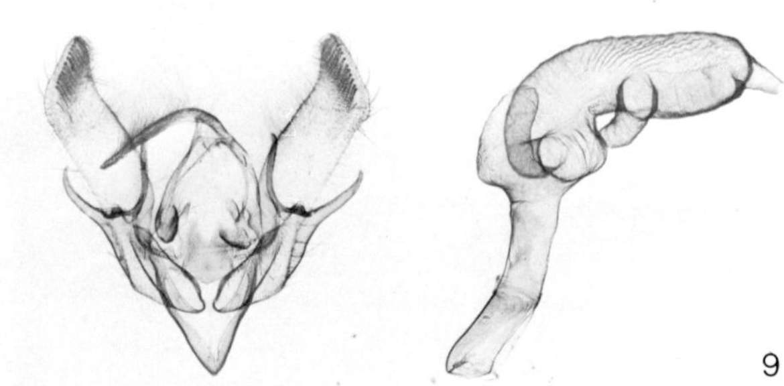
9

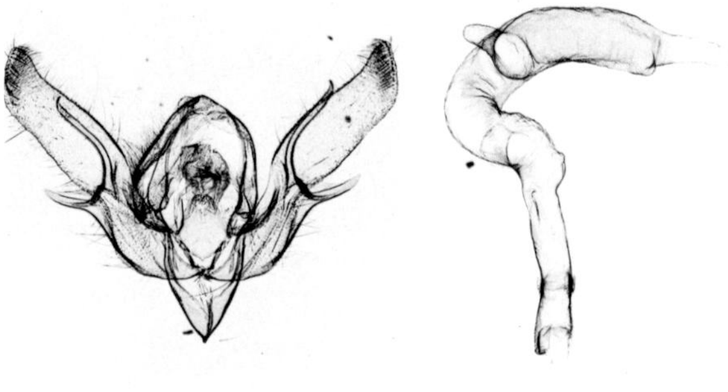
10

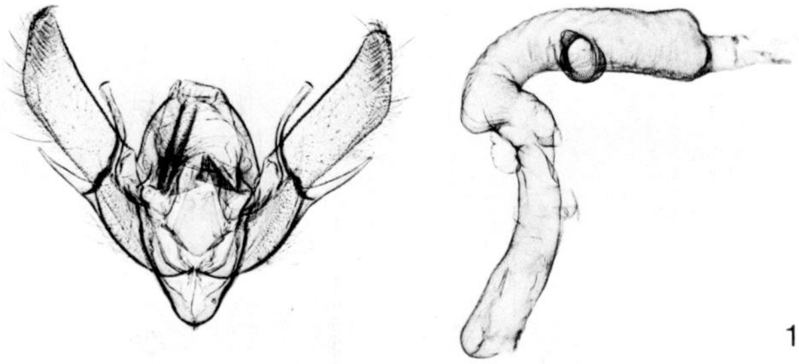

1

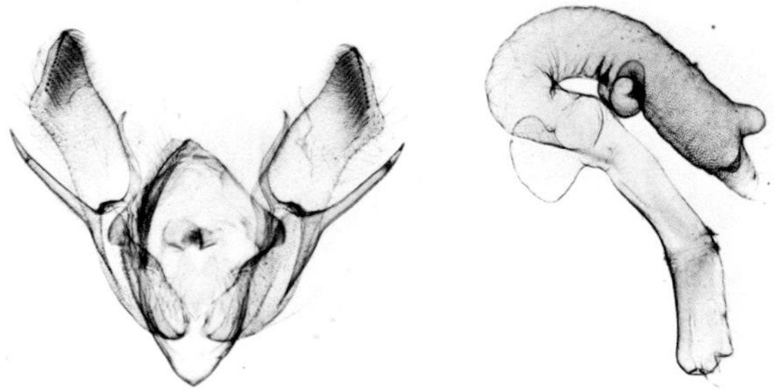

2

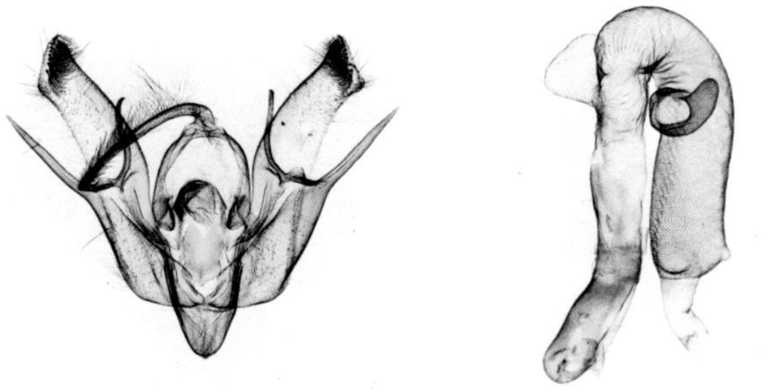

3

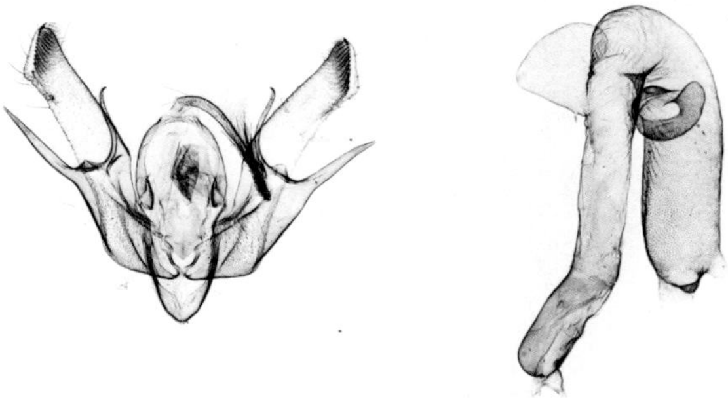

4

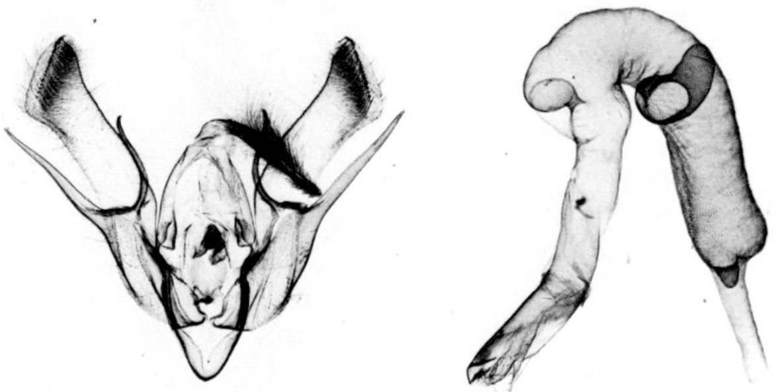

5

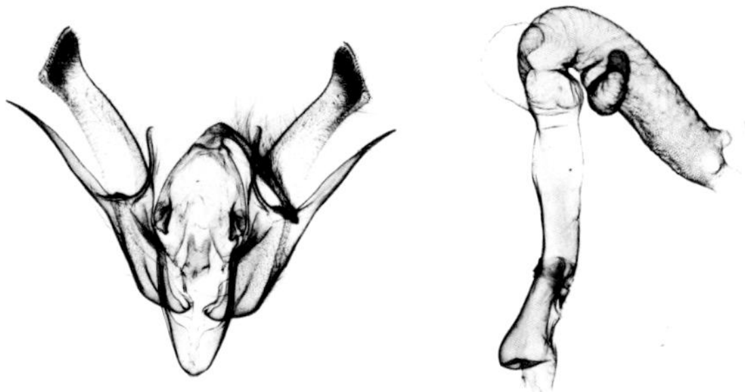

6

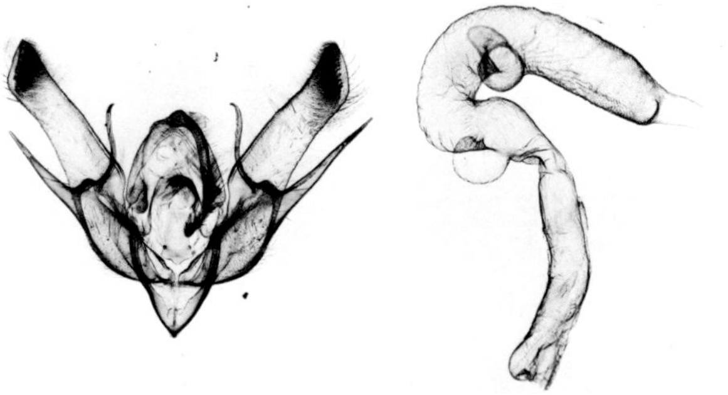

7

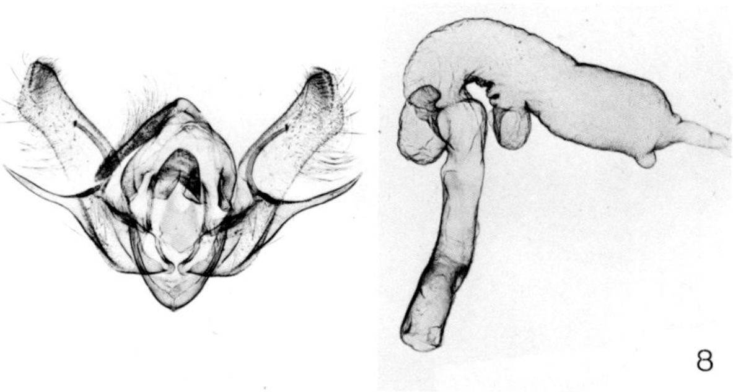

8

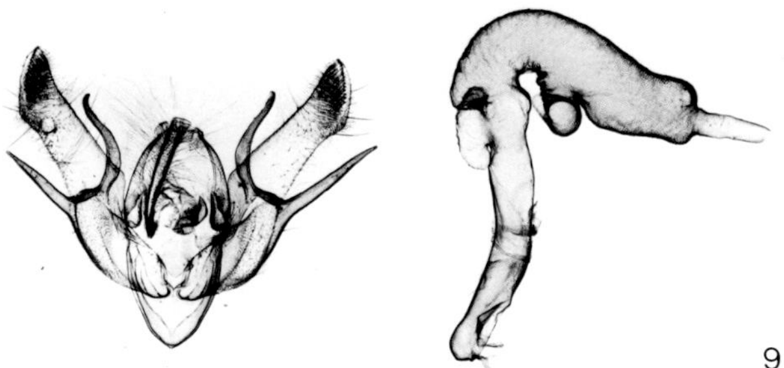

9

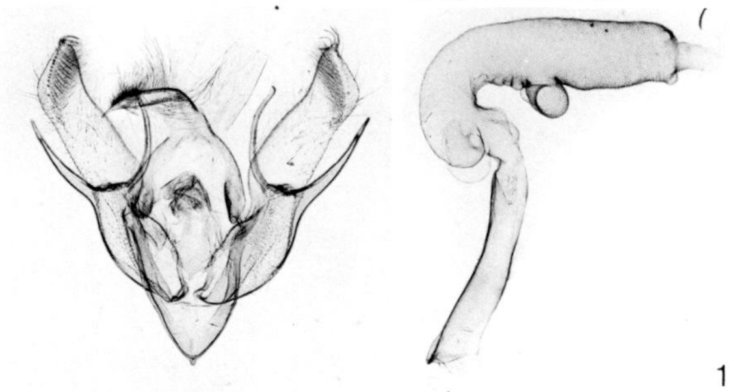

10

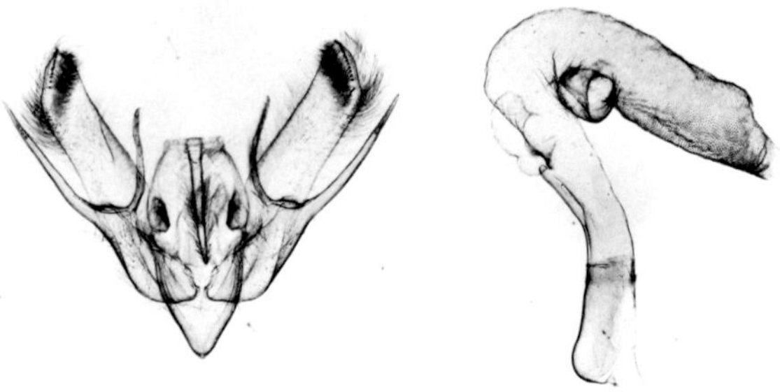
1

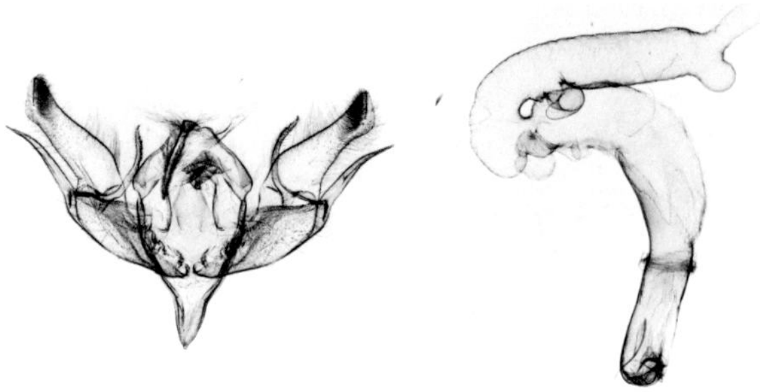
2

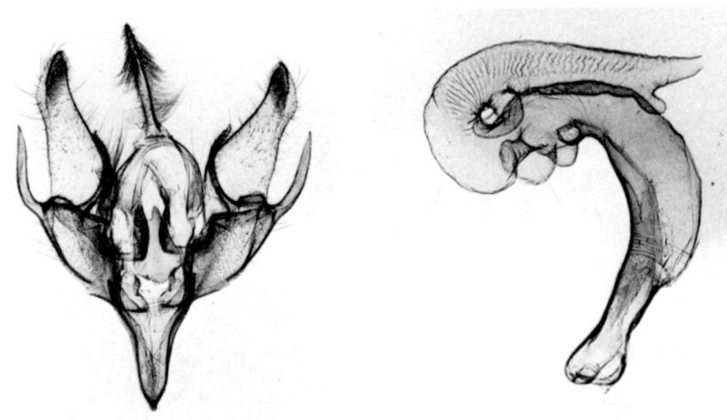
3

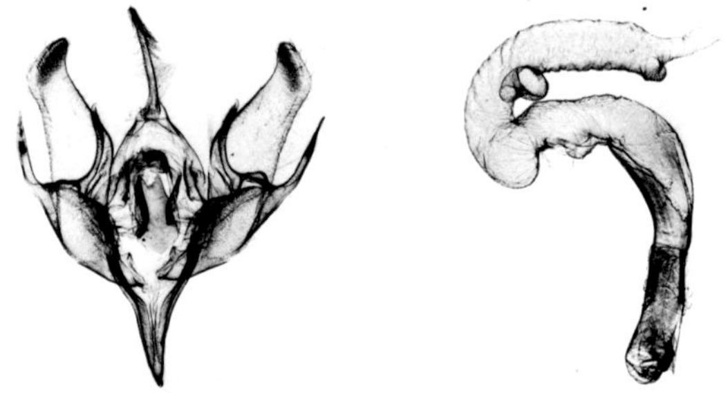
4

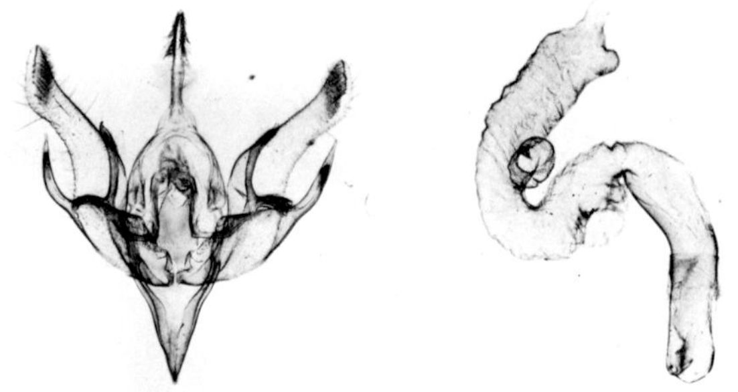
5

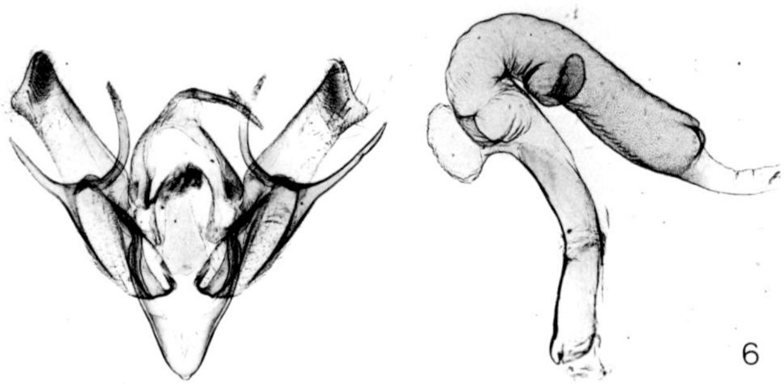
6

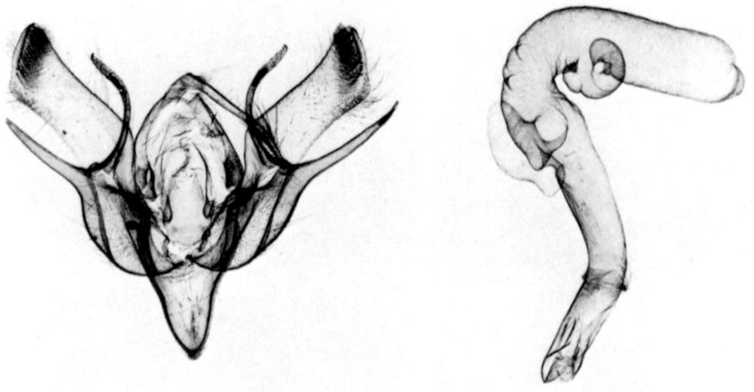
7

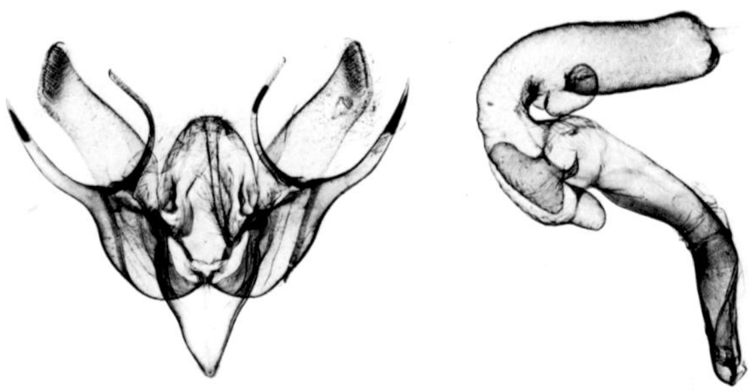
8

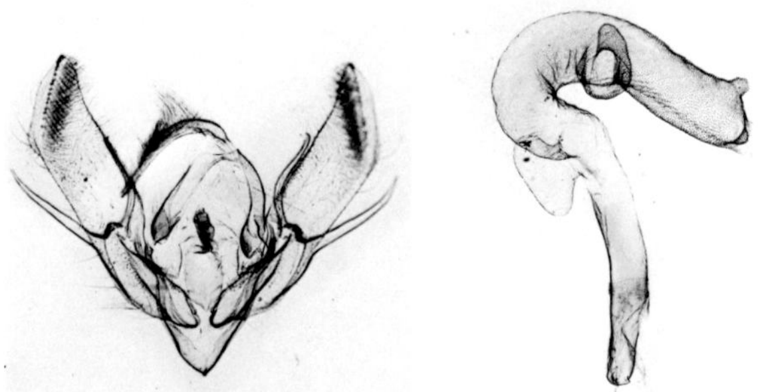
9

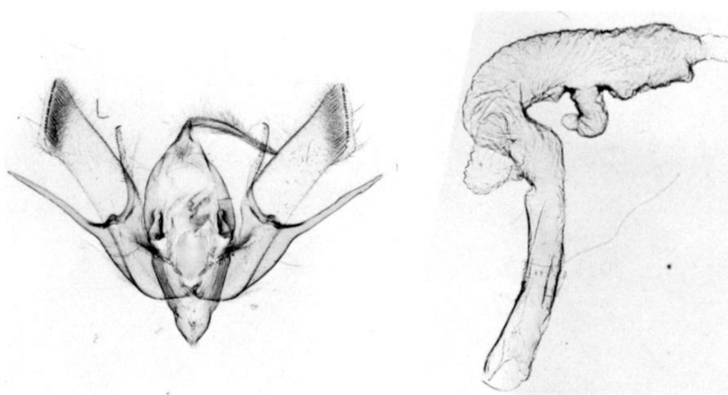
10

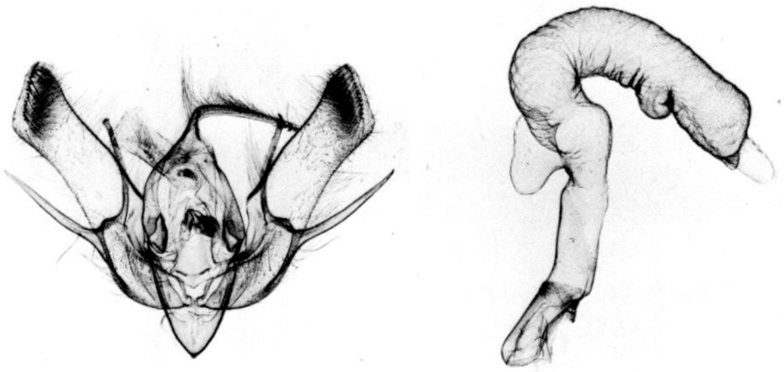

1

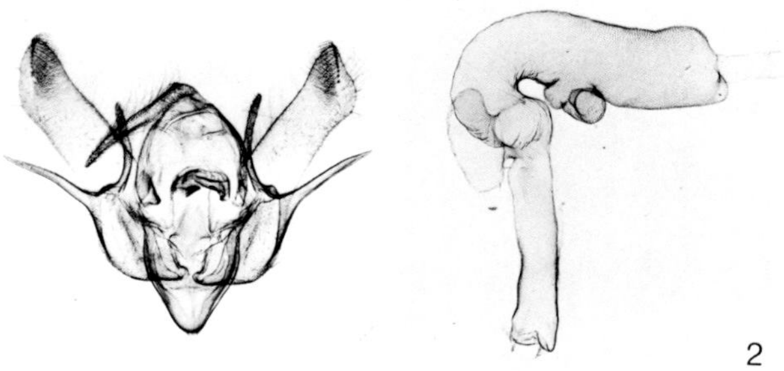

2

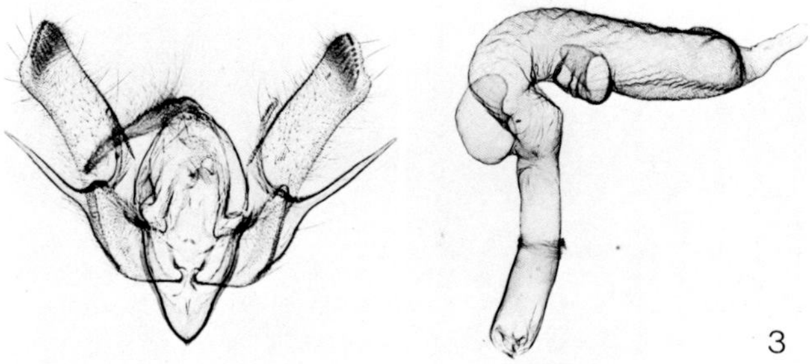

3

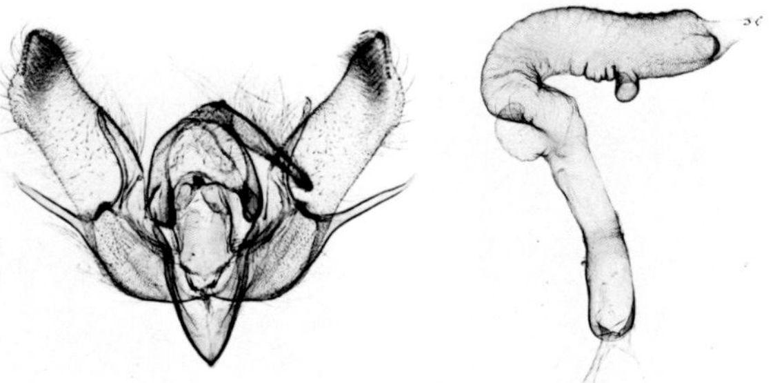

4

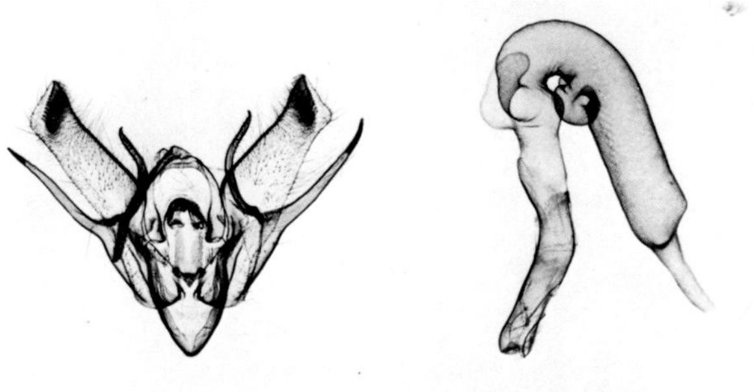

5

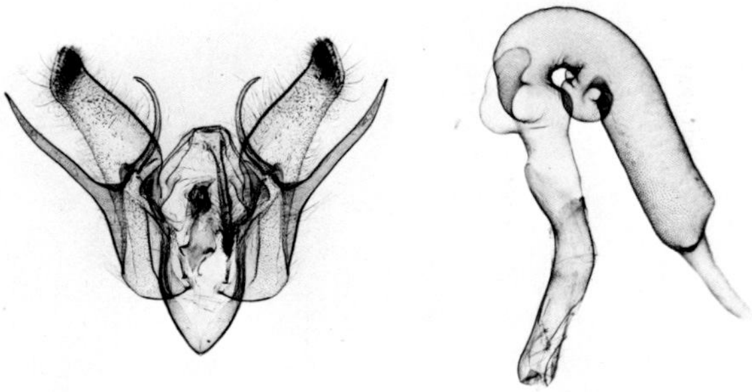

6

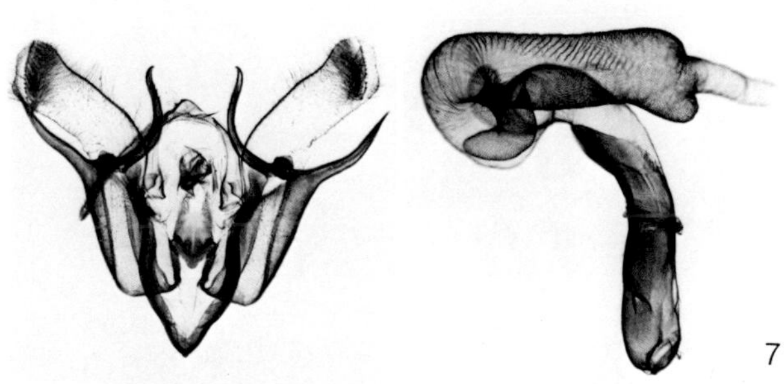

7

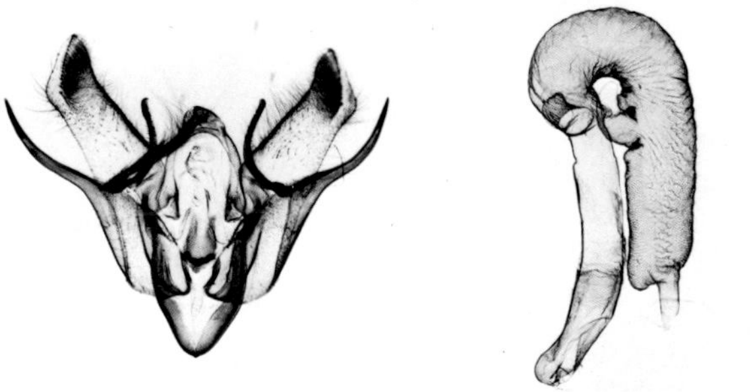

8

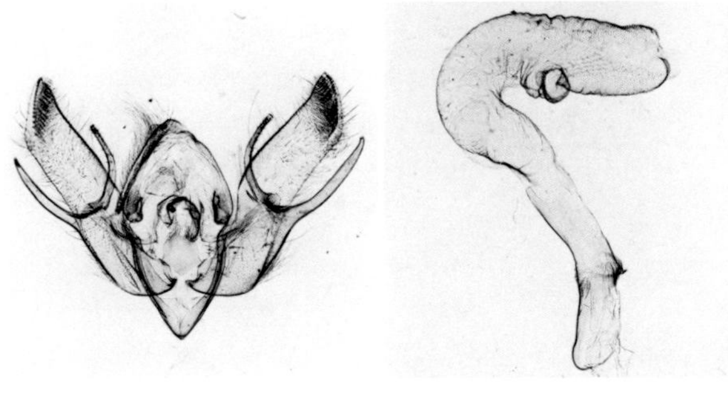

9

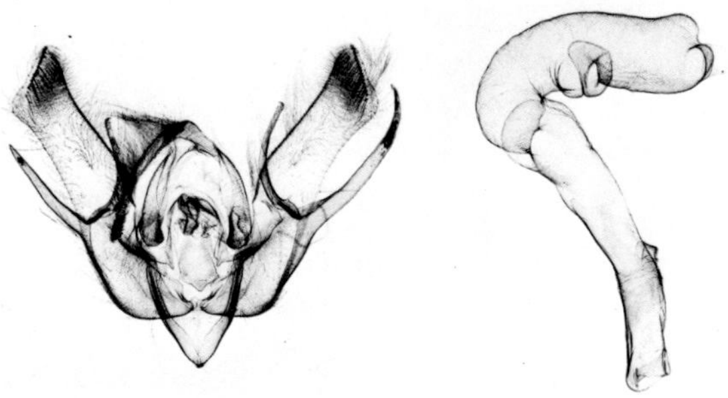

10

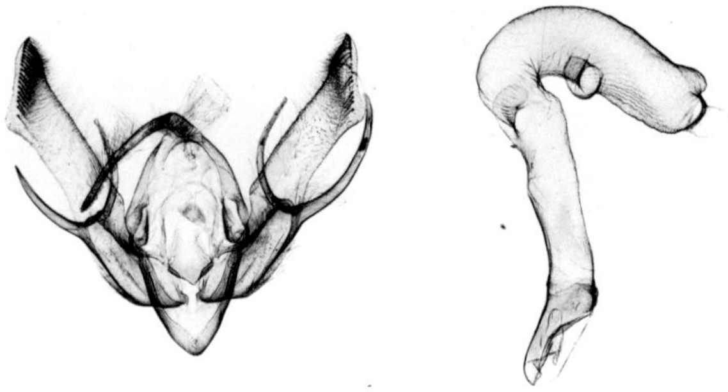
1

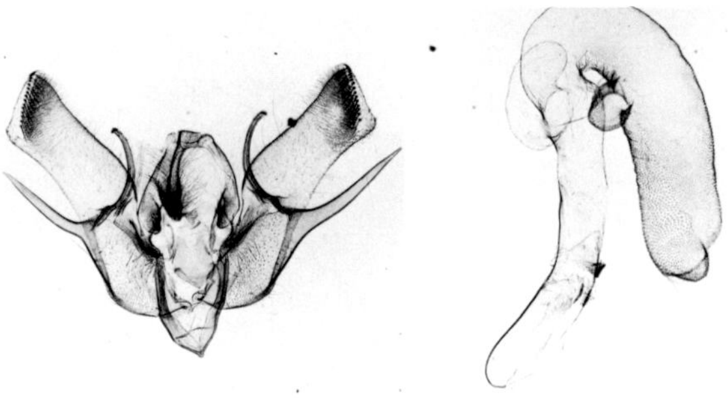
2

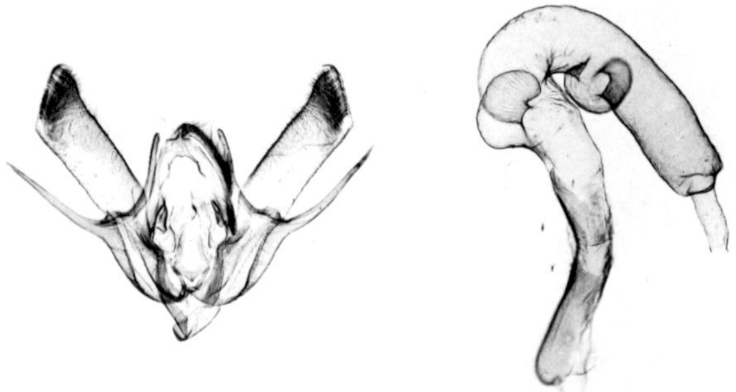
3

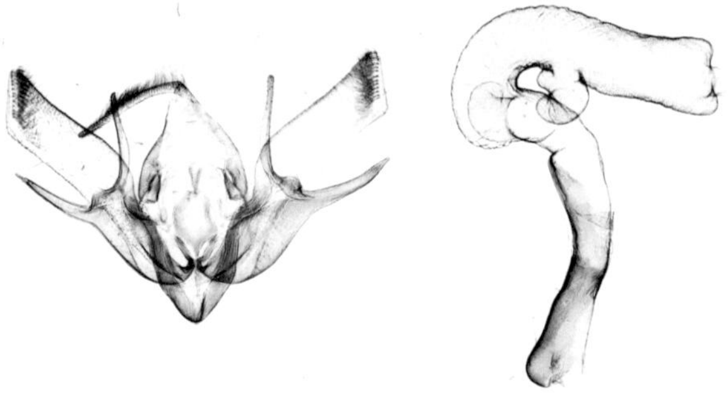
4

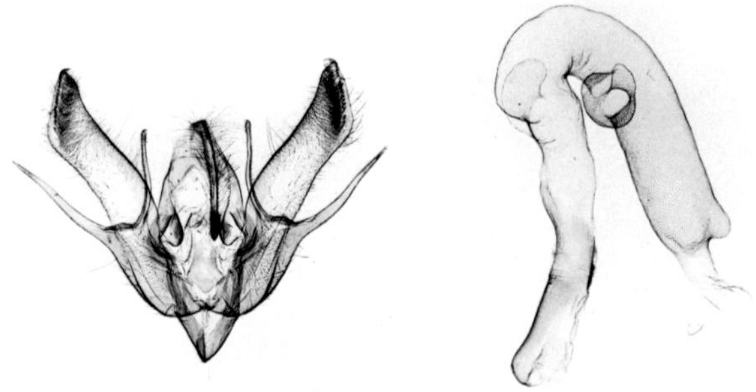
5

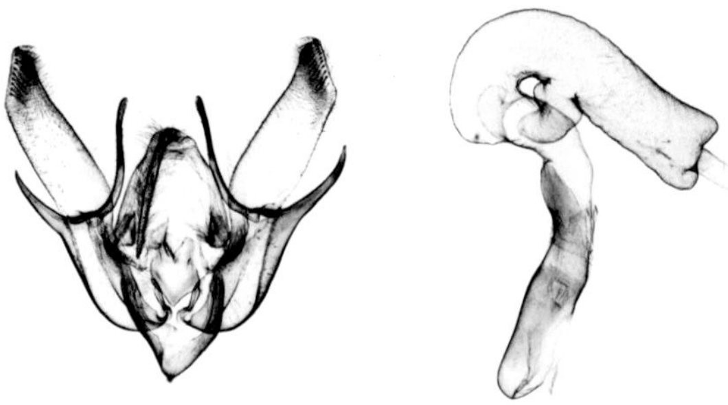
6

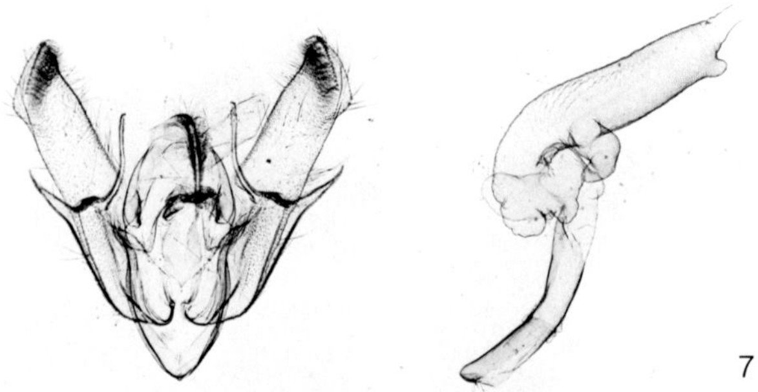
7

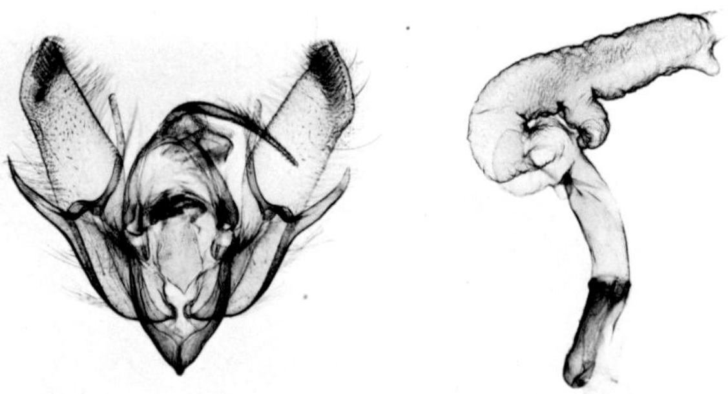
8

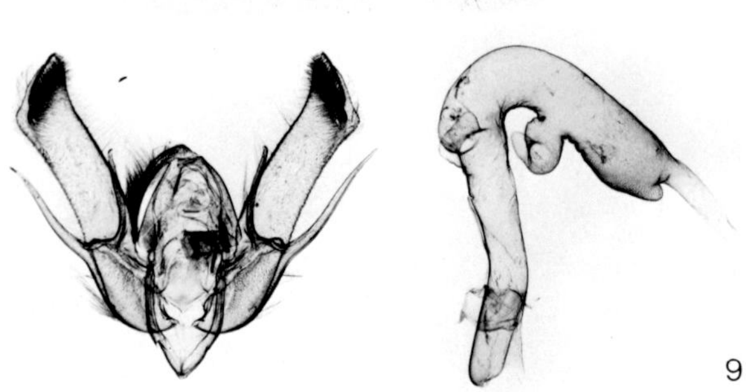
9

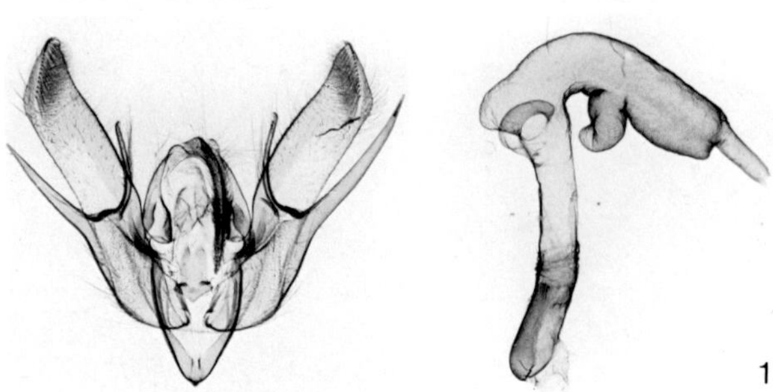
10

PLATE I: MALE GENITALIA OF *EUXOA* SPECIES

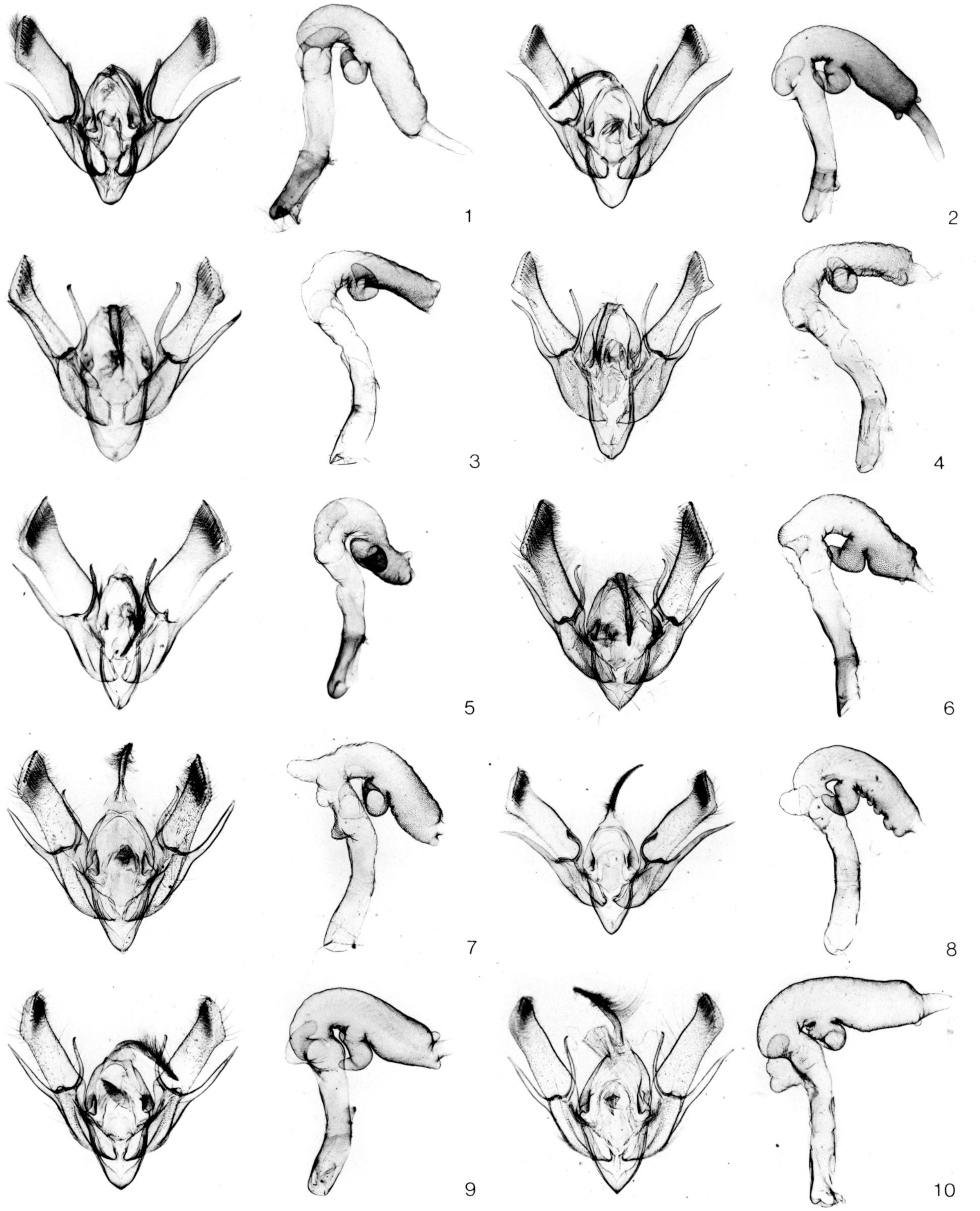

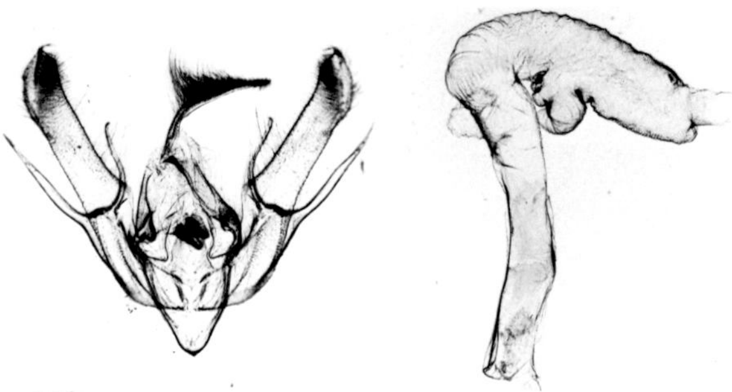
1

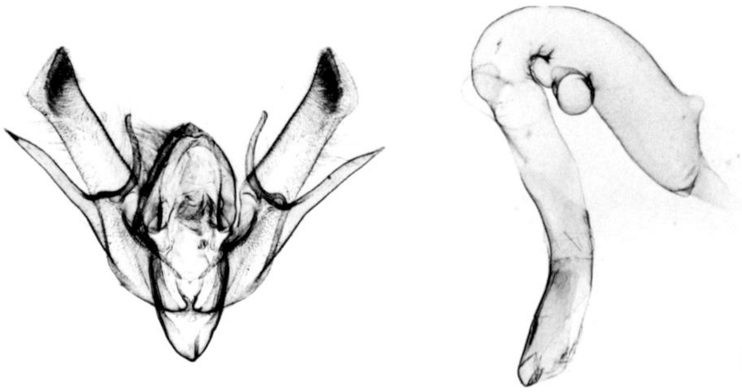
2

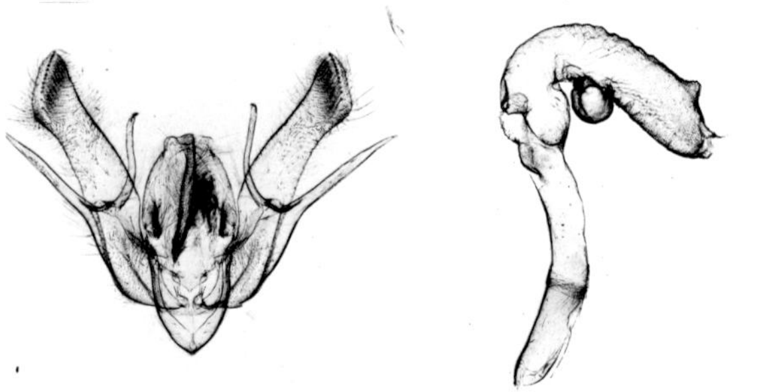
3

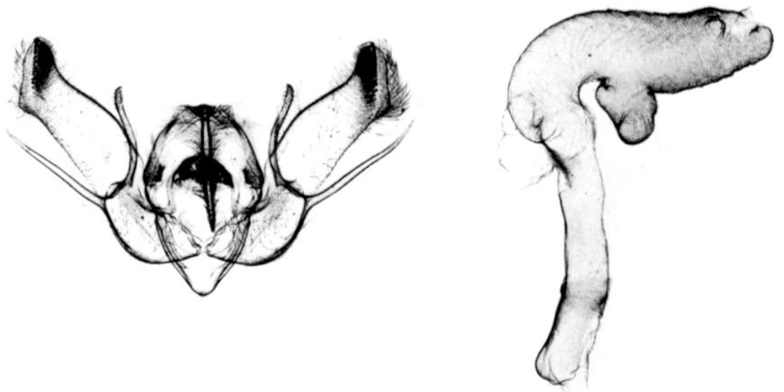
4

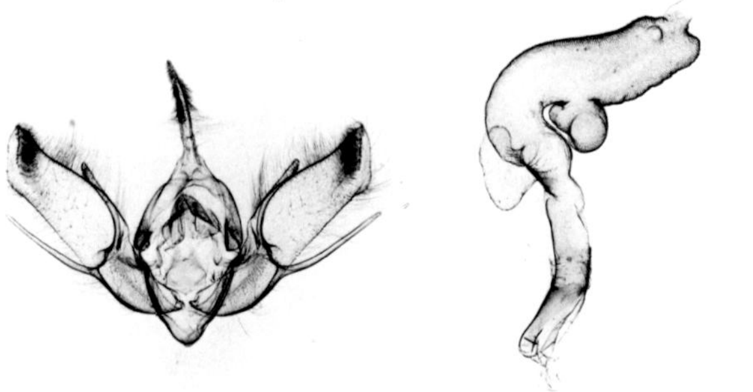
5

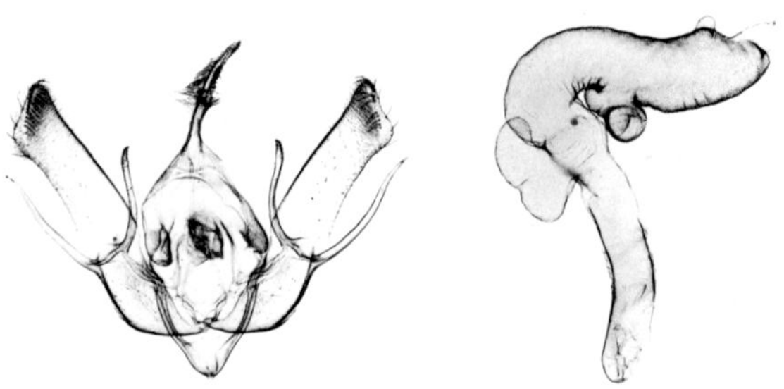
6

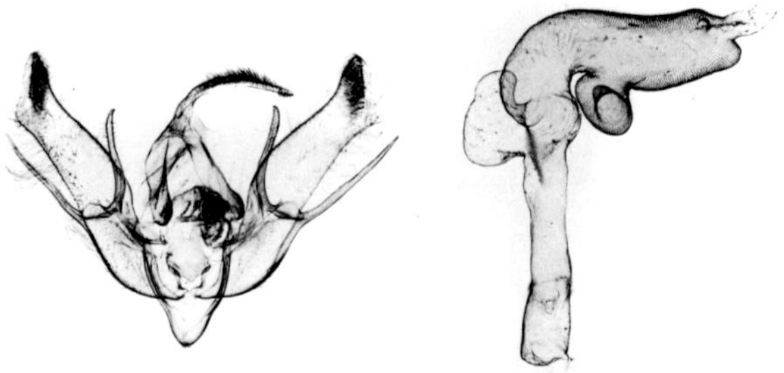
7

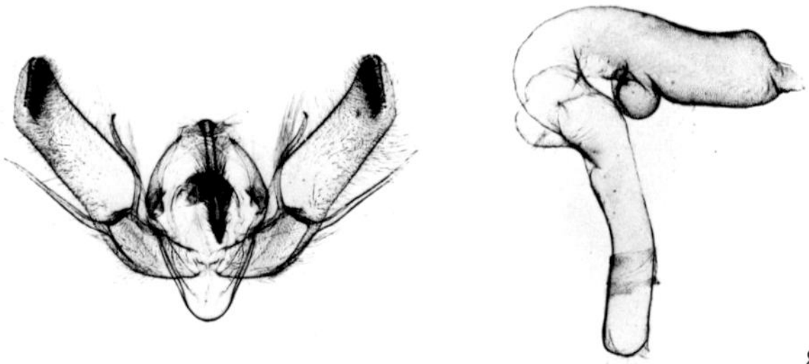
8

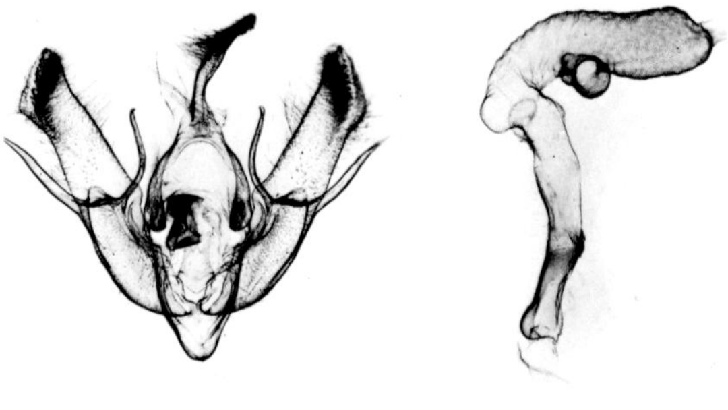
9

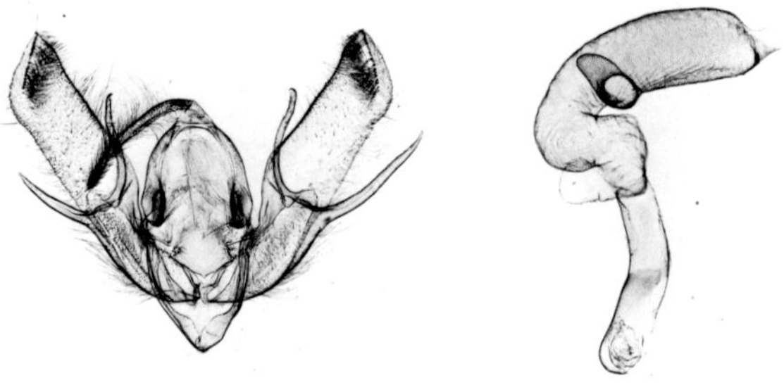
10

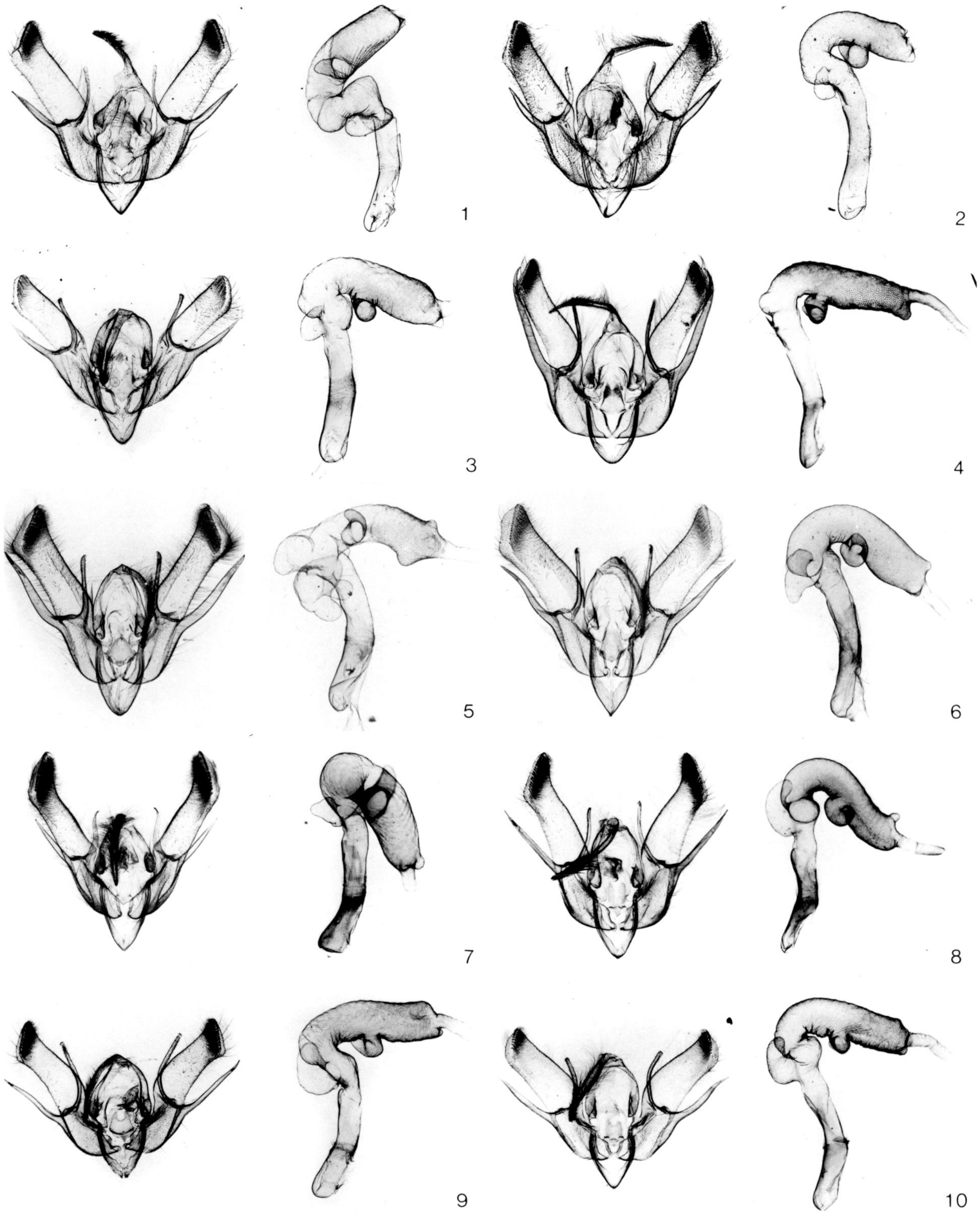
1
2
3
4
5
6
7
8
9
10

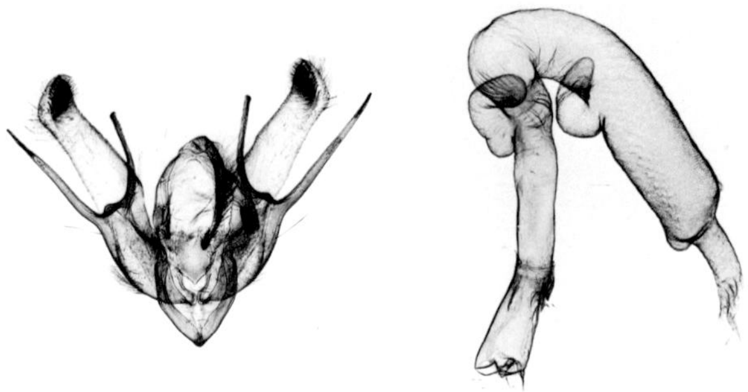

1

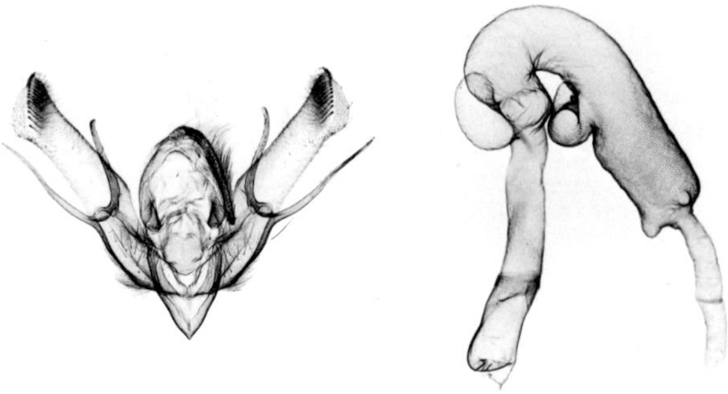

2

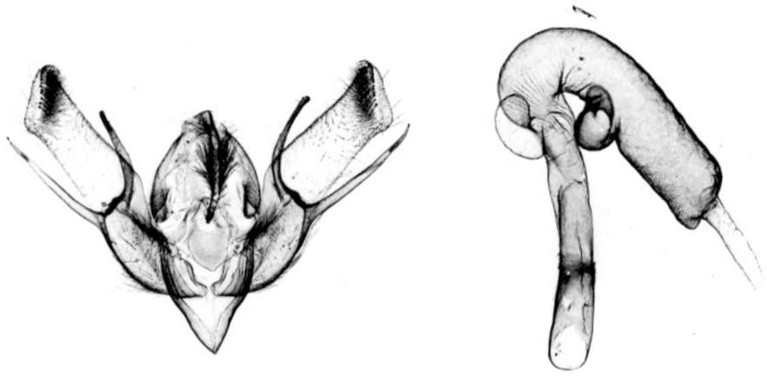

3

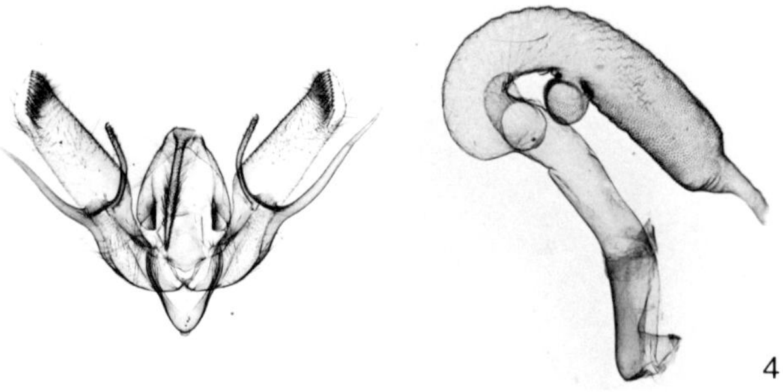

4

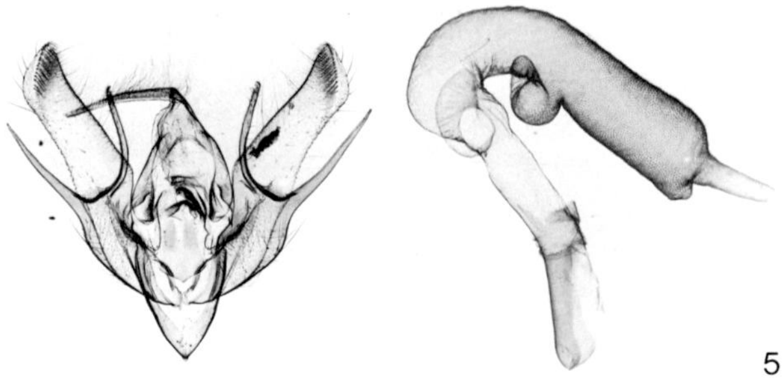

5

6

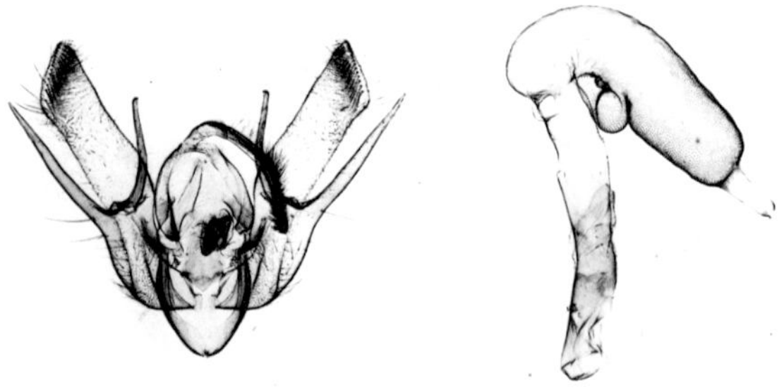

7

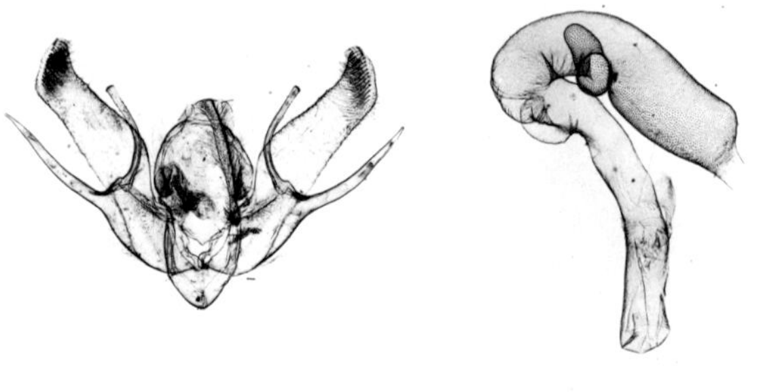

8

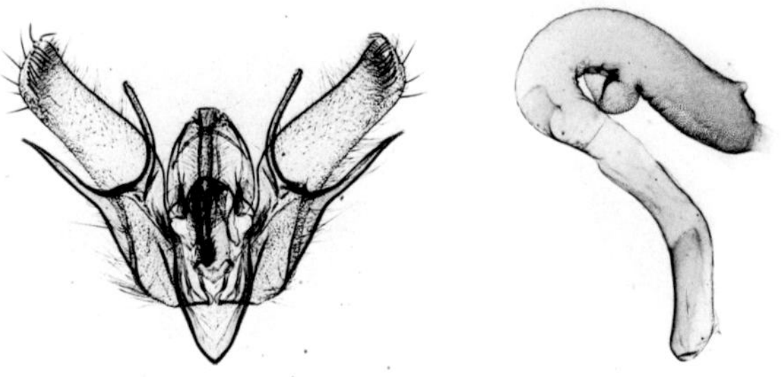

9

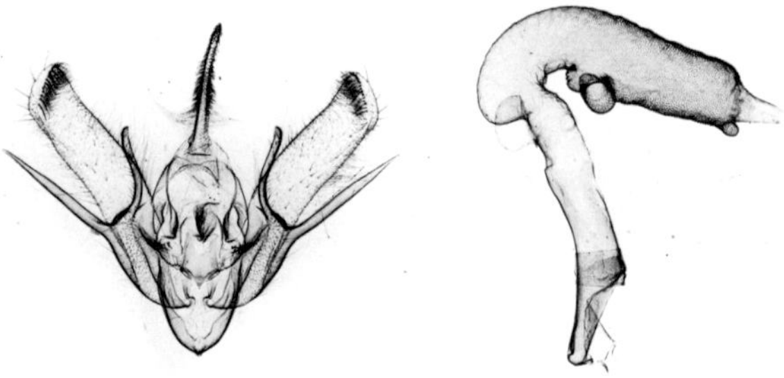

10

PLATE M: MALE GENITALIA OF *EUXOA* SPECIES

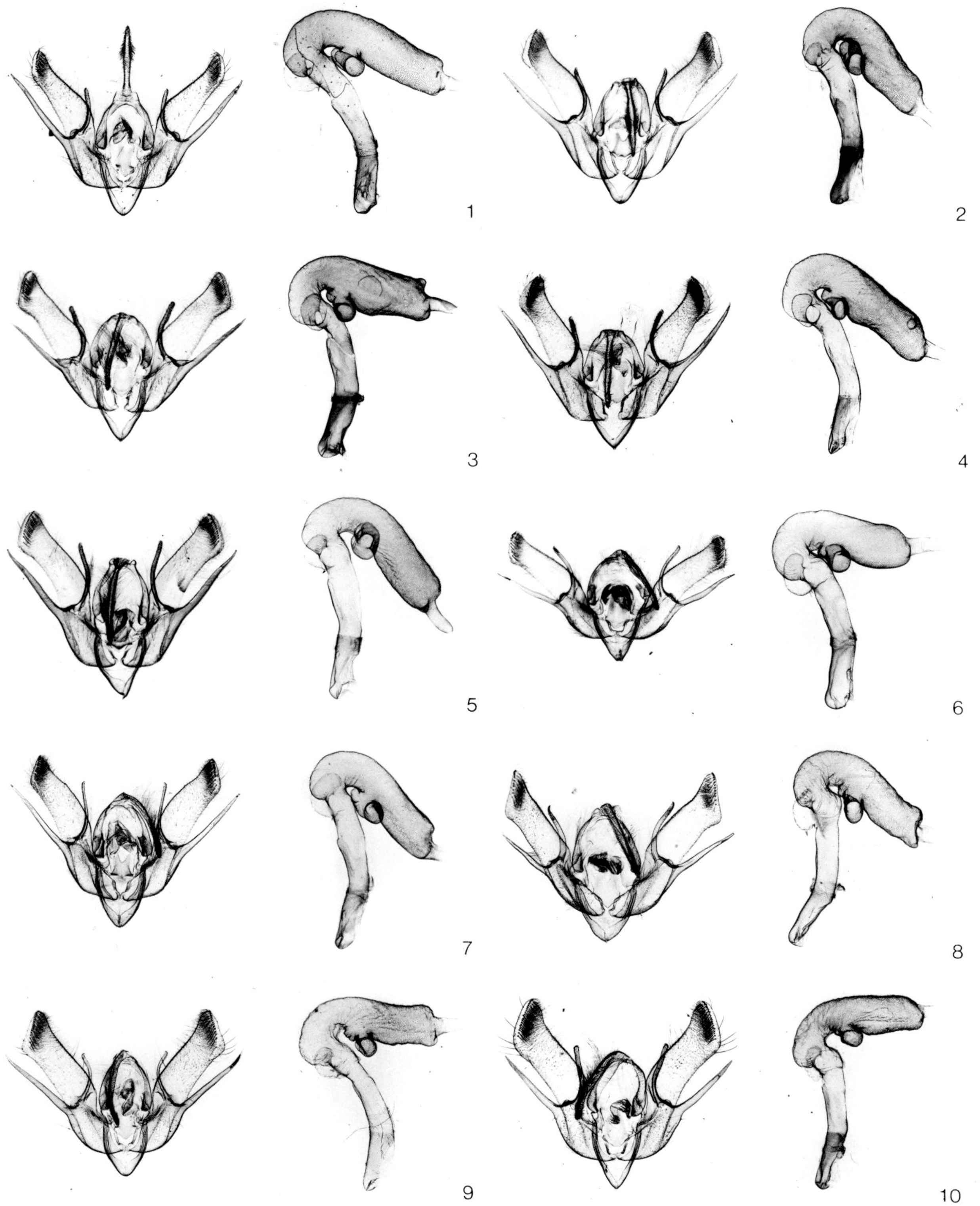

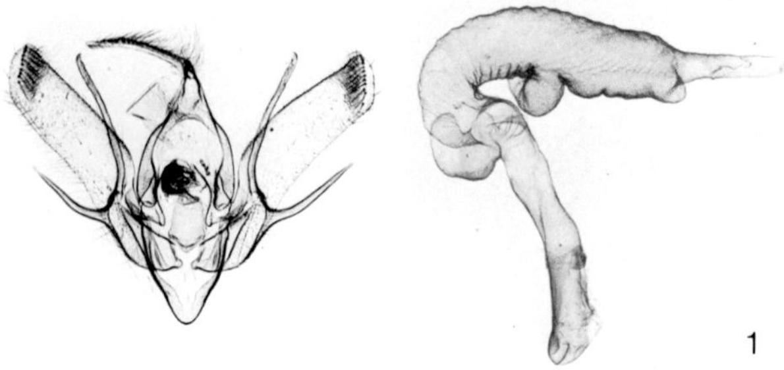

1

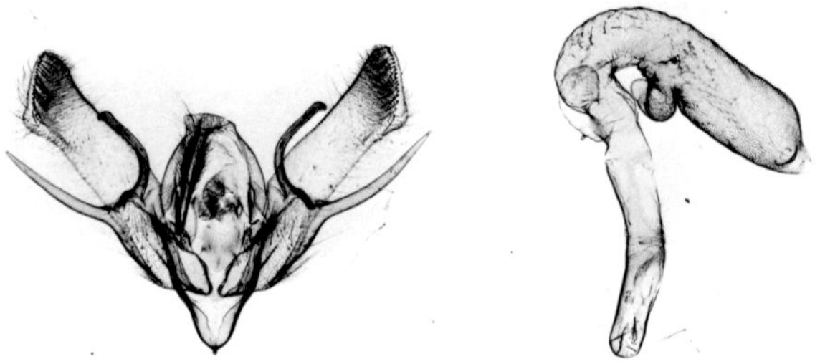

2

3

4

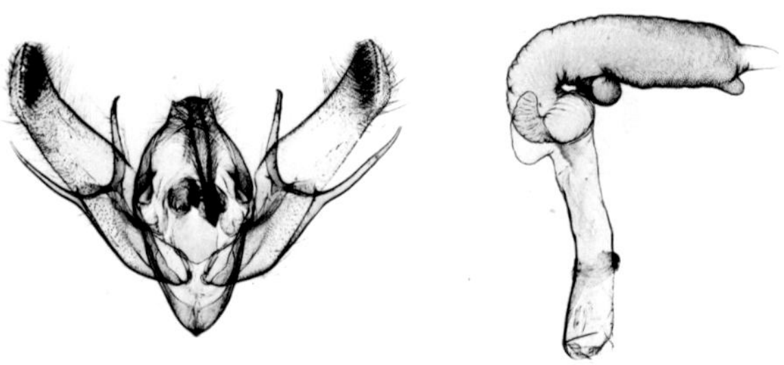

5

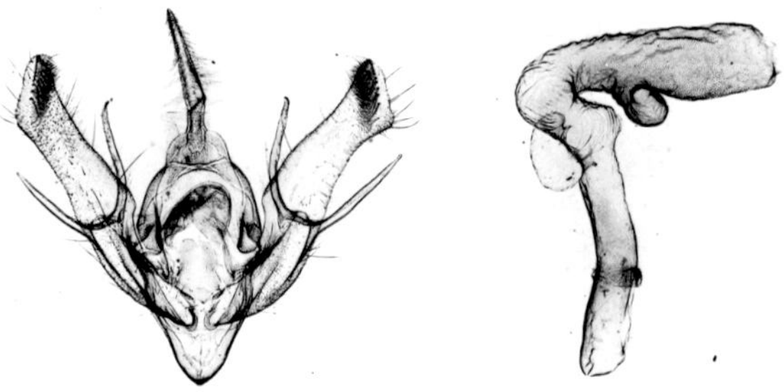

6

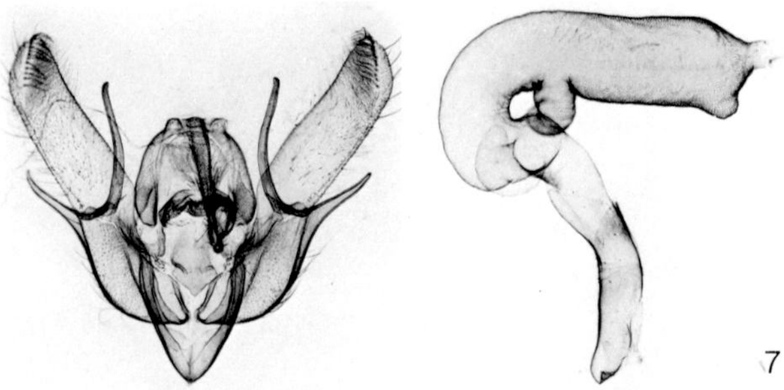

7

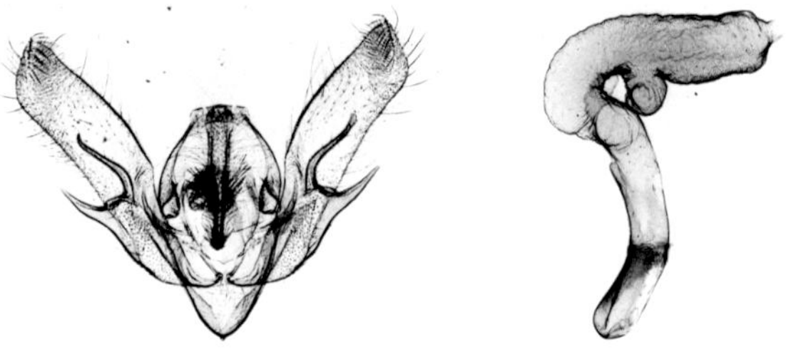

8

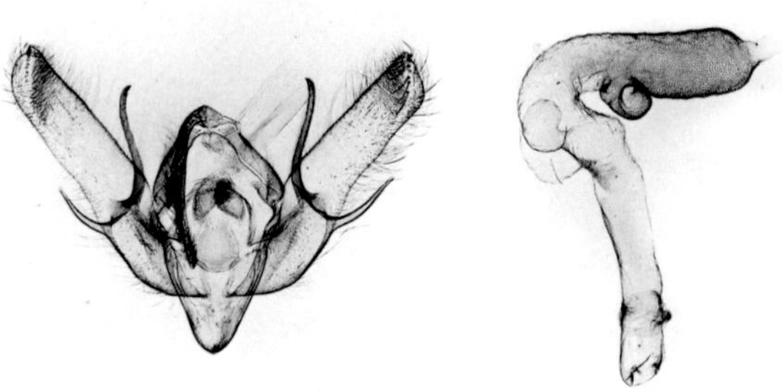

9

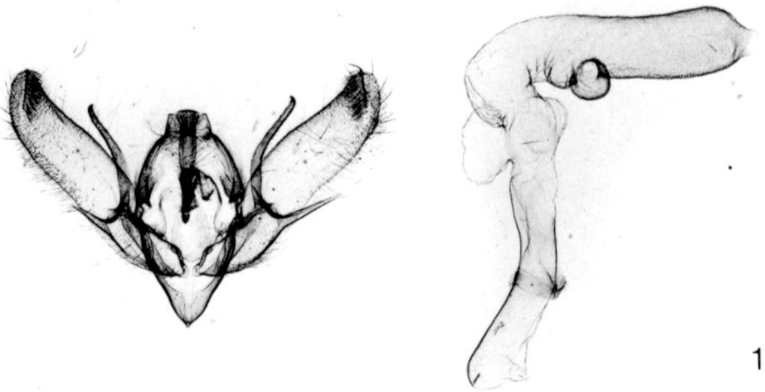

10

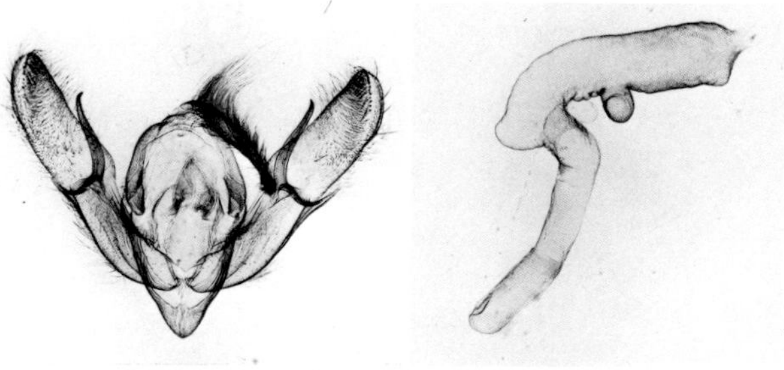
1

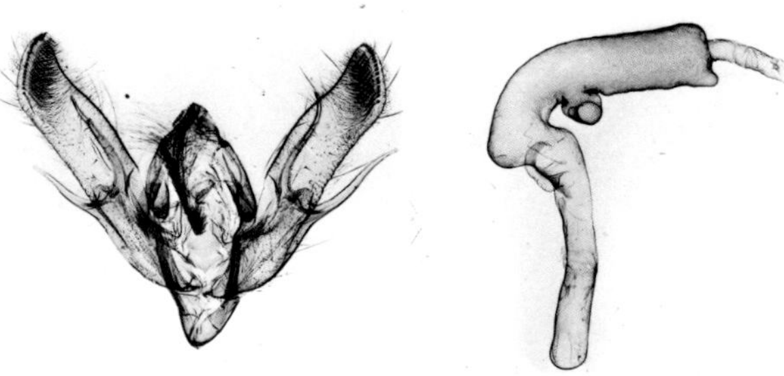
2

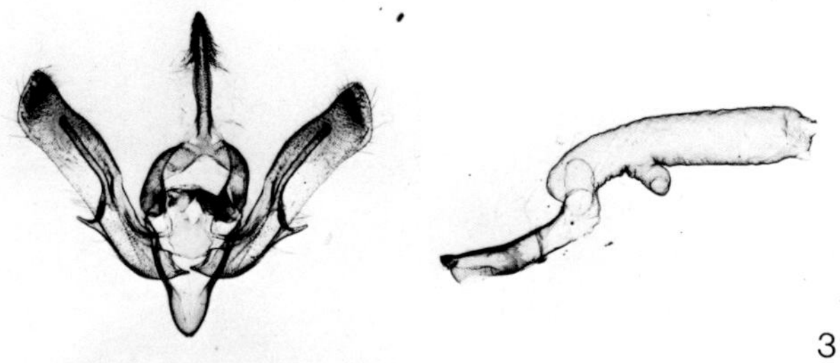
3

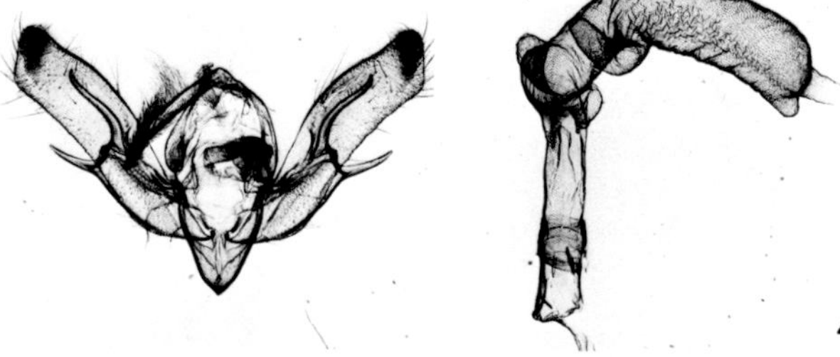
4

5

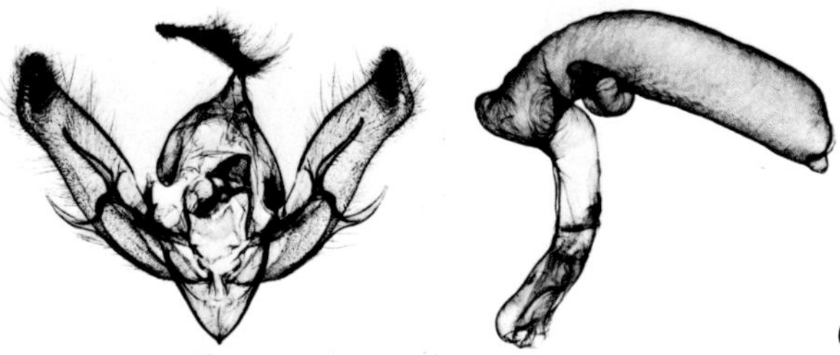
6

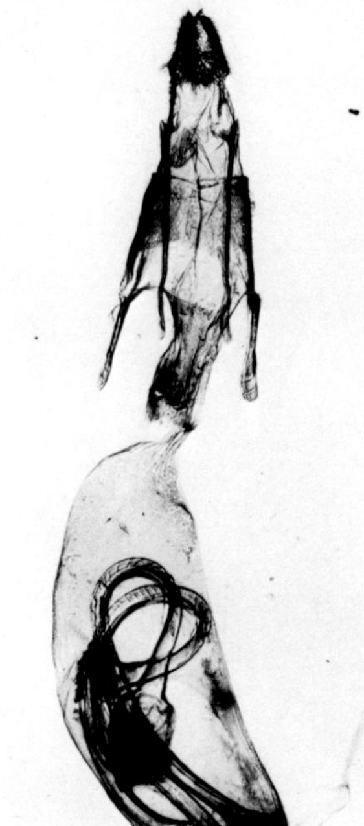
7

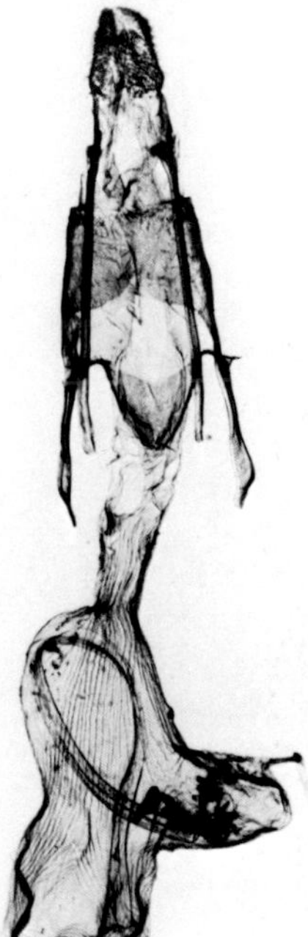
8

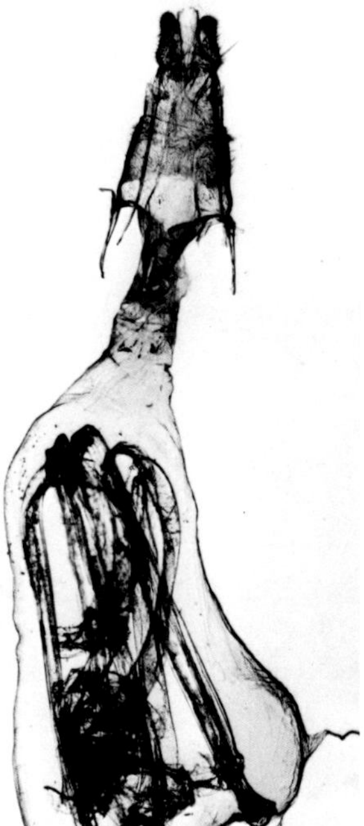
9

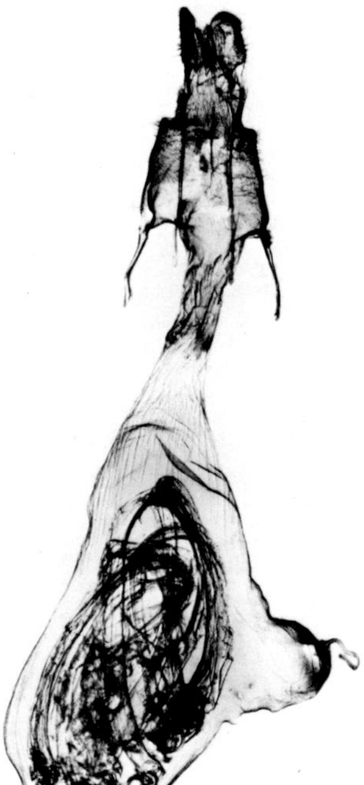
10

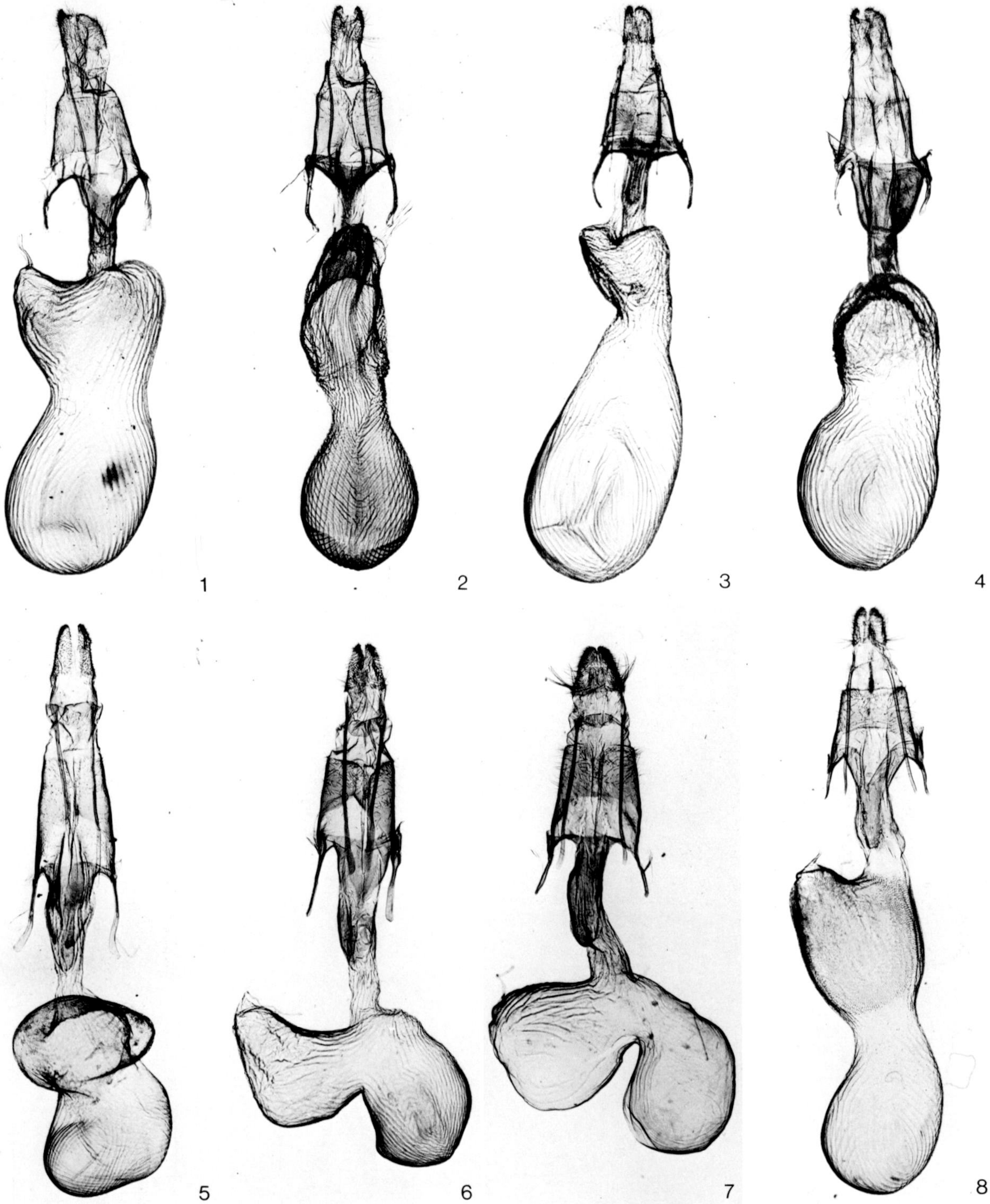
1
2
3
4
5
6
7
8

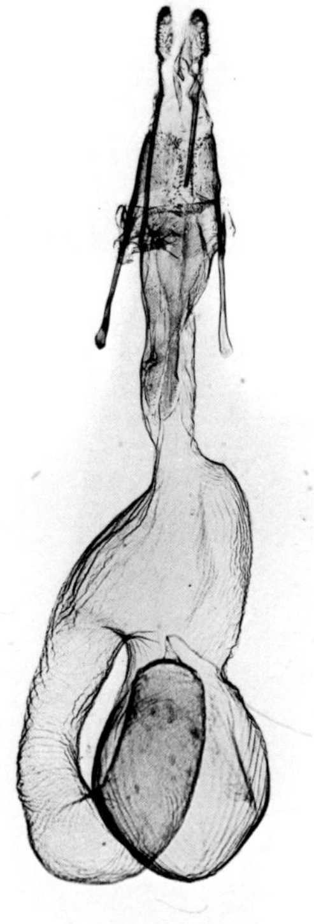

1

2

3

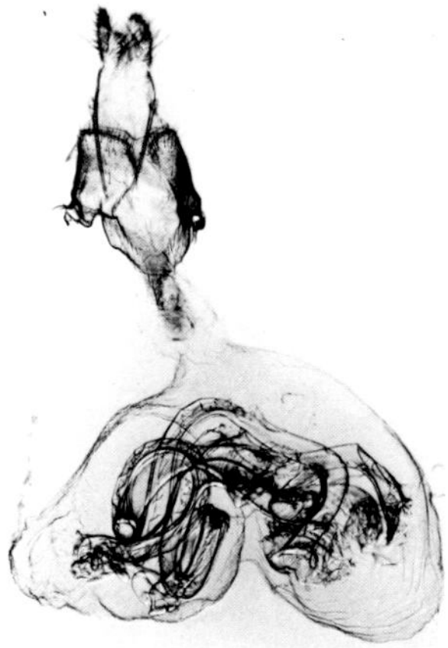

4

5

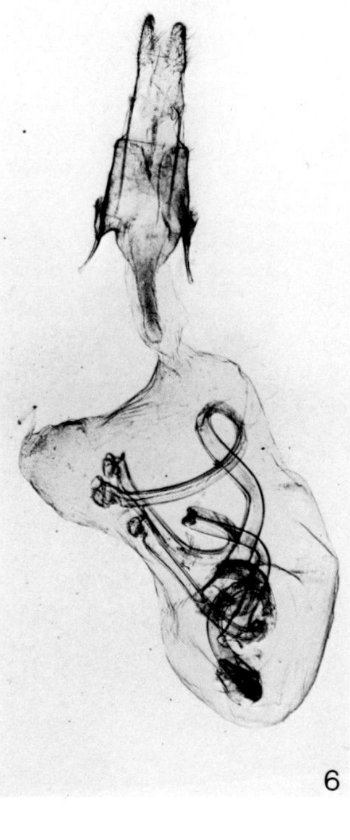

6

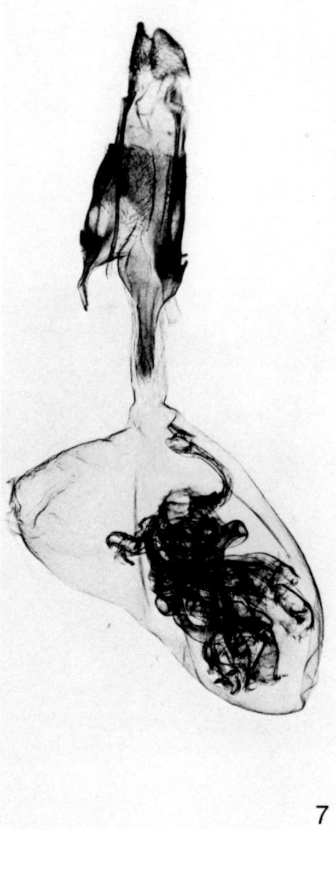

7

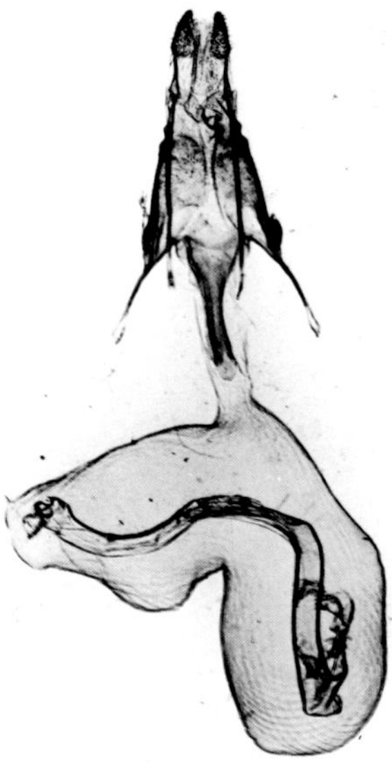

8

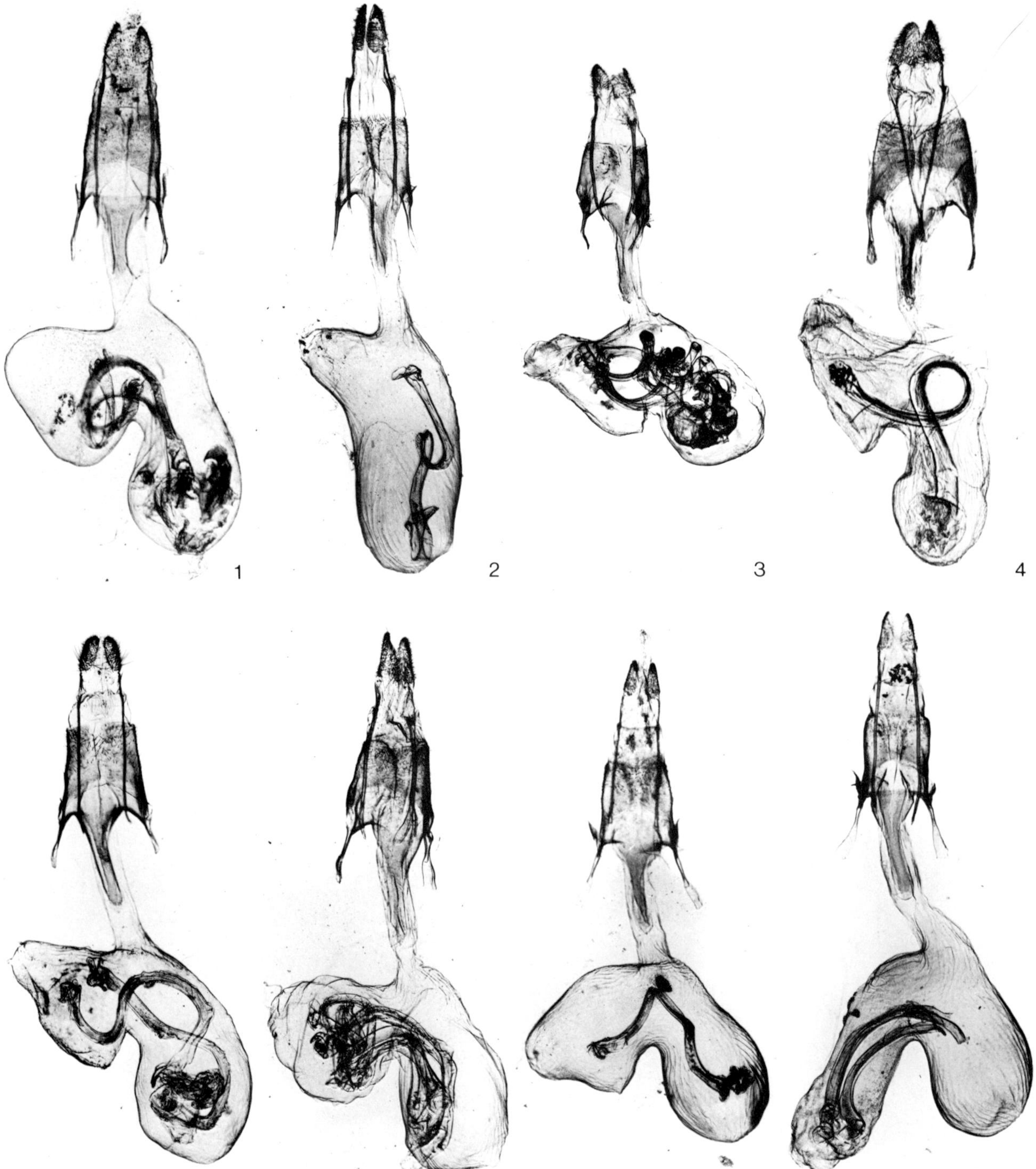
1
2
3
4
5
6
7
8

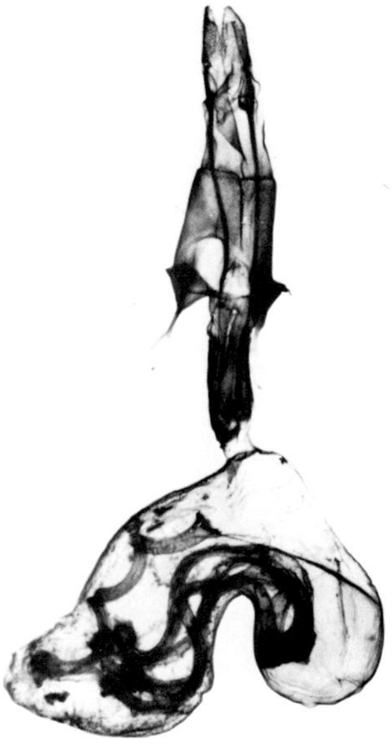

1

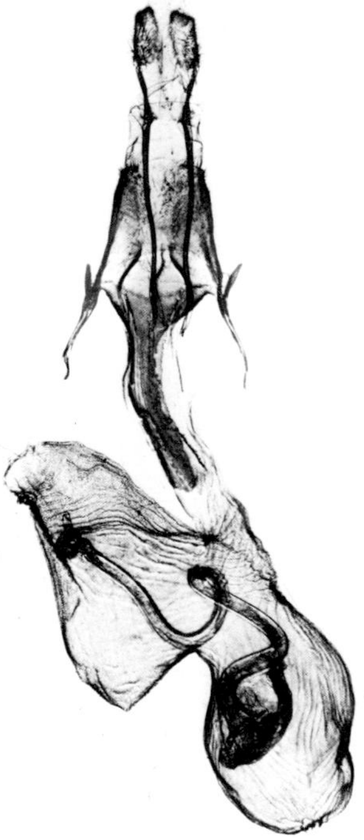

2

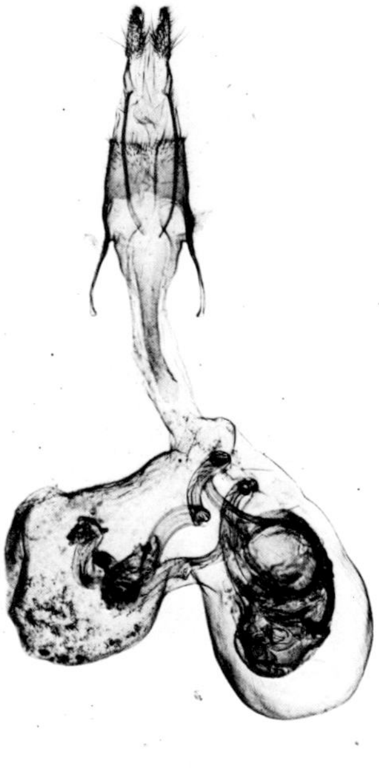

3

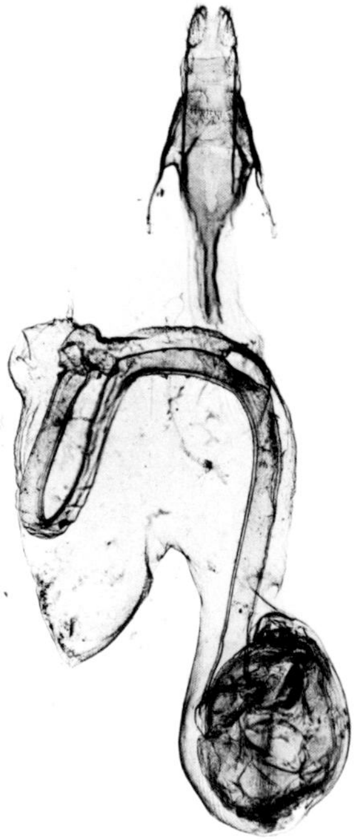

4

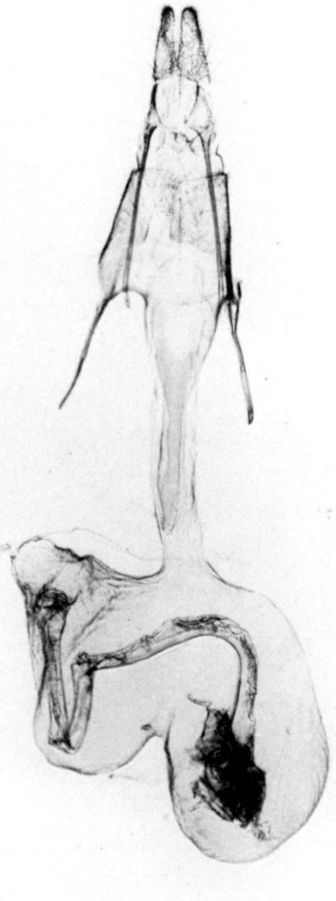

5

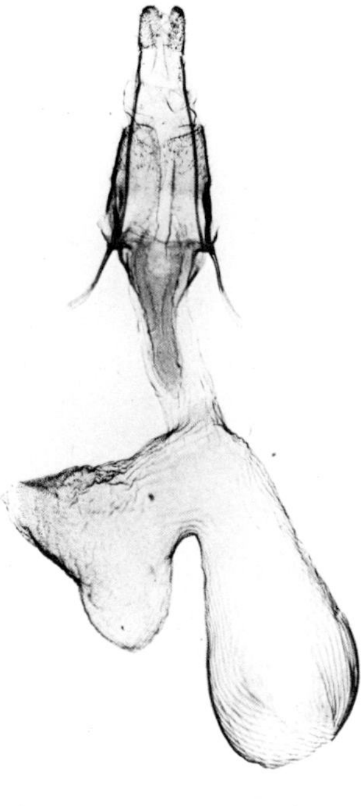

6

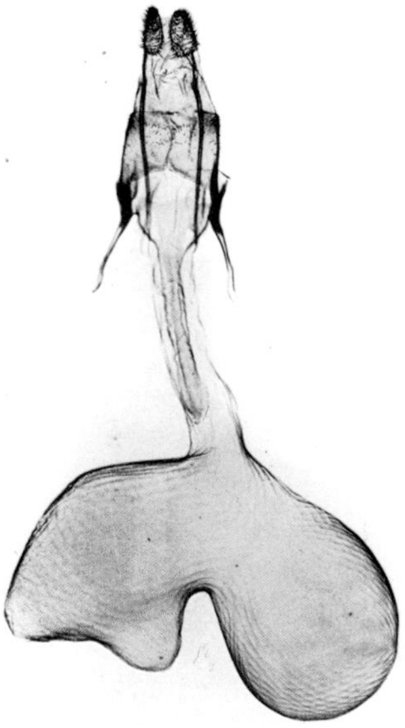

7

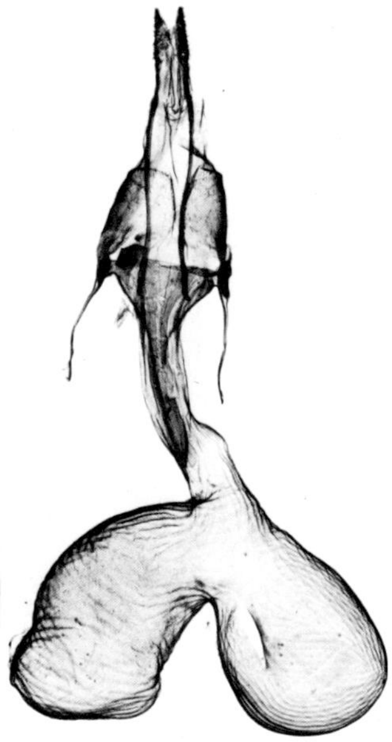

8

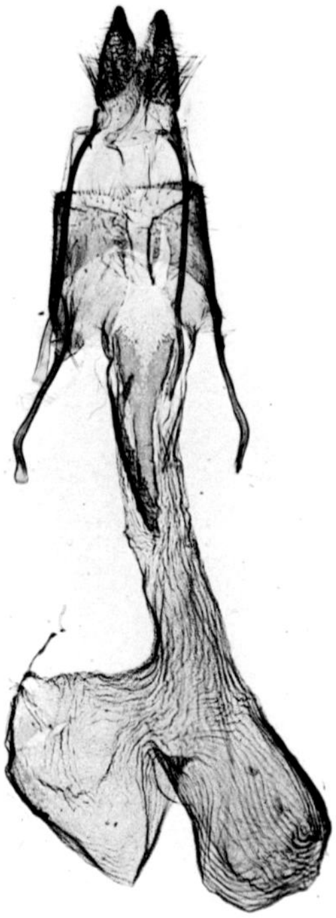

1

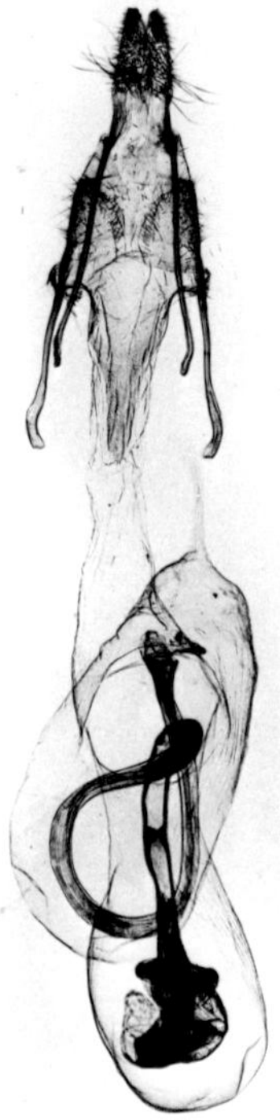

2

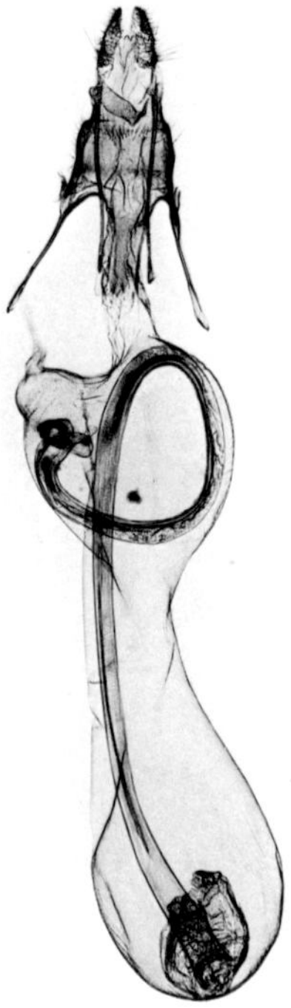

3

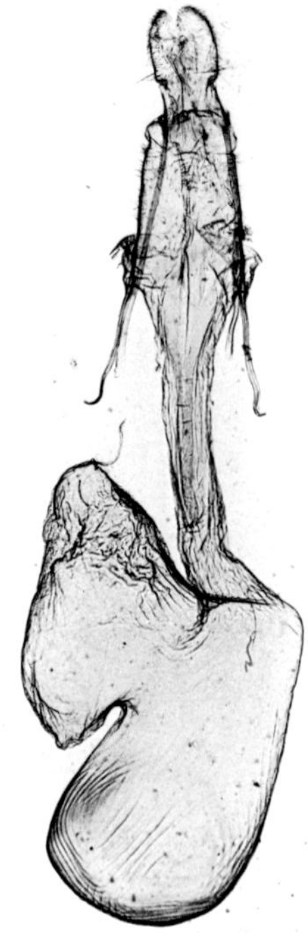

4

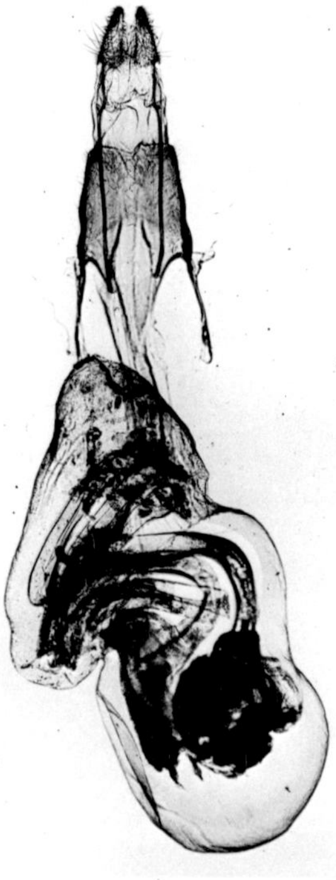

5

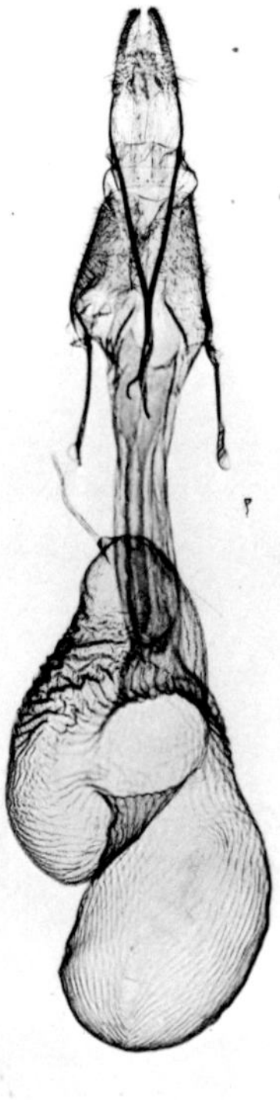

6

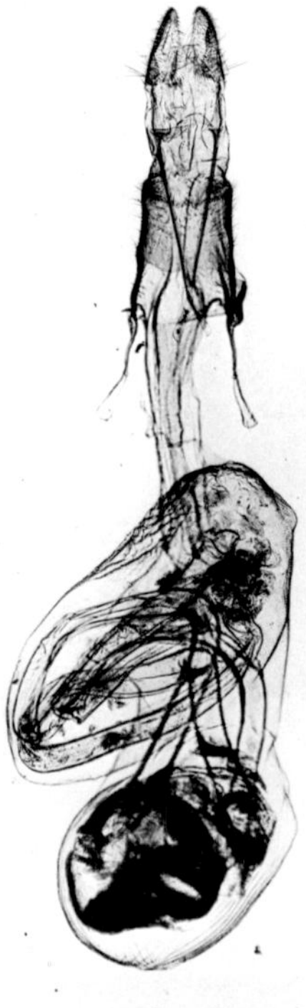

7

8

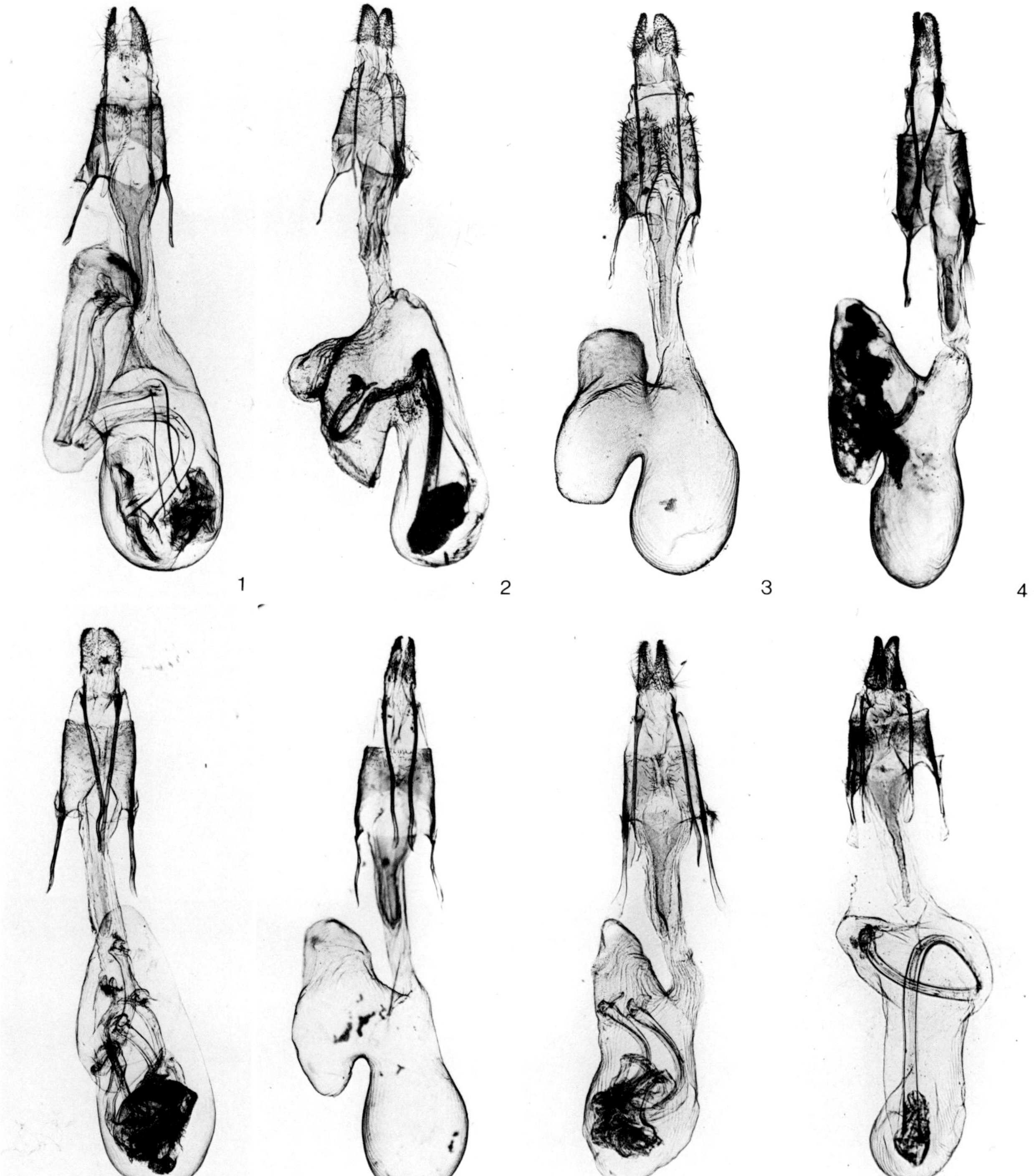
1
2
3
4
5
6
7
8

1

2

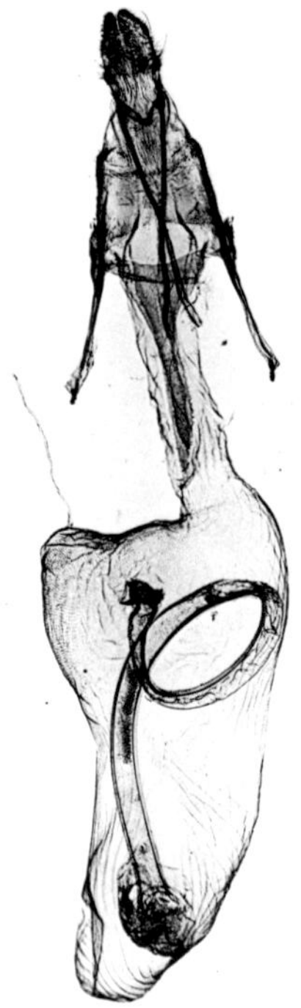
3

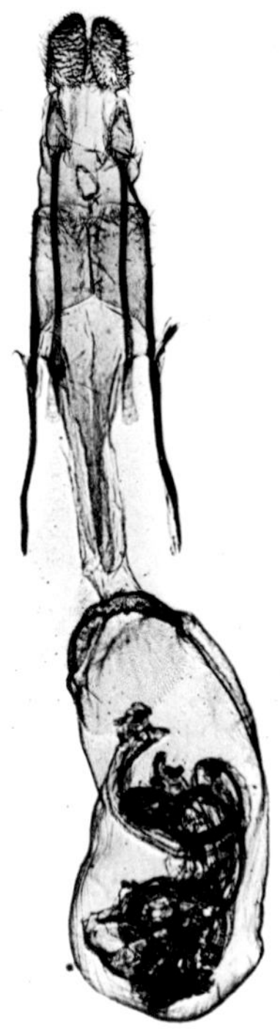
4

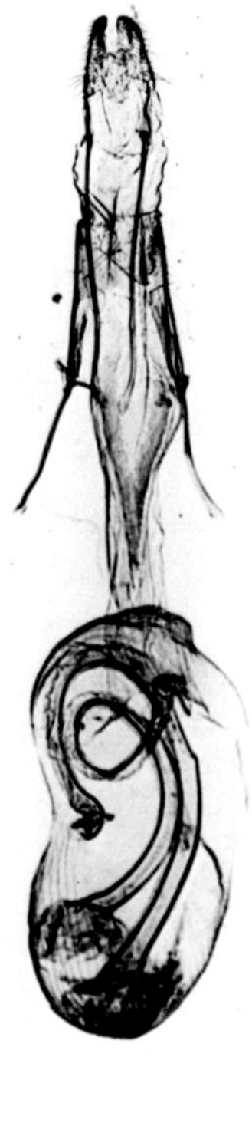
5

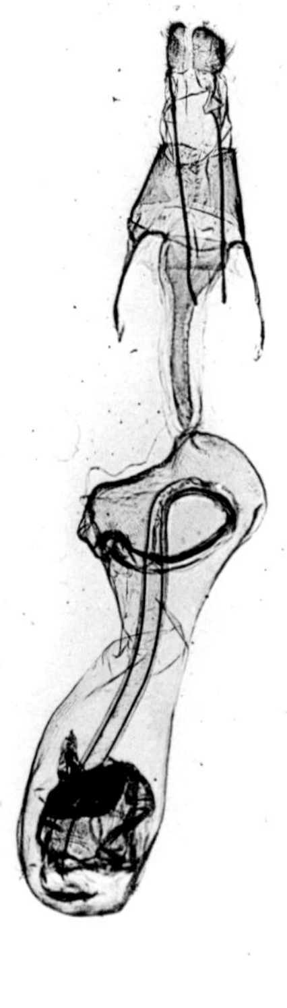
6

7

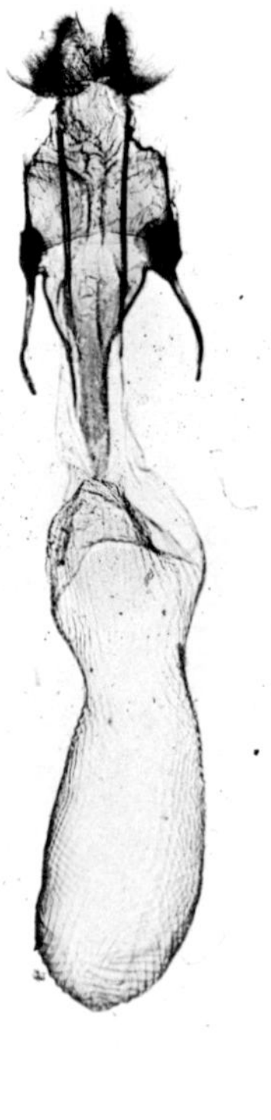
8

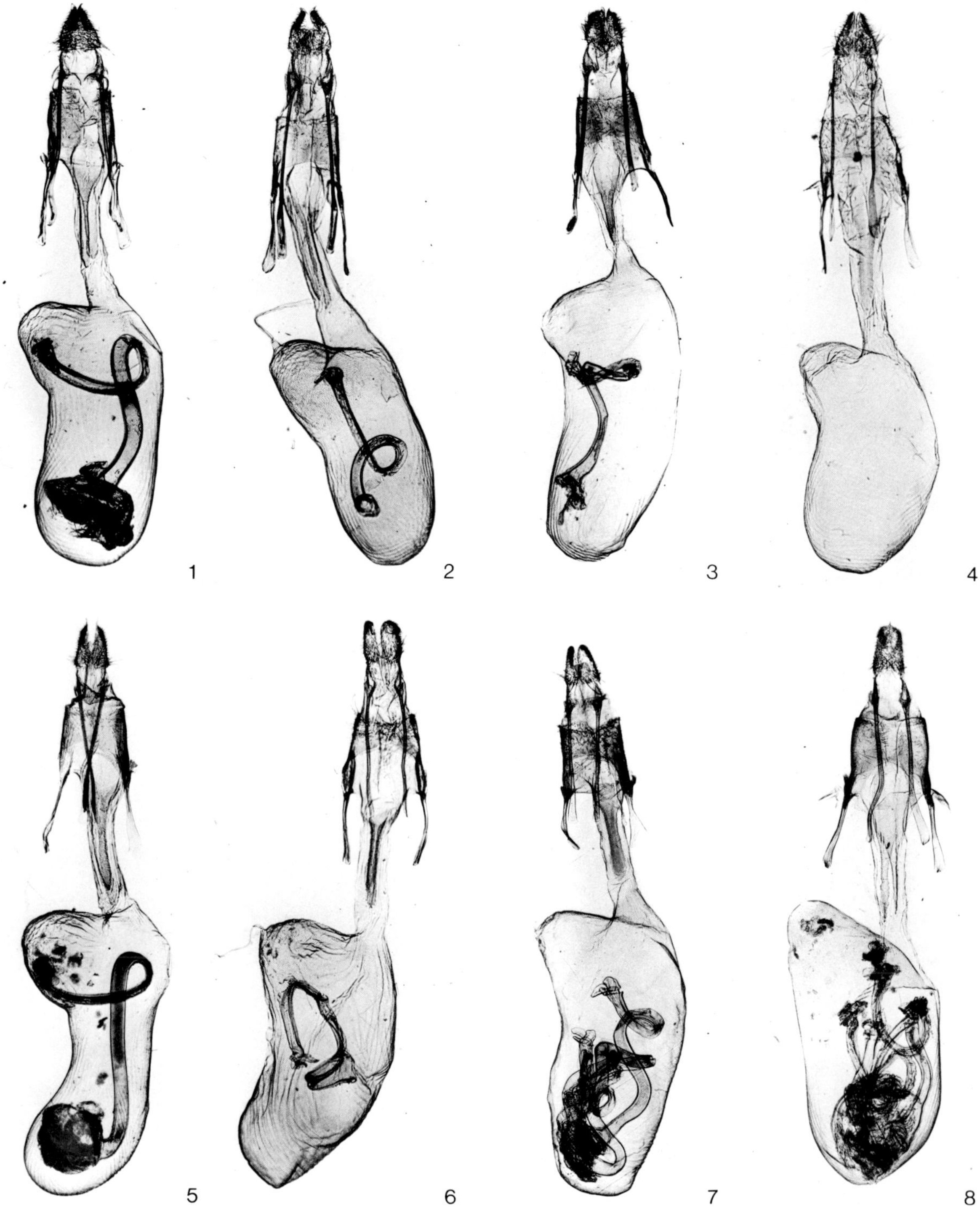
1
2
3
4
5
6
7
8

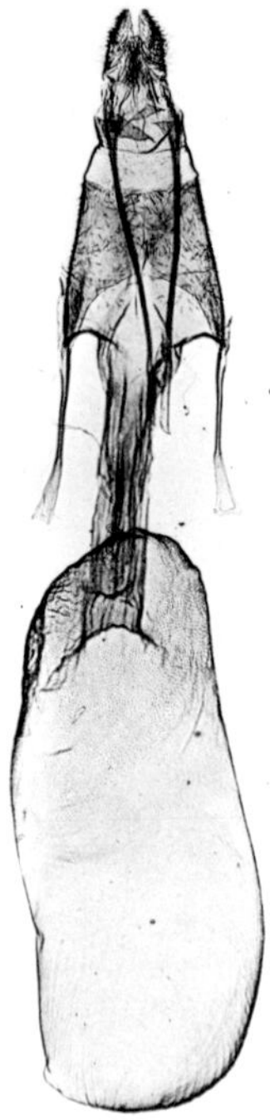

1

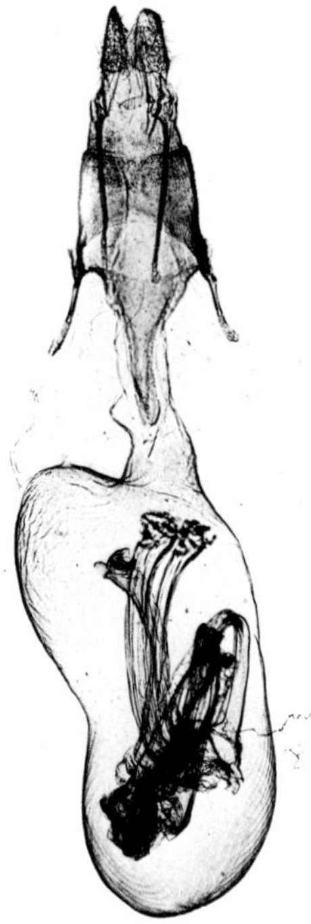

2

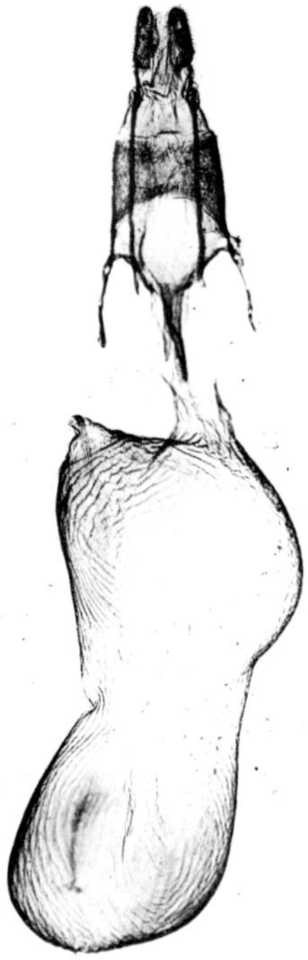

3

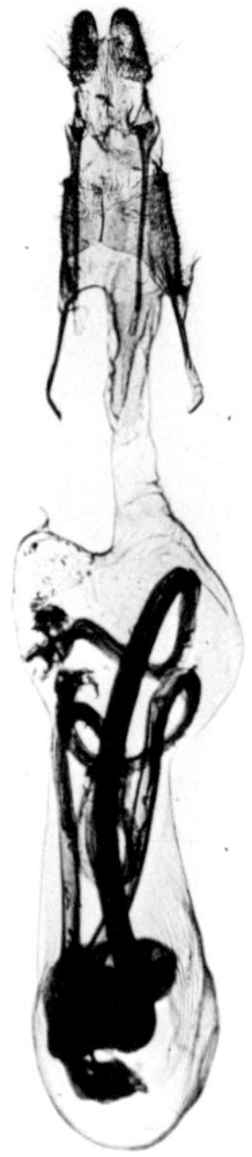

4

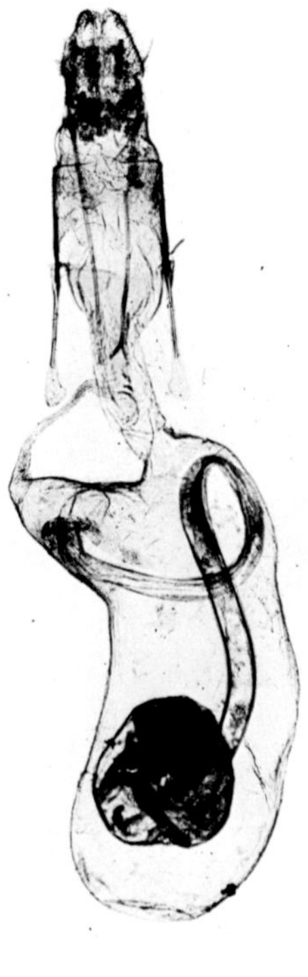

5

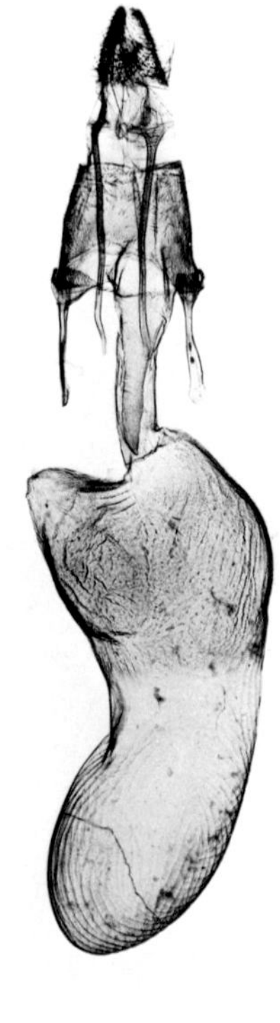

6

7

8

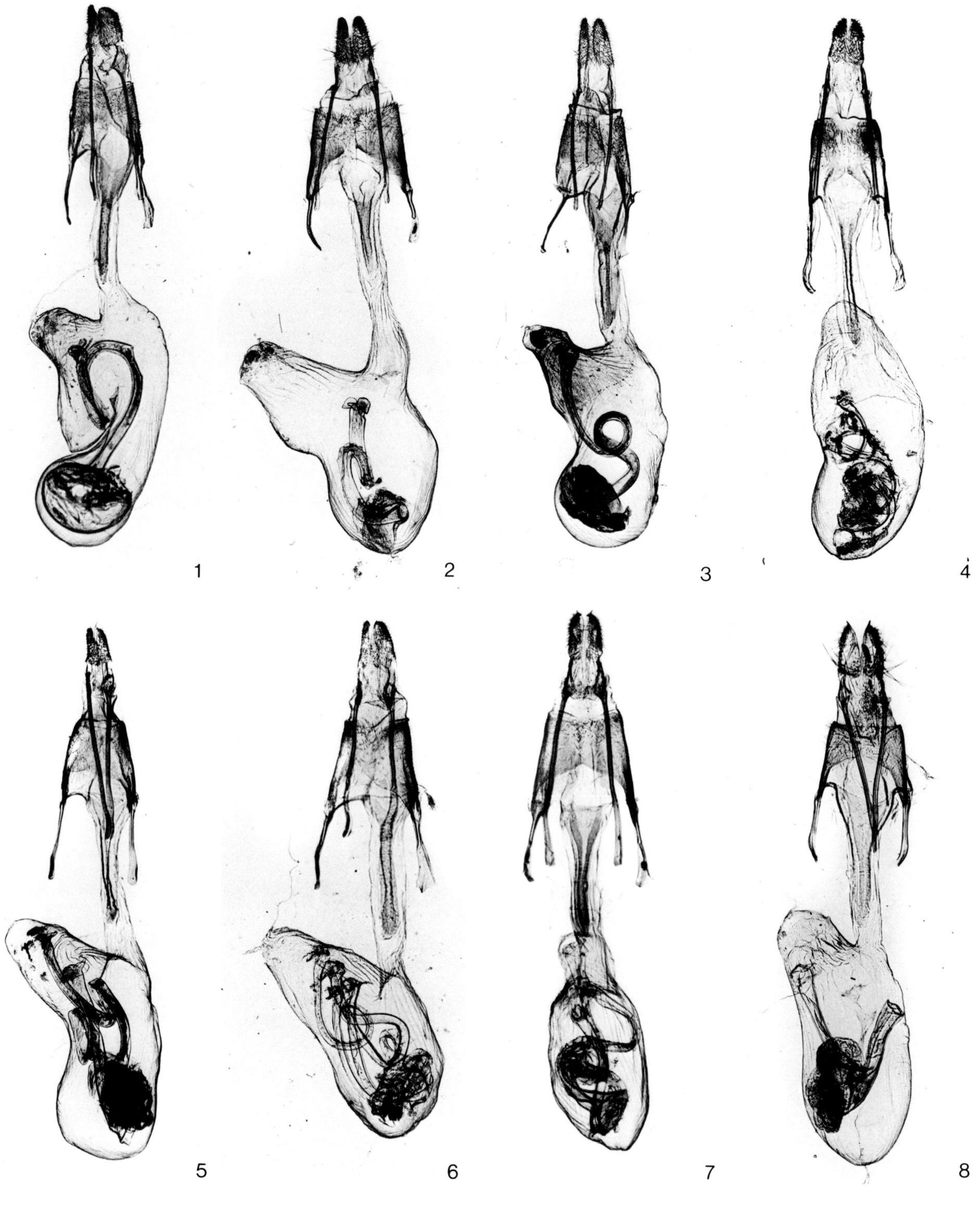
1
2
3
4
5
6
7
8

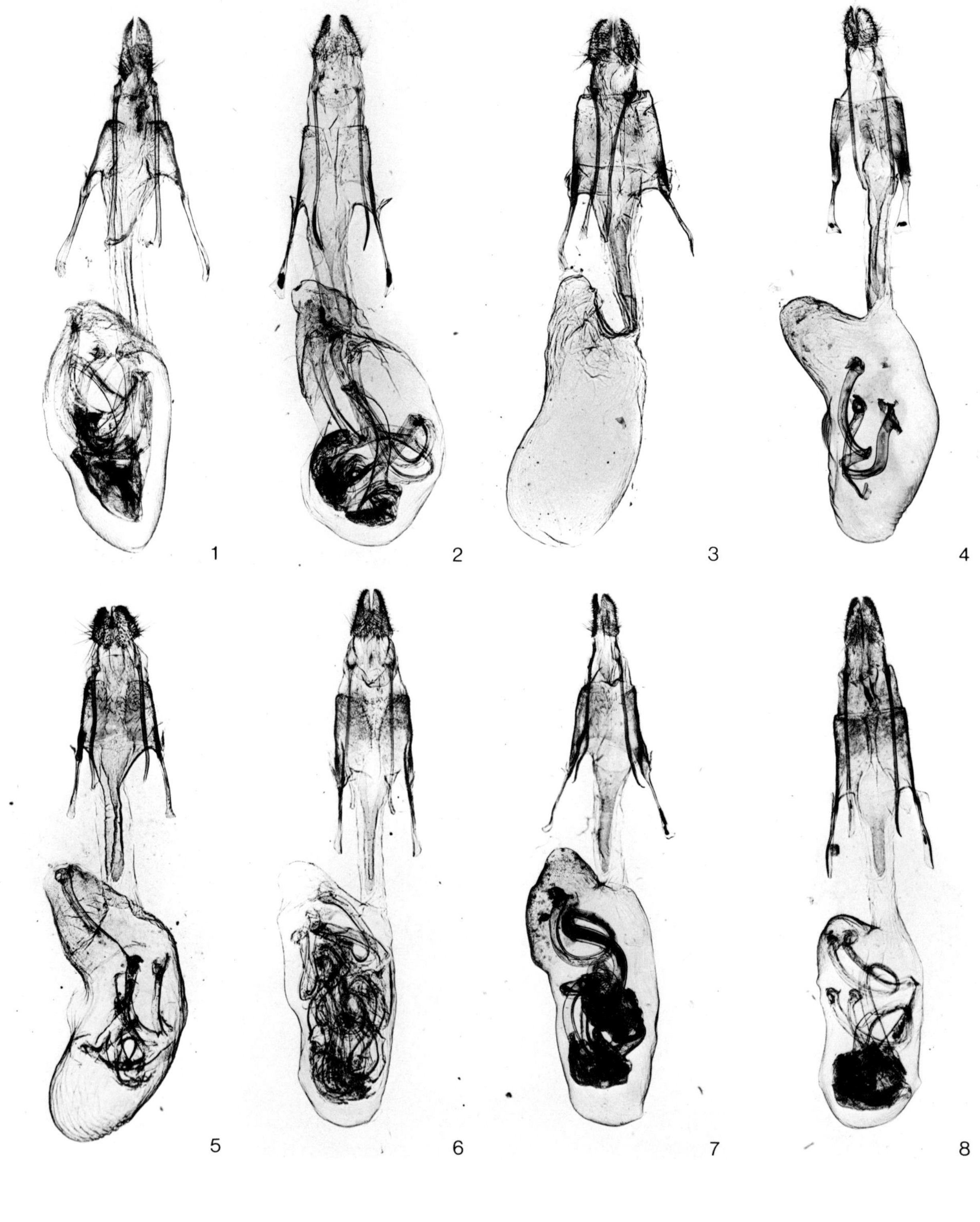
1
2
3
4
5
6
7
8

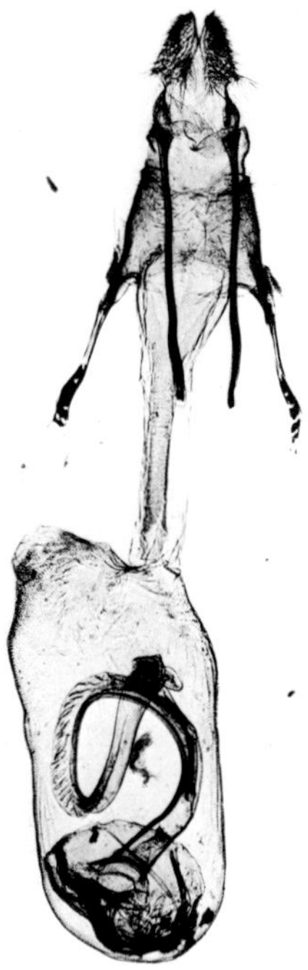

1

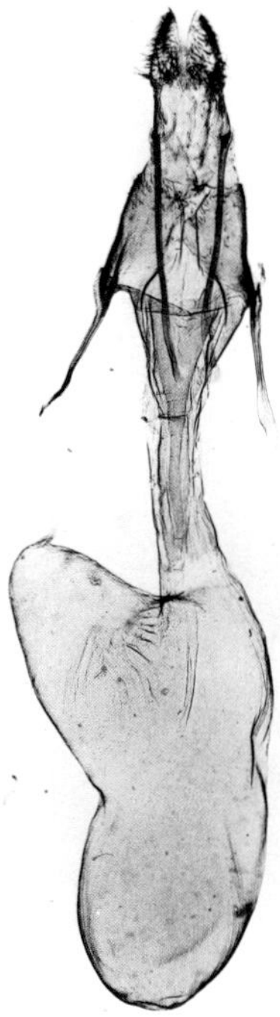

2

3

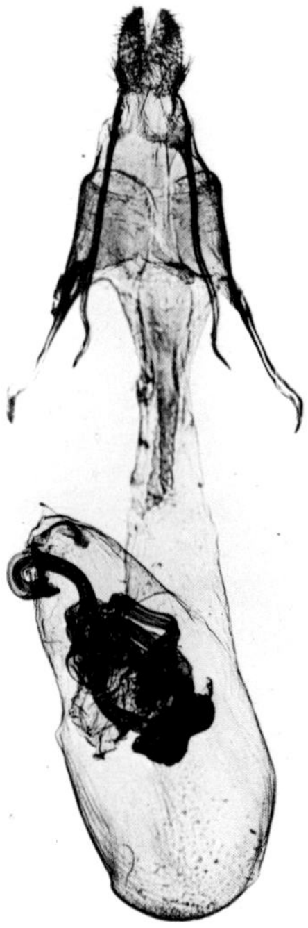

4

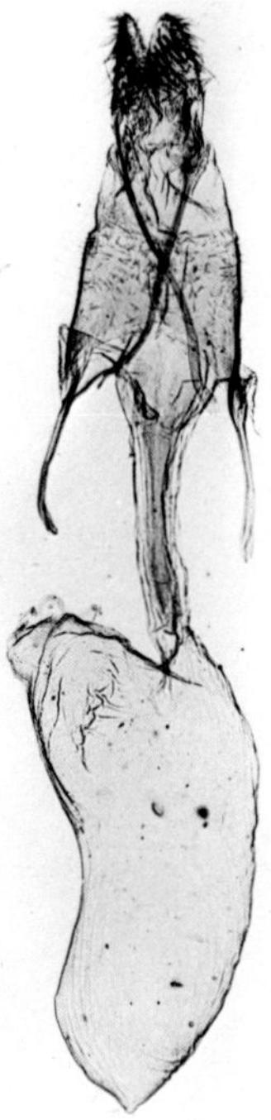

5

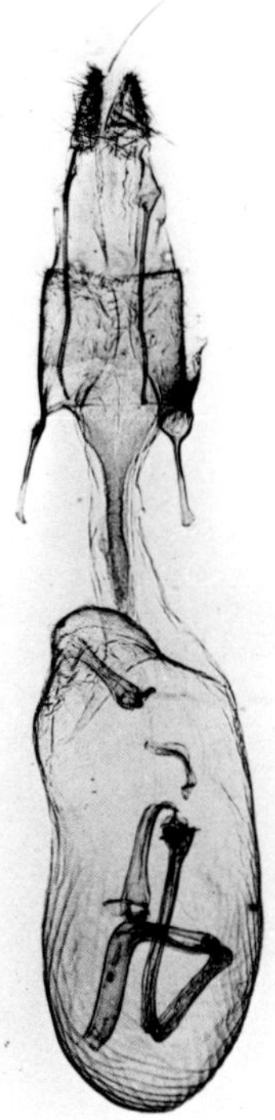

6

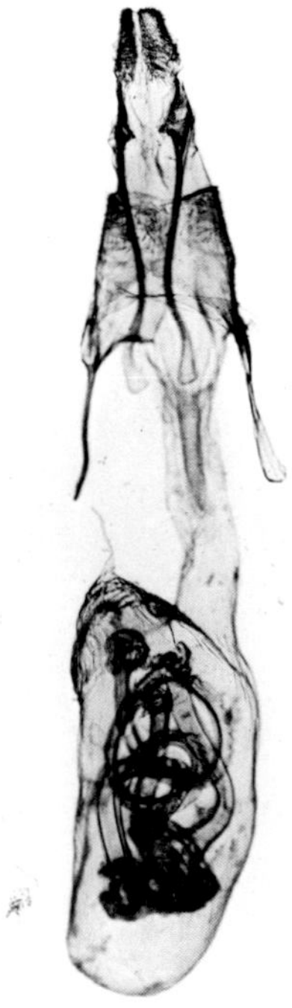

7

8

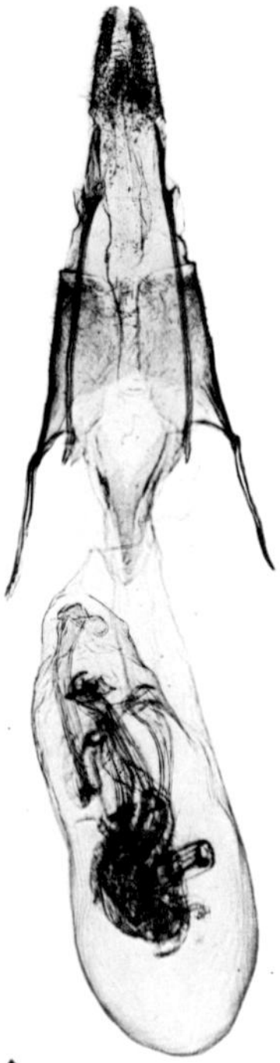

1

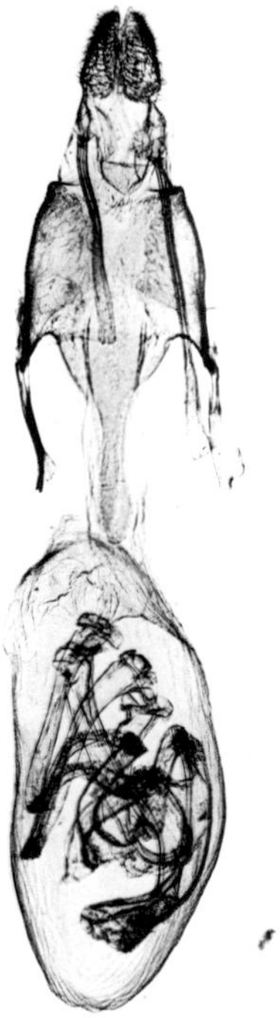

2

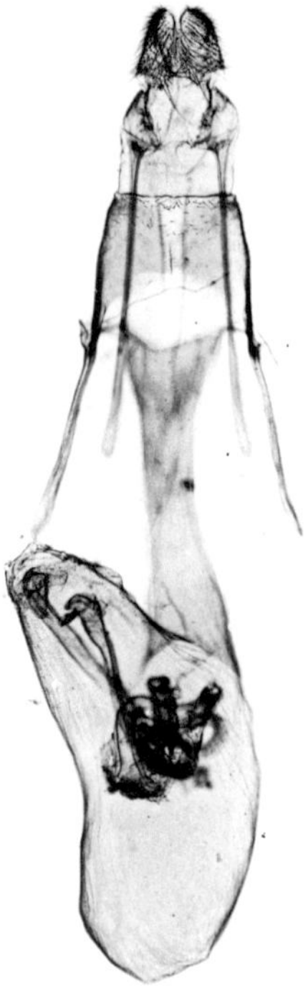

3

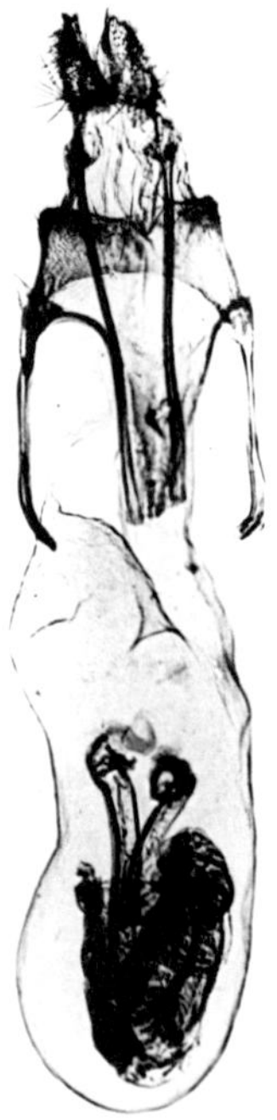

4

5

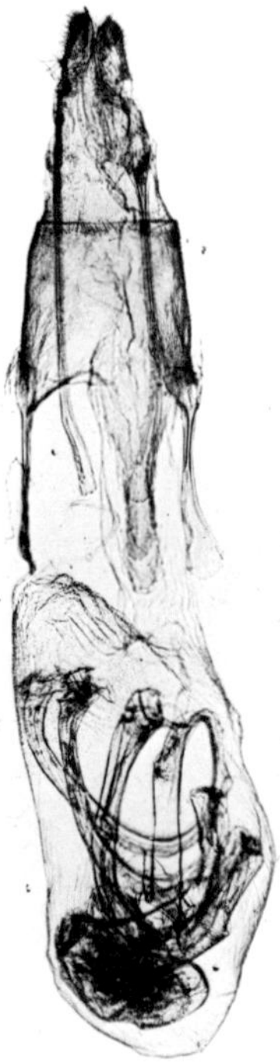

6

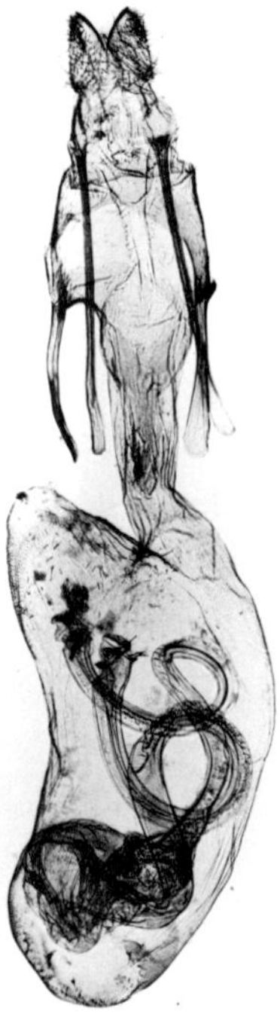

7

8

PLATE EE: FEMALE GENITALIA OF *EUXOA* SPECIES

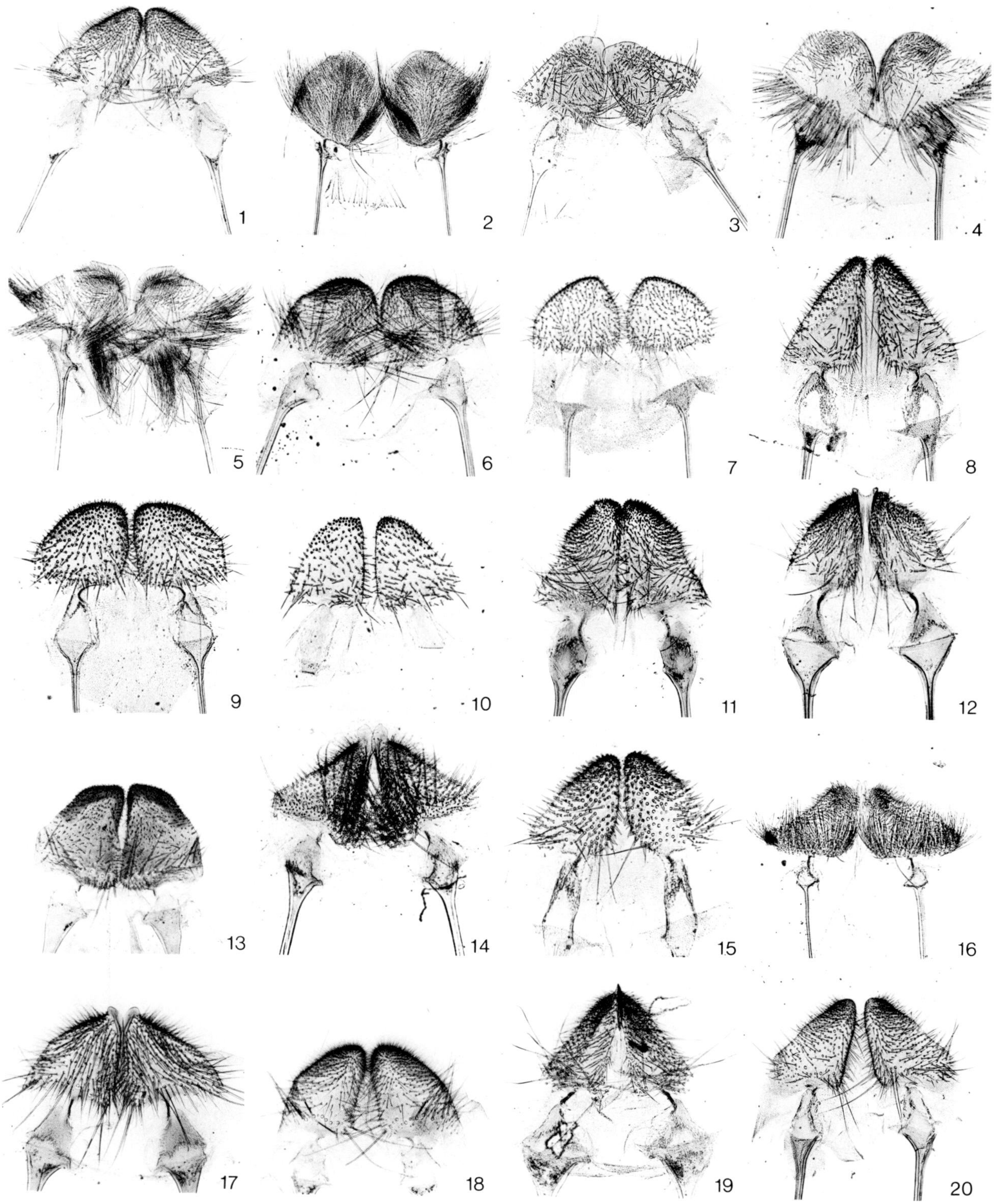

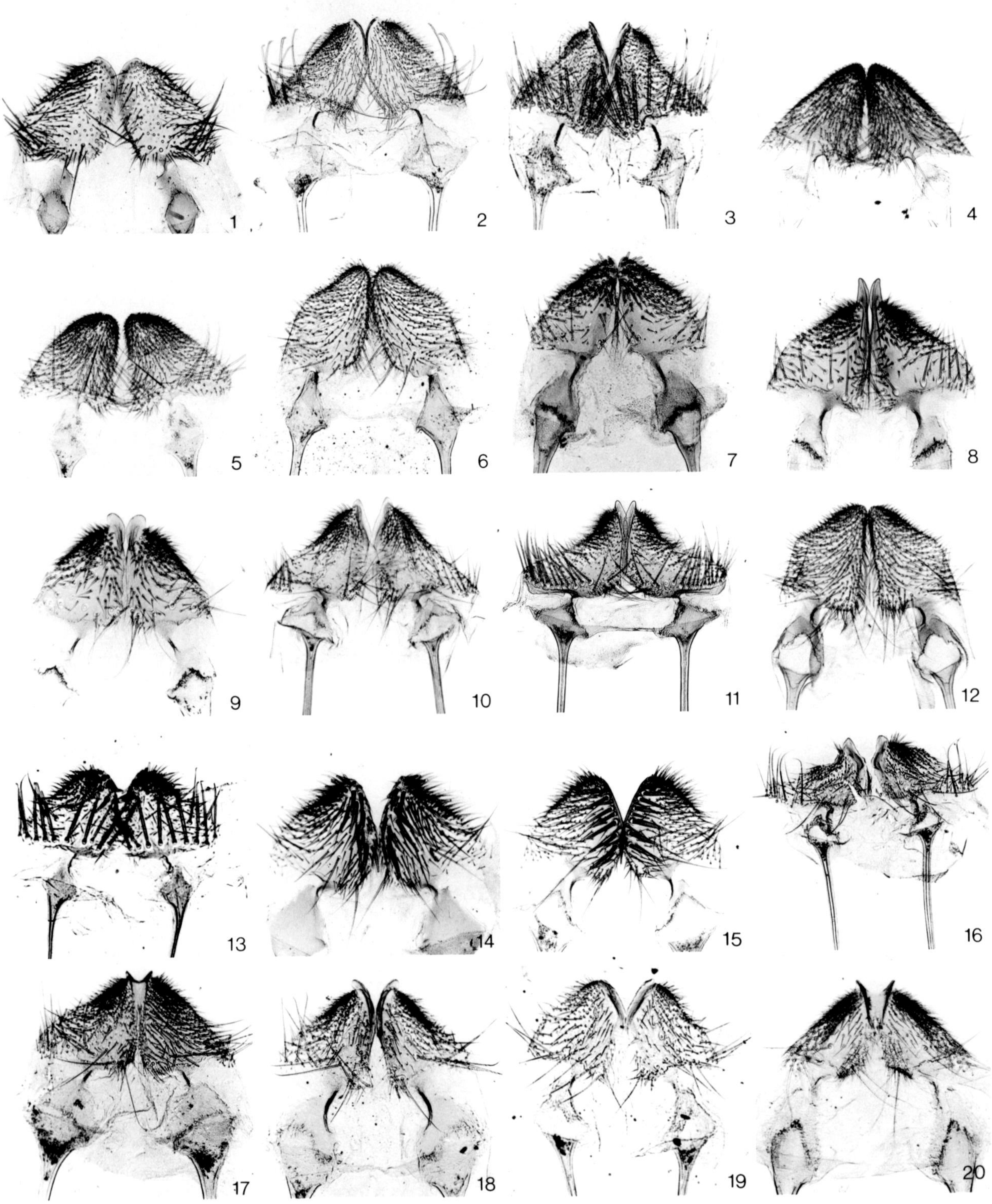
1
2
3
4
5
6
7
8
9
10
11
12
13
14
15
16
17
18
19
20

COLOR PLATES

Noctuoidea

PLATE 1

Noctuoidea

NOCTUIDAE

figs. 1–55

NATURAL SIZE 1:1

1. *Euxoa lidia drewseni* (Stgr.), ♂. Narssarssuaq, Greenland, August 1962, N. L. Wolff (CNC). (p. 29).
2. *Euxoa lidia thanatologia* (Dyar), ♂. Saskatoon, Saskatchewan, emerged 12 December 1976, J. R. Byers (CNC). (p. 30).
3. *Euxoa lidia thanatologia* (Dyar), ♂. Saskatoon, Saskatchewan, emerged 2 December 1978, J. R. Byers (CNC). (p. 30).
4. *Euxoa lidia thanatologia* (Dyar), ♀. Kaslo, British Columbia, 26 August 1902, J. W. Cockle (CNC). (p. 30).

5. *Euxoa lidia thanatologia* (Dyar), ♀. Saskatoon, Saskatchewan, emerged 2 April 1979, J. R. Byers (CNC). (p. 30).
6. *Euxoa auxiliaris* (Grt.), ♀. Buena Vista 7,800′, Colorado, 21 June 1961, M. R. MacKay (CNC). (p. 30).
7. *Euxoa auxiliaris* (Grt.), ♂. Boulder 5,500′, Colorado, 10 June 1961, M. R. MacKay (CNC). (p. 30).
8. *Euxoa auxiliaris* (Grt.), ♂. Boulder 5,500′, Colorado, 11 June 1961, M. R. MacKay (CNC). (p. 30).
9. *Euxoa inconcinna* (Harv.), ♀. Big Bend Natl. Park, Texas, 6 May 1959, L. Bottimer (CNC). (p. 31).
10. *Euxoa inconcinna* (Harv.), ♀. Ramsay Canyon, Cochise Co., Arizona, 15 May 1967, R. Sternitsky (CNC). (p. 31).
11. *Euxoa terrealis* (Grt.), ♀. Fort Valley 7,350′, 7½ mi NW Flagstaff, Coconino Co., Arizona, 16 July 1961, J. G. Franclemont (CNC). (p. 31).
12. *Euxoa terrealis* (Grt.), ♂. Hart Prairie 8,500′, 10 mi NNW Flagstaff, Coconino Co., Arizona, 24 June 1964, J. G. Franclemont (JGF). (p. 31).
13. *Euxoa terrealis* (Grt.), ♂. 5 mi N Prescott 5,450′, Yavapai Co., Arizona, 5 June 1972, Lloyd M. Martin (CNC). (p. 31).
14. *Euxoa shasta shasta* Lafontaine, ♂. Paratype. McBride Springs 4,800′, Siskiyou Co., California, 20 July 1965, E. & I. Munroe (CNC). (p. 32).
15. *Euxoa biformata* Sm., ♂. Nelson Creek, Plumas Co., California, 17 August 1940, W. R. Bauer (CNC). (p. 33).
16. *Euxoa biformata* Sm., ♀. Hat Creek 4,300′, Shasta Co., California, 29 August 1948, W. R. Bauer (CNC). (p. 33).
17. *Euxoa intermontana* Lafontaine, ♂. Paratype. 15 mi S Kanosh 6,600′, Utah, 27 August 1965, D. F. Hardwick (CNC). (p. 33).
18. *Euxoa intermontana* Lafontaine, ♀. Paratype. 5 mi SW Ely 7,400′, Nevada, 6 September 1969, D. F. Hardwick (CNC). (p. 33).
19. *Euxoa mimallonis mimallonis* (Grt.), ♂. Forestville, Quebec, 9 August 1950, R. de Rouette (CNC). (p. 34).
20. *Euxoa mimallonis gagates* (Grt.), ♀. Capulin Natl. Monument, 6 mi W Folsom 7,300′, New Mexico, 13 September 1968, D. F. Hardwick (CNC). (p. 34).
21. *Euxoa mimallonis gagates* (Grt.), ♂. Blue Ridge 7,386′, San Gabriel Mts., Los Angeles Co., California, 8 September 1963, C. Henne (LACM). (p. 34).
22. *Euxoa vernalis* Lafontaine, ♀. Paratype. Fort Valley 7,350′, 7½ mi NW Flagstaff, Coconino Co., Arizona, J. G. Franclemont (CNC). (p. 35).
23. *Euxoa septentrionalis* (Wlk.), ♂. 1 mi SW Bass Lake 3,400′, California, 1 September 1966, D. F. Hardwick (CNC). (p. 36).
24. *Euxoa septentrionalis* (Wlk.), ♀. 6 mi E Mayfield 6,500′, Utah, 25 August 1965, D. F. Hardwick (CNC). (p. 36).
25. *Euxoa olivia* (Morr.), ♀. Juniper Hills 3,500′, Mojave Desert, Los Angeles Co., California, 15 October 1962, C. Henne (LACM). (p. 36).
26. *Euxoa olivia* (Morr.), ♂. Juniper Hills 3,600′, Mojave Desert, Los Angeles Co., California, 5 October 1961, C. Henne (LACM). (p. 36).
27. *Euxoa olivia* (Morr.), ♀. Juniper Hills 3,500′, Mojave Desert, Los Angeles Co., California, 15 October 1961, C. Henne (LACM). (p. 36).
28. *Euxoa olivia* (Morr.), ♂. 6 mi NW Hugo 5,200′, Colorado, 9 September 1968, D. F. Hardwick (CNC). (p. 36).
29. *Euxoa messoria* (Harr.), ♂. Buckhorn Flats 6,800′, San Gabriel Mts., Los Angeles Co., California, 25 July 1947, C. Henne (LACM). (p. 37).
30. *Euxoa messoria* (Harr.), ♀. Mt. Pinos 8,500′, Los Padres Natl. Forest, Kern Co., California, 19 September 1963, C. Henne (LACM). (p. 37).
31. *Euxoa divergens* (Wlk.), ♂. 10 mi WSW Georgetown 11,000′, Colorado, 1 August 1967, D. F. Hardwick (CNC). (p. 39).
32. *Euxoa sinelinea* Hdwk., ♂. T29N R2W, Sec. 18, Otsego Co., Michigan, 5 July 1980, M. C. Nielsen (CNC). (p. 39).
33. *Euxoa sinelinea* Hdwk., ♀. Lukinto Lake, Hwy 11, Ontario, 15–21 July 1978, A. Mutuura (CNC). (p. 39).
34. *Euxoa edictalis* (Sm.), ♂. White Mts. 10,000′, Mono Co., California, 28 June 1962, P. Manis & D. Mathias (CNC). (p. 40).
35. *Euxoa edictalis* (Sm.), ♂. Summerland, British Columbia, 17 August 1930, A. N. Gartrell (CNC). (p. 40).
36. *Euxoa westermanni* (Stgr.), ♂. Cameron Bay, Great Bear Lake, Northwest Territories, 12 July 1937, T. N. Freeman (CNC). (p. 44).
37. *Euxoa dissona* (Mösch.), ♂. Hopedale, Labrador, 20 July 1922 (CNC). (p. 46).
38. *Euxoa dissona* (Mösch.), ♂. Mt. Katahdin 4,600′–5,200′, Maine, 8 July 1950, A. E. Brower (CNC). (p. 46).
39. *Euxoa dissona* (Mösch.), ♀. Lake of the Clouds 5,000′, Mt. Washington, New Hampshire, 31 July 1954, Becker, Munroe & Mason (CNC). (p. 46).
40. *Euxoa hyperborea* Lafontaine, ♂. Holotype. Gubik Gas Field 200′, 1 mi S Colville River, Alaska, 30 June 1970, J. Hok (USNM). (p. 46).
41. *Euxoa quebecensis* (Sm.), ♂. 4 mi SE Cowley 3,800′, Alberta, 11 June 1979, J. D. Lafontaine (CNC). (p. 47).
42. *Euxoa quebecensis* (Sm.), ♂. 4 mi SE Cowley 3,800′, Alberta, 11 June 1979, J. D. Lafontaine (CNC). (p. 47).
43. *Euxoa scandens* (Riley), ♀. Lethbridge, Alberta, 3 August 1927, H. L. Seamans (CNC). (p. 47).
44. *Euxoa scandens* (Riley), ♂. 12 km SW Ottawa, Ontario, 1 July 1986, J. D. Lafontaine (CNC). (p. 47).
45. *Euxoa aurulenta* (Sm.), ♂. Wallula, Washington, 7 May 1949, W. C. Cook (CNC). (p. 48).
46. *Euxoa aurulenta* (Sm.), ♂. Port Colborne, Ontario, 2 June 1934, J. J. de Gryse (CNC). (p. 48).
47. *Euxoa vallus vallus* (Sm.), ♂. Slate Peak 7,400′, Okanogan Co., Washington, 3 August 1963, J. Shepherd (CNC). (p. 49).
48. *Euxoa vallus luteosita* (Sm.), ♂. 12 mi WNW Estes Park 11,600′, Colorado, 29 July 1967, D. F. Hardwick (CNC). (p. 49).
49. *Euxoa vallus luteosita* (Sm.), ♂. Pennsylvania Mtn. 3,550 m, Park Co., Colorado, D. Ford (CNC). (p. 49).
50. *Euxoa vallus bivittata* Lafontaine, ♂. Holotype. 7 mi WSW Lee Vining 9,600′, California, 12 August 1967, D. F. Hardwick (CNC). (p. 49).
51. *Euxoa c. churchillensis* (McD.), ♀. 21 mi E Tuktoyaktuk, 20–25 June 1971, D. M. Wood (CNC). (p. 45).
52. *Euxoa churchillensis alpina* Lafontaine, ♂. Holotype. Pennsylvania Mtn., Park Co., Colorado, 21 July 1980, D. Ford (CNC). (p. 45).
53. *Euxoa churchillensis alpina* Lafontaine, ♀. Paratype. Wheeler Park near Twining 12,000′, Taos Co., New Mexico, 3 July 1966, M. Toliver (LACM). (p. 45).
54. *Euxoa m. macleani* McD., ♂. Slate Peak 7,400′, Okanogan Co., Washington, 3 August 1963, J. Shepherd (CNC). (p. 50).
55. *Euxoa macleani chimoensis* Hdwk., ♂. Holotype. Ft. Chimo, Quebec, 25 June 1948, H. N. Smith (CNC). (p. 50).

1
2
3
4
5
6
7
8
9
10
11
12
13
14
15
16
17
18
19
20
21
22
23
24
25
26
27
28
29
30
31
32
33
34
35
36
37
38
39
40
41
42
43
44
45
46
47
48
49
50
51
52
53
54
55

PLATE 2

Noctuoidea

NOCTUIDAE

figs. 1–55

NATURAL SIZE 1:1

1. *Euxoa trifasciata* (Sm.), ♀. 3 mi NE Mt. Shasta 5,300′, California, 7 September 1974, J. D. Lafontaine (CNC). (p. 50).
2. *Euxoa l. lewisi* (Grt.), ♂. 12 mi WNW Estes Park 11,600′, Colorado, 29 July 1967, D. F. Hardwick (CNC). (p. 51).
3. *Euxoa lewisi juliae* Hdwk., ♂. Paratype. 7 mi WSW Lee Vining 9,600′, California, 13 August 1967, D. F. Hardwick (CNC). (p. 51).
4. *Euxoa altens* McD., ♂. 10 mi NE Mineral 7,400′, California, 24 August 1967, D. F. Hardwick (CNC). (p. 51).

5. *Euxoa altens* McD., ♀. 10 mi NE Mineral 7,400′, California, 23 August 1967, D. F. Hardwick (CNC). (p. 51).

6. *Euxoa austrina* Hdwk., ♂. Paratype. Buckhorn Flats 6,500′, San Gabriel Mts., Los Angeles Co., California, 1 July 1948, Claude I. Smith (CNC). (p. 52).

7. *Euxoa austrina* Hdwk., ♀. Mt. Pinos 8,500′, Los Padres Natl. Forest, Kern Co., California, 10 June 1961, C. Henne (LACM). (p. 52).

8. *Euxoa cryptica* Hdwk., ♂. Paratype. 20 mi W Bishop 10,000′, California, 11 August 1967, D. F. Hardwick (CNC). (p. 52).

9. *Euxoa leuschneri* Lafontaine, ♂. Holotype. Barton Flats 6,300′, San Bernardino Mts., California, 15 June 1985, R. H. Leuschner (LACM). (p. 52).

10. *Euxoa extranea* (Sm.), ♂. 10 mi NE Mineral 7,400′, California, 24 August 1967, D. F. Hardwick (CNC). (p. 50).

11. *Euxoa tristicula* (Morr.), ♀. Cypress Hills, Saskatchewan, 9 June 1939, A. R. Brooks (CNC). (p. 54).

12. *Euxoa tristicula* (Morr.), ♂. Manyberries, Alberta, 12 July 1951, D. F. Hardwick (CNC). (p. 54).

13. *Euxoa vetusta* (Wlk.), ♀. Stimson Creek, Mason Co., Washington, 1 June 1948, D. Frechin (CNC). (p. 54).

14. *Euxoa fuscigera* (Harv.), ♂. 6 mi WSW Stoneyford 1,600′, California, 27 September 1966, D. F. Hardwick (CNC). (p. 55).

15. *Euxoa fuscigera* (Harv.), ♀. Petaluma, California, 12 October 1946, W. R. Bauer (CNC). (p. 55).

16. *Euxoa atomaris* (Sm.), ♂. Red Rover Mine Canyon SW Acton 3,200′, Los Angeles Co., California, 24 September 1963, C. Henne (LACM). (p. 55).

17. *Euxoa atomaris* (Sm.), ♀. Vincent Gap 6,500′, San Gabriel Mts., Los Angeles Co., California, 20 September 1966, C. Henne (LACM). (p. 55).

18. *Euxoa atomaris* (Sm.), ♂. 5 mi SE Follyfarm 4,250′, Oregon, 15 September 1961, D. F. Hardwick (CNC). (p. 55).

19. *Euxoa atomaris* (Sm.), ♂. 4 mi N Sequim 200′, Washington, 17 August 1960, D. F. Hardwick (CNC). (p. 55).

20. *Euxoa pleuritica* (Grt.), ♂. Manyberries, Alberta, 13 July 1951, D. F. Hardwick (CNC). (p. 56).

21. *Euxoa pleuritica* (Grt.), ♀. Manyberries, Alberta, 14 July 1951, D. F. Hardwick (CNC). (p. 56).

22. *Euxoa pestula* Sm., ♂. Lethbridge, Alberta, July 1983, J. R. Byers (CNC). (p. 57).

23. *Euxoa pestula* Sm., ♂. Manyberries, Alberta, 12 July 1951, D. F. Hardwick (CNC). (p. 57).

24. *Euxoa simona* McD., ♂. 5 mi W Lee Vining 8,900′, California, 7 August 1967, D. F. Hardwick (CNC). (p. 58).

25. *Euxoa simona* McD., ♀. 12 mi WNW Estes Park 11,600′, Colorado, 21 July 1967, D. F. Hardwick (CNC). (p. 58).

26. *Euxoa hardwicki* Lafontaine, ♂. Holotype. Walla Walla, Washington, 6 June 1955, W. C. Cook (CNC). (p. 63).

27. *Euxoa hardwicki* Lafontaine, ♀. 7¾ mi N Big Timber near Big Timber Creek, Sweet Grass Co., Montana, 2 July 1969, John G. Franclemont (JGF). (p. 63).

28. *Euxoa camalpa* (Dyar), ♂. Limpia Canyon, Jefferson Davis Co., Texas, 20 May 1950, E. C. Johnston (CNC). (p. 64).

29. *Euxoa camalpa* (Dyar), ♀. Alpine, Texas, 7–15 June 1926 (CNC). (p. 64).

30. *Euxoa maderensis* Lafontaine, ♀. Paratype. Madera Canyon 4,880′, Santa Rita Mts., Santa Cruz Co., Arizona, 20 June 1963, J. G. Franclemont (CNC). (p. 65).

31. *Euxoa serotina* Lafontaine, ♂. Sinton, Welder Wildlife Foundation (San Patricio Co.), Texas, 27 October 1964, A. & M. E. Blanchard (CNC). (p. 65).

32. *Euxoa bostoniensis* (Grt.), ♂. East Wareham, Massachusetts, 27 September 1972, W. E. Tomlinson (CNC). (p. 66).

33. *Euxoa bostoniensis* (Grt.), ♀. East Wareham, Massachusetts, 27 September 1972, W. E. Tomlinson (CNC). (p. 66).

34. *Euxoa medialis* (Sm.), ♂. Granite Dells 4 mi N Prescott, Yavapai Co., Arizona, 21 October 1971, Lloyd M. Martin (CNC). (p. 66).

35. *Euxoa medialis* (Sm.), ♂. 12 mi N Scott City 2,800′, Kansas, 3 October 1968, D. F. Hardwick (CNC). (p. 66).

36. *Euxoa medialis* (Sm.), ♀. Granite Dells 4 mi N Prescott, Yavapai Co., Arizona, 22 October 1971, Lloyd M. Martin (CNC). (p. 66).

37. *Euxoa medialis* (Sm.), ♂. 4 mi SE Cowley 3,800′, Alberta, 12 September 1973, J. D. Lafontaine (CNC). (p. 66).

38. *Euxoa medialis* (Sm.), ♂. Granite Dells 4 mi N Prescott, Yavapai Co., Arizona, 20 November 1970, Lloyd M. Martin (CNC). (p. 66).

39. *Euxoa medialis* (Sm.), ♂. Cruthers Creek SW Valyermo 4,400′, San Gabriel Mts., Los Angeles Co., California, 2 November 1962, C. Henne (LACM). (p. 66).

40. *Euxoa bifasciata* (Sm.), ♂. Upper Santa Ana River, San Bernardino Co., California, 18 July 1948, Grace H. & John L. Sperry (CNC). (p. 67).

41. *Euxoa bifasciata* (Sm.), ♀. East Verde River N of Payson 5,000′, Gila Co., Arizona, 8 July 1959, C. Henne (LACM). (p. 67).

42. *Euxoa bifasciata* (Sm.), ♀. SW of Valyermo 5,000′, north slope San Gabriel Mts., Los Angeles Co., California, 11 July 1960, C. Henne (LACM). (p. 67).

43. *Euxoa perexcellens* (Grt.), ♂. 9 mi NW Leavenworth 2,200′, Washington, 24 August 1974, J. D. Lafontaine (CNC). (p. 67).

44. *Euxoa perexcellens* (Grt.), ♂. Robson, British Columbia, 31 August 1943, H. R. Foxlee (CNC). (p. 67).

45. *Euxoa perexcellens* (Grt.), ♀. 9 mi NW Leavenworth 2,200′, Washington, 24 August 1974, J. D. Lafontaine (CNC). (p. 67).

46. *Euxoa perexcellens* (Grt.), ♀. 9 mi NW Leavenworth 2,200′, Washington, 24 August 1974, J. D. Lafontaine (CNC). (p. 67).

47. *Euxoa sculptilis* (Harv.), ♀. 5 mi SW Portal 5,400′, Arizona, 8 October 1969, D. F. Hardwick (CNC). (p. 67).

48. *Euxoa ustulata* Lafontaine, ♀. Paratype. Nelson Creek, Plumas Co., California, 18 August 1940, W. R. Bauer (CNC). (p. 68).

49. *Euxoa r. rufula* (Sm.), ♂. Doolittle Ranch 9,800′, Mt. Evans, Colorado, 7 August 1961, E. W. Rockburne (CNC). (p. 69).

50. *Euxoa rufula basiflava* (Sm.), ♀. Lillooet, British Columbia, 18 September 1938, J. K. Jacob (CNC). (p. 69).

51. *Euxoa rufula basiflava* (Sm.), ♀. 10 mi NE Mineral 7,400′, California, 24 August 1967, D. F. Hardwick (CNC). (p. 69).

52. *Euxoa intrita* (Morr.), ♂. Saskatoon, Saskatchewan, emerged 23 February 1972, J. R. Byers (CNC). (p. 69).

53. *Euxoa intrita* (Morr.), ♀. 3 mi W Brocket 3,400′, Alberta, 7 August 1961, D. F. Hardwick (CNC). (p. 69).

54. *Euxoa intrita* (Morr.), ♂. Manning Park 3,900′, British Columbia, 5 August 1960, D. F. Hardwick (CNC). (p. 69).

55. *Euxoa intrita* (Morr.), ♂. 10 mi NE Davis Creek 5,800′, California, 29 August 1967, D. F. Hardwick (CNC). (p. 69).

1 2 3 4 5
6 7 8 9 10
11 12 13 14 15
16 17 18 19 20
21 22 23 24 25
26 27 28 29 30
31 32 33 34 35
36 37 38 39 40
41 42 43 44 45
46 47 48 49 50
51 52 53 54 55

PLATE 3

Noctuoidea

NOCTUIDAE

figs. 1–55

NATURAL SIZE 1:1

1. *Euxoa a. annulipes* (Sm.), ♂. Walnut Canyon 6,500′, 6⅓ mi EESE Flagstaff, Coconino Co., Arizona, 2 July 1965, J. G. Franclemont (JGF). (p. 70).
2. *Euxoa a. annulipes* (Sm.), ♀. Mineral King, Tulare Co., California, 10 August 1960, W. E. Simonds (California Department Agriculture, Sacramento). (p. 70).
3. *Euxoa a. annulipes* (Sm.), ♀. Colorado Natl. Monument, Colorado, 15 July 1968, J. E. H. Martin (CNC). (p. 70).
4. *Euxoa annulipes oncocnemoides* (B. & Benj.), ♀. La Tuna Canyon, Los Angeles Co., California, 8 June 1946, W. H. Evans (CNC). (p. 70).
5. *Euxoa scholastica* McD., ♀. Boulderwood, Halifax, Nova Scotia, 15 August 1959, D. C. Ferguson (CNC). (p. 71).
6. *Euxoa terrena* (Sm.), ♀. Shirley Meadows 6,000′, Greenhorn Mts., Kern Co., California, 10 July 1972, C. Henne (LACM). (p. 71).
7. *Euxoa terrena* (Sm.), ♀. Cruthers Creek SW Valyermo 4,400′, San Gabriel Mts., Los Angeles Co., California, 3 July 1962, C. Henne (LACM). (p. 71).
8. *Euxoa terrena* (Sm.), ♂. NW Jackson Lake 5,500′, San Gabriel Mts., Los Angeles Co., California, 14 July 1967, C. Henne (LACM). (p. 71).
9. *Euxoa antica* Lafontaine, ♀. Walnut Canyon 6,500′, 6⅓ mi EESE Flag-

staff, Coconino Co., Arizona, 10 July 1965, J. G. Franclemont (JGF). (p. 72).

10. *Euxoa antica* Lafontaine, ♀. Paratype. Big Timber Creek, 7 mi N Big Timber, Sweetgrass Co., Montana, 4 July 1966, Douglas C. Ferguson (CNC). (p. 72).
11. *Euxoa franclemonti* Lafontaine, ♂. Holotype. Walnut Canyon 6,500′, 6⅓ mi EESE Flagstaff, Coconino Co., Arizona, 5 July 1965, J. G. Franclemont (JGF). (p. 72).
12. *Euxoa franclemonti* Lafontaine, ♀. Paratype. Walnut Canyon 6,500′, 6⅓ mi EESE Flagstaff, Coconino Co., Arizona, 6 July 1965, J. G. Franclemont (JGF). (p. 72).
13. *Euxoa absona* Lafontaine, ♂. Holotype. Angel Creek 7,000′, E. Humboldt Mts., SSW of Wells, Elko Co., Nevada, 18 July 1971, D. C. Ferguson (USNM). (p. 73).
14. *Euxoa absona* Lafontaine, ♀. Paratype. Clark Co., Nevada, 24–30 June (USNM). (p. 73).
15. *Euxoa serricornis* (Sm.), ♀. Vasquez Rocks 2,500′, Los Angeles Co., California, 21 February 1961, C. Henne (LACM). (p. 74).
16. *Euxoa serricornis* (Sm.), ♀. Apple Valley, California, 17 May 1955, J. E. H. Martin (CNC). (p. 74).
17. *Euxoa tocoyae* (Sm.), ♂. Pine Flat, Sonoma Co., California, 28 April 1940, W. R. Bauer (CNC). (p. 74).
18. *Euxoa tocoyae* (Sm.), ♀. Pinnacles, San Benito Co., California, 13 May 1937, E. C. Johnston (CNC). (p. 74).
19. *Euxoa scotogrammoides* McD., ♀. Paratype. Jefferson Co., Montana, W. C. Cook (CNC). (p. 74).
20. *Euxoa scotogrammoides* McD., ♀. Hermiston, Oregon, 22 May 1961, Kenneth Goeden (CNC). (p. 74).
21. *Euxoa pluralis* (Grt.), ♂. Mt. Pinos 8,200′, Los Padres National Forest, Kern Co., California, 11 July 1961, C. Henne (LACM). (p. 75).
22. *Euxoa pluralis* (Grt.), ♀. Vaseaux Lake 1,200′, 13 mi SSE Penticton, British Columbia, 18 June 1979, J. D. Lafontaine (CNC). (p. 75).
23. *Euxoa cinnabarina* B. & McD., ♀. Buckhorn Flats 6,500′, San Gabriel Mts., Los Angeles Co., California, 3 August 1948, Claude I. Smith (UCB). (p. 75).
24. *Euxoa permixta* McD., ♀. Telluride 8,745′, San Miguel Co., Colorado, 11 July 1977, D. C. Ferguson (USNM). (p. 76).
25. *Euxoa permixta* McD., ♂. 11 mi SE Cedar City 8,300′, Utah, 29 August 1965, D. F. Hardwick (CNC). (p. 76).
26. *Euxoa setonia* McD., ♂. Horseshoe Canyon 2,750′, Drumheller, Alberta, 17 July 1960, D. F. Hardwick (CNC). (p. 76).
27. *Euxoa setonia* McD., ♀. Walnut Canyon 6,500′, 6⅓ mi EESE Flagstaff, Coconino Co., Arizona, 29 June 1965, J. G. Franclemont (JGF). (p. 76).
28. *Euxoa setonia* McD., ♀. 6 mi W Telluride 8,000′, San Miguel Co., Colorado, 15 July 1977, D. C. Ferguson (USNM). (p. 76).
29. *Euxoa pallidimacula* Lafontaine, ♂. Holotype. Eureka, Utah, 17 July 1911, Tom Spalding (CNC). (p. 77).
30. *Euxoa pallidimacula* Lafontaine, ♂. Paratype. Angel Creek 7,000′, E. Humboldt Mts., SSW of Wells, Elko Co., Nevada, 25 July 1971, D. C. Ferguson (USNM). (p. 77).
31. *Euxoa mojave* Lafontaine, ♂. Holotype. Apple Valley, California, 22 May 1955, W. R. Richards (CNC). (p. 77).
32. *Euxoa mojave* Lafontaine, ♂. Paratype. Wildrose Canyon, Panamint Mts., California, 29 May 1960, D. S. Verity (LACM). (p. 77).
33. *Euxoa declarata* (Wlk.), ♂. 12 km SW Ottawa, Ontario, 6 September 1985, J. D. Lafontaine (CNC). (p. 79).
34. *Euxoa declarata* (Wlk.), ♂. Big Pines Area 6,700′, San Gabriel Mts., Los Angeles Co., California, 25 July 1963, C. Henne (LACM). (p. 79).
35. *Euxoa declarata* (Wlk.), ♀. Big Pines Area 6,800′, San Gabriel Mts., Los Angeles Co., California, 16 August 1963, C. Henne (LACM). (p. 79).
36. *Euxoa declarata* (Wlk.), ♂. 6 mi SW Folsom 7,300′, New Mexico, 13 September 1968, D. F. Hardwick (CNC). (p. 79).
37. *Euxoa campestris* (Grt.), ♂. Forestville, Quebec, 10 August 1950, R. de Ruette (CNC). (p. 79).
38. *Euxoa campestris* (Grt.), ♀. 7 mi E Tuxford 1,700′, Saskatchewan, 17 August 1968, D. F. Hardwick (CNC). (p. 79).
39. *Euxoa rockburnei* Hdwk., ♂. Paratype. 4 mi NW Cedarville 6,200′, California, 30 August 1967, D. F. Hardwick (CNC). (p. 80).
40. *Euxoa rockburnei* Hdwk., ♀. Paratype. Truckee, California, Ximena McGlashan (CNC). (p. 80).
41. *Euxoa flavidens* (Sm.), ♂. 2 mi SE Greens Peak 9,500′, White Mts., Apache Co., Arizona, 6 August 1962, E. & I. Munroe (CNC). (p. 80).
42. *Euxoa flavidens* (Sm.), ♀. Greer 8,500′, White Mts., Apache Co., Arizona, 6 August 1962, E. & I. Munroe (CNC). (p. 80).
43. *Euxoa silens* (Grt.), ♂. Juniper Hills 3,500′, Mojave Desert, Los Angeles Co., California, 29 April 1962, C. Henne (LACM). (p. 81).
44. *Euxoa silens* (Grt.), ♀. Juniper Hills 3,500′, Mojave Desert, Los Angeles Co., California, 5 May 1962, C. Henne (LACM). (p. 81).
45. *Euxoa pimensis* B. & McD., ♂. La Tuna Canyon, Los Angeles Co., California, 24 April 1949, William H. Evans (CNC). (p. 81).
46. *Euxoa pimensis* B. & McD., ♀. Madera Canyon 4,880′, Santa Rita Mts., Santa Cruz Co., Arizona, 9 May 1963, J. G. Franclemont (CNC). (p. 81).
47. *Euxoa pimensis* B. & McD., ♀. Madera Canyon 4,880′, Santa Rita Mts., Santa Cruz Co., Arizona, 3 May 1963, J. G. Franclemont (CNC). (p. 81).
48. *Euxoa immixta* (Grt.), ♂. Lake Brownwood State Park, Brown Co., Texas, 23 April 1965, A. & M. E. Blanchard (LACM). (p. 81).
49. *Euxoa immixta* (Grt.), ♀. Sioux Falls, South Dakota (CNC). (p. 81).
50. *Euxoa simulata* McD., ♂. Big Pines Area 6,800′, San Gabriel Mts., Los Angeles Co., California, 9 August 1963, C. Henne (LACM). (p. 82).
51. *Euxoa simulata* McD., ♂. Almanor 4,500′, Plumas Co., California, 1 August 1965, E. & I. Munroe (CNC). (p. 82).
52. *Euxoa punctigera* (Wlk.), ♂. 11 mi E Cle Elum 2,300′, Washington, 25 August 1960, D. F. Hardwick (CNC). (p. 83).
53. *Euxoa punctigera* (Wlk.), ♀. 19 mi N Leavenworth 2,000′, Washington, 26 August 1960, D. F. Hardwick (CNC). (p. 83).
54. *Euxoa punctigera* (Wlk.), ♀. Big Pines Area 6,800′, San Gabriel Mts., Los Angeles Co., California, 14 August 1963, C. Henne (LACM). (p. 83).
55. *Euxoa punctigera* (Wlk.), ♀. Big Pines Area 6,800′, San Gabriel Mts., Los Angeles Co., California, 14 August 1963, C. Henne (LACM). (p. 83).

1 2 3 4 5
6 7 8 9 10
11 12 13 14 15
16 17 18 19 20
21 22 23 24 25
26 27 28 29 30
31 32 33 34 35
36 37 38 39 40
41 42 43 44 45
46 47 48 49 50
51 52 53 54 55

PLATE 4

Noctuoidea

NOCTUIDAE

figs. 1–60

NATURAL SIZE 1:1

1. *Euxoa cana* Lafontaine, ♂. Holotype. Sumpter 4,430′, 28 mi W Baker, Oregon, 10–11 September 1963, W. C. Cook (CNC). (p. 84).
2. *Euxoa cana* Lafontaine, ♀. Paratype. 6 mi SSW Grasmere 5,400′, Idaho, 24 August 1969, D. F. Hardwick (CNC). (p. 84).
3. *Euxoa aurantiaca* Lafontaine, ♂. Holotype. 4 mi SW Alcova 5,500′, Wyoming, 14 August 1965, D. F. Hardwick (CNC). (p. 84).
4. *Euxoa aurantiaca* Lafontaine, ♂. Walnut Canyon 6,500′, 6⅓ mi EESE Flagstaff, Coconino Co., Arizona, 29 June 1965, J. G. Franclemont (JGF). (p. 84).
5. *Euxoa aurantiaca* Lafontaine, ♀. 10 mi E Panaca 6,500′, Nevada, 1 September 1965, D. F. Hardwick (CNC). (p. 84).
6. *Euxoa stygialis* (B. & McD.), ♂. Onion Saddle 7,600′, Chiricahua Mts., Cochise Co., Arizona, 11 July 1966, J. G. Franclemont (JGF). (p. 84).
7. *Euxoa stygialis* (B. & McD.), ♀. East Turkey Creek 6,400′, Chiricahua Mts., Cochise Co., Arizona, 6 July 1966, J. G. Franclemont (JGF). (p. 84).
8. *Euxoa stigmatalis* (Sm.), ♂. McGaffey 7,500′, Zuni Mts., McKinley Co., New Mexico, 20 July 1962, E. & I. Munroe (CNC). (p. 85).
9. *Euxoa stigmatalis* (Sm.), ♀. 17 mi NE Paragonah 7,800′, Utah, 29 August 1971, D. F. Hardwick (CNC). (p. 85).
10. *Euxoa stigmatalis* (Sm.), ♀. McGaffey 7,500′, Zuni Mts., McKinley Co., New Mexico, 20 July 1962, E. & I. Munroe (CNC). (p. 85).

11. *Euxoa spumata* McD., ♂. 13 mi N Wolf Creek 3,800′, Montana, 13 August 1961, D. F. Hardwick (CNC). (p. 85).
12. *Euxoa spumata* McD., ♂. 4 mi SW Alcova 5,500′, Wyoming, 14 August 1965, D. F. Hardwick (CNC). (p. 85).
13. *Euxoa spumata* McD., ♀. 11 mi E Milo 3,100′, Alberta, 2 August 1961, D. F. Hardwick (CNC). (p. 85).
14. *Euxoa pallipennis* (Sm.), ♂. Manyberries, Alberta, 10 August 1951, D. F. Hardwick (CNC). (p. 86).
15. *Euxoa pallipennis* (Sm.), ♀. Swift Current, Saskatchewan, 30 August 1939, A. R. Brooks (CNC). (p. 86).
16. *Euxoa baja* Lafontaine, ♂. Holotype. Long Beach, California, 26 September 1939 (USNM). (p. 86).
17. *Euxoa tessellata* (Harr.), ♂. 12 km SW Ottawa, Ontario, 5 July 1986, J. D. Lafontaine (CNC). (p. 87).
18. *Euxoa tessellata* (Harr.), ♀. 12 km SW Ottawa, Ontario, 8 July 1986, J. D. Lafontaine (CNC). (p. 87).
19. *Euxoa tessellata* (Harr.), ♀. 12 km SW Ottawa, Ontario, 5 July 1986, J. D. Lafontaine (CNC). (p. 87).
20. *Euxoa tessellata* (Harr.), ♀. 12 km SW Ottawa, Ontario, 1 August 1985, J. D. Lafontaine (CNC). (p. 87).
21. *Euxoa tessellata* (Harr.), ♂. Katherine Lake 2,200′, Riding Mountain National Park, Manitoba, 21 July 1979, J. D. Lafontaine (CNC). (p. 87).
22. *Euxoa tessellata* (Harr.), ♀. Walla Walla, Washington, 21 June 1949, W. C. Cook (CNC). (p. 87).
23. *Euxoa tessellata* (Harr.), ♀. McBride Springs 4,800′, Mt. Shasta, Siskiyou Co., California, 20 July 1965, E. & I. Munroe (CNC). (p. 87).
24. *Euxoa tessellata* (Harr.), ♂. Walla Walla, Washington, 9 July 1948, W. C. Cook (CNC). (p. 87).
25. *Euxoa tessellata* (Harr.), ♂. 5 mi W Lee Vining 8,900′, California, 7 August 1967, D. F. Hardwick (CNC). (p. 87).
26. *Euxoa tessellata* (Harr.), ♀. Mt. Pinos 8,200′, Los Padres Natl. Forest, Kern Co., California, 11 July 1961, C. Henne (LACM). (p. 87).
27. *Euxoa tessellata* (Harr.), ♂. Camp Ozena, Upper Cuyama, Ventura Co., California, 8 June 1963, C. W. Kirkwood (CNC). (p. 87).
28. *Euxoa tessellata* (Harr.), ♂. Dillon Beach, Marin Co., California, 10 July 1962, M. G. Tunzi (LACM). (p. 87).
29. *Euxoa plagigera* (Morr.), ♂. 1 mi S Beverley 2,400′, Saskatchewan, 22 August 1968, D. F. Hardwick (CNC). (p. 89).
30. *Euxoa plagigera* (Morr.), ♀. 8 mi SE Laramie 8,400′, Wyoming, 28 August 1964, D. F. Hardwick (CNC). (p. 89).
31. *Euxoa plagigera* (Morr.), ♀. Barton Flats 6,600′, San Bernardino Mts., San Bernardino Co., California, 2 August 1964, C. Henne (LACM). (p. 89).
32. *Euxoa albipennis* (Grt.), ♂. Mt. Pinos 7,600′, Los Padres Natl. Forest, Kern Co., California, 24 September 1965, C. Henne (LACM). (p. 90).
33. *Euxoa albipennis* (Grt.), ♀. Mt. Pinos 7,600′, Los Padres Natl. Forest, Kern Co., California, 26 September 1963, C. Henne (LACM). (p. 90).
34. *Euxoa albipennis* (Grt.), ♀. 3 mi NE Dixon 5,700′, New Mexico, 16 September 1968, D. F. Hardwick (CNC). (p. 90).
35. *Euxoa albipennis* (Grt.), ♀. Swift Current, Saskatchewan, 30 August 1939, A. R. Brooks (CNC). (p. 90).
36. *Euxoa albipennis* (Grt.), ♀. 12 km SW Ottawa, Ontario, 6 September 1985, J. D. Lafontaine (CNC). (p. 90).
37. *Euxoa henrietta* (Sm.), ♂. 25 mi NW Canby 4,200′, California, 1 September 1967, D. F. Hardwick (CNC). (p. 91).
38. *Euxoa henrietta* (Sm.), ♂. Gorman, Ridge Route, California, 30 September 1957, F. P. Sala (CNC). (p. 91).
39. *Euxoa henrietta* (Sm.), ♂. 4 mi N Lyle 1,500′, Washington, 21 August 1960, D. F. Hardwick (CNC). (p. 91).
40. *Euxoa henrietta* (Sm.), ♂. 7 mi WNW Orland 700′, California, 30 September 1970, D. F. Hardwick (CNC). (p. 91).
41. *Euxoa cincta* B. & Benj., ♂. Onion Saddle 7,600′, Chiricahua Mts., Cochise Co., Arizona, 8 August 1966, J. G. Franclemont (JGF). (p. 92).
42. *Euxoa cincta* B. & Benj., ♀. Sierra Vista, Arizona, 15 August 1966, R. F. Sternitzky (CNC). (p. 92).
43. *Euxoa coconino* Lafontaine, ♂. Holotype. Kaibab Natl. Forest 7,000′, Coconino Plateau, Arizona, 20 August 1960, E. G. Munroe (CNC). (p. 92).
44. *Euxoa hollemani* (Grt.), ♂. 4 mi NW Cedarville 6,200′, California, 30 August 1967, D. F. Hardwick (CNC). (p. 93).
45. *Euxoa hollemani* (Grt.), ♀. 13 mi SE Gardnerville 5,900′, Nevada, 18 September 1969, D. F. Hardwick (CNC). (p. 93).
46. *Euxoa subandera* Lafontaine, ♀. Holotype. La Tuna Canyon, Los Angeles Co., California, 1 September 1948, William H. Evans (CNC). (p. 94).
47. *Euxoa xasta* B. & McD., ♂. White City, Eddy Co., New Mexico, 16 May 1950, E. C. Johnston (CNC). (p. 94).
48. *Euxoa xasta* B. & McD., ♀. Manzanita Spring 5,500′, Guadalupe Mts., Culberson Co., Texas, 24 May 1973, D. C. Ferguson (USNM). (p. 94).
49. *Euxoa catenula* (Grt.), ♂. 4 mi NW Cedarville 6,200′, California, 30 August 1967, D. F. Hardwick (CNC). (p. 95).
50. *Euxoa catenula* (Grt.), ♂. Juniper Hills 3,500′, Mojave Desert, Los Angeles Co., California, 8 October 1966, C. Henne (LACM). (p. 95).
51. *Euxoa catenula* (Grt.), ♀. 6 mi S Big Bear City 6,600′, California, 7 September 1966, D. F. Hardwick (CNC). (p. 95).
52. *Euxoa comosa altera* McD., ♀. Paratype. Cartwright, Manitoba, 30 August 1903, E. F. Heath (CNC). (p. 98).
53. *Euxoa comosa altera* McD., ♂. Saskatoon, Saskatchewan, 25 July 1923, K. M. King (CNC). (p. 98).
54. *Euxoa comosa altera* McD., ♀. 3 mi W Brocket 3,400′, Alberta, 7 August 1961, D. F. Hardwick (CNC). (p. 98).
55. *Euxoa comosa altera* McD., ♂. St. Paul, Minnesota, 13 August 1928, Carl T. Schmidt (CNC). (p. 98).
56. *Euxoa comosa annir* (Stkr.), ♂. 17 mi W Dupuyer 4,900′, Montana, 11 August 1961, D. F. Hardwick (CNC). (p. 99).
57. *Euxoa comosa annir* (Stkr.), ♂. 2 mi N Scout Lake 2,600′, Saskatchewan, 27 August 1968, D. F. Hardwick (CNC). (p. 99).
58. *Euxoa comosa annir* (Stkr.), ♂. Big Timber Creek 7 mi N Big Timber, Sweetgrass Co., Montana, 9 August 1966, Douglas C. Ferguson (CNC). (p. 99).
59. *Euxoa comosa annir* (Stkr.), ♀. 9 mi SW Grantsville 6,000′, Utah, 14 September 1965, D. F. Hardwick (CNC). (p. 99).
60. *Euxoa comosa annir* (Stkr.), ♂. 5 mi N Pueblo 5,100′, Colorado, 22 August 1975, J. D. Lafontaine (CNC). (p. 99).

1 2 3 4 5
6 7 8 9 10
11 12 13 14 15
16 17 18 19 20
21 22 23 24 25
26 27 28 29 30
31 32 33 34 35
36 37 38 39 40
41 42 43 44 45
46 47 48 49 50
51 52 53 54 55
56 57 58 59 60

PLATE 5

Noctuoidea

NOCTUIDAE

figs. 1–60

NATURAL SIZE 1:1

1. *Euxoa c. comosa* (Morr.), ♀. 3 mi SW Ute Park 7,300′, New Mexico, 14 September 1968, D. F. Hardwick (CNC). (p. 99).
2. *Euxoa c. comosa* (Morr.), ♀. Greer 8,500′, White Mts., Apache Co., Arizona, 7 August 1962, E. & I. Munroe (CNC). (p. 99).
3. *Euxoa c. comosa* (Morr.), ♀. 11 mi SE Cedar City 8,300′, Utah, 29 August 1965, D. F. Hardwick (CNC). (p. 99).
4. *Euxoa c. comosa* (Morr.), ♀. Diamond Rock, White Mts., Arizona, 6 September 1947, Grace H. & John L. Sperry (CNC). (p. 99).
5. *Euxoa c. comosa* (Morr.), ♂. 16 mi W Kayenta 7,300′, Arizona, 4 September 1971, D. F. Hardwick (CNC). (p. 99).
6. *Euxoa comosa lutulenta* (Sm.), ♀. Dayton 1,620′, Washington, 13 August 1963, R. E. Miller (CNC). (p. 99).
7. *Euxoa comosa lutulenta* (Sm.), ♂. km 140.5 Dempster Highway 900 m, Yukon, 21–23 July 1980, G. & M. Wood (CNC). (p. 99).
8. *Euxoa comosa lutulenta* (Sm.), ♀. 7 mi NNW Brooks 4,600′, Montana, 28 August 1970, D. F. Hardwick (CNC). (p. 99).
9. *Euxoa comosa lutulenta* (Sm.), ♀. Bluejay near Lake Arrowhead, San Bernardino Co., California, 9–10 August 1956, Noel McFarland (LACM). (p. 99).
10. *Euxoa comosa lutulenta* (Sm.), ♂. Buckhorn Flats 6,500′, San Gabriel Mts., Los Angeles Co., California, 3 August 1948, Claude I. Smith (UCB). (p. 99).
11. *Euxoa comosa lutulenta* (Sm.), ♂. Big Pines Area 6,800′, San Gabriel Mts., Los Angeles Co., California, 25 July 1963, C. Henne (LACM). (p. 99).

12. *Euxoa comosa ontario* (Sm.), ♂. Boulderwood, Halifax, Nova Scotia, 5 August 1963, D. C. Ferguson (USNM). (p. 100).

13. *Euxoa comosa ontario* (Sm.), ♂. Peggy's Cove, Halifax Co., Nova Scotia, 20 August 1952, D. C. Ferguson (USNM). (p. 100).

14. *Euxoa comosa ontario* (Sm.), ♂. West Dover, Halifax Co., Nova Scotia, 1 September 1957, D. C. Ferguson (USNM). (p. 100).

15. *Euxoa comosa ontario* (Sm.), ♀. Keweenaw Co., Michigan, 28 July 1977, M. C. Nielsen (CNC). (p. 100).

16. *Euxoa lineifrons* (Sm.), ♂. Buena Vista, Chaffee Co., Colorado, 28 August 1957, Margot May (CNC). (p. 100).

17. *Euxoa lineifrons* (Sm.), ♂. Gibbonsville, Lemhi Co., Idaho, 8 August 1958, Margot May (CNC). (p. 100).

18. *Euxoa guadalupensis* Lafontaine & Byers, ♂. Holotype. Bear Canyon 5,700′, Guadalupe Mts., Texas, 2 October 1969, A. & M. E. Blanchard (USNM). (p. 100).

19. *Euxoa guadalupensis* Lafontaine & Byers, ♀. Paratype. Bear Canyon 5,700′, Guadalupe Mts., Texas, 4 September 1969, A. & M. E. Blanchard (USNM). (p. 100).

20. *Euxoa lucida* B. & McD., ♂. 8 mi S Pioche 4,900′, Nevada, 3 September 1965, D. F. Hardwick (CNC). (p. 101).

21. *Euxoa lucida* B. & McD., ♀. 9 mi W Caliente 6,200′, Nevada, 2 September 1965, D. F. Hardwick (CNC). (p. 101).

22. *Euxoa lucida* B. & McD., ♀. Garfield Co. 6,000′, Colorado, D. Bruce (CNC). (p. 101).

23. *Euxoa lucida* B. & McD., ♂. Walnut Canyon 6,500′, 6⅓ mi EESE Flagstaff, Coconino Co., Arizona, 2 September 1964, J. G. Franclemont (JGF). (p. 101).

24. *Euxoa fumalis* (Grt.), ♂. Ithaca, New York, 25 August 1939, J. G. Franclemont (JGF). (p. 101).

25. *Euxoa fumalis* (Grt.), ♀. Ithaca, New York, 13 September 1936, J. G. Franclemont (JGF). (p. 101).

26. *Euxoa occidentalis* Lafontaine & Byers, ♂. Mt. Pinos 7,600′, Los Padres Natl. Forest, Kern Co., California, 24 September 1963, C. Henne (LACM). (p. 102).

27. *Euxoa occidentalis* Lafontaine & Byers, ♀. Paratype. Sumpter 4,430′, 28 mi W Baker, Oregon, 10–11 September 1963, W. C. Cook (CNC). (p. 102).

28. *Euxoa velleripennis* (Grt.), ♂. 14 km S Sharbot Lake, Ontario, 2 August 1981, J. D. Lafontaine (CNC). (p. 102).

29. *Euxoa velleripennis* (Grt.), ♀. Petite Riviere, Lunenburg Co., Nova Scotia, 3 September 1954, D. C. Ferguson (USNM). (p. 102).

30. *Euxoa infausta* (Wlk.), ♀. Pullman, Washington, 23 July 1933, J. F. Clarke (CNC). (p. 104).

31. *Euxoa infausta* (Wlk.), ♂. Pullman, Washington, 27 July 1932, J. F. Clarke (CNC). (p. 104).

32. *Euxoa infausta* (Wlk.), ♂. La Mesa, San Diego Co., California, 15 April 1950, E. C. Johnston (CNC). (p. 104).

33. *Euxoa infausta* (Wlk.), ♂. Big Timber Creek 7 mi N Big Timber, Sweetgrass Co., Montana, 10 July 1966, Douglas C. Ferguson (CNC). (p. 104).

34. *Euxoa infausta* (Wlk.), ♀. Big Timber Creek 7 mi N Big Timber, Sweetgrass Co., Montana, 10 July 1966, Douglas C. Ferguson (CNC). (p. 104).

35. *Euxoa infausta* (Wlk.), ♂. 9 mi N Union Creek 4,200′, Oregon, 3 September 1974, J. D. Lafontaine (CNC). (p. 104).

36. *Euxoa satis* (Harv.), ♀. Almanor 4,500′, Plumas Co., California, 1 August 1965, E. & I. Munroe (CNC). (p. 105).

37. *Euxoa satis* (Harv.), ♀. Salmon Arm, British Columbia, 19 July 1926, W. R. Buckell (CNC). (p. 105).

38. *Euxoa satis* (Harv.), ♀. 28 mi N Squamish 1,800′, British Columbia, 16 August 1974, J. D. Lafontaine (CNC). (p. 105).

39. *Euxoa satis* (Harv.), ♀. Boulder 5,400′, Colorado, 20 July 1961, E. W. Rockburne (CNC). (p. 105).

40. *Euxoa satis* (Harv.), ♀. McBride Springs 4,800′, Siskiyou Co., California, 24 July 1965, E. & I. Munroe (CNC). (p. 105).

41. *Euxoa b. brunneigera* (Grt.), ♀. Brooks Memorial State Park 3,100′, Satus Pass, Washington, 24–26 July 1964, W. C. Cook (CNC). (p. 106).

42. *Euxoa b. brunneigera* (Grt.), ♀. La Sierra Canyon 1,000′, Santa Monica Mts., Los Angeles Co., California, 20 May 1961, C. Henne (LACM). (p. 106).

43. *Euxoa b. brunneigera* (Grt.), ♀. Petaluma, Sonoma Co., California, 10 June 1936, W. R. Bauer (CNC). (p. 106).

44. *Euxoa b. brunneigera* (Grt.), ♀. Camp Ozena, Upper Cuyama, Ventura Co., California, 16 June 1963, C. W. Kirkwood (CNC). (p. 106).

45. *Euxoa brunneigera excogita* (Sm.), ♂. Big Timber Canyon ca. 6,500′, (Half Moon Park), Crazy Mts., Montana, 10 August 1966, Douglas C. Ferguson (CNC). (p. 107).

46. *Euxoa brunneigera excogita* (Sm.), ♂. Big Timber Creek 7 mi N Big Timber, Sweetgrass Co., Montana, 10 July 1966, Douglas C. Ferguson (CNC). (p. 107).

47. *Euxoa bicollaris* (Grt.), ♀. Smokey Valley 6,500′, Tulare Co., California, 20 June 1945 (LACM). (p. 107).

48. *Euxoa bicollaris* (Grt.), ♀. 5 mi N Beverly Hills 1,100′, Oak Pass Road, Santa Monica Mts., Los Angeles Co., California, Noel McFarland (LACM). (p. 107).

49. *Euxoa bicollaris* (Grt.), ♀. Summerland, Santa Barbara Co., California, 19 May 1949, C. W. Kirkwood (LACM). (p. 107).

50. *Euxoa inyoca* Benj., ♂. Rock Creek, Mono Co., 7 May 1934, M. L. Walton (LACM). (p. 108).

51. *Euxoa selenis* (Sm.), ♀. La Mesa, San Diego Co., California, 15 April 1950, E. C. Johnston (CNC). (p. 108).

52. *Euxoa selenis* (Sm.), ♂. San Diego, California, 21 April 1909, Geo. H. Field (CNC). (p. 108).

53. *Euxoa selenis* (Sm.), ♀. Juniper Hills 3,500′, Mojave Desert, Los Angeles Co., California, 5 May 1962, C. Henne (LACM). (p. 108).

54. *Euxoa piniae* Buckett & Bauer, ♂. Almanor 4,500′, Plumas Co., California, 31 July 1965, E. & I. Munroe (CNC). (p. 108).

55. *Euxoa piniae* Buckett & Bauer, ♀. 2 mi NW Almanor 4,500′, Plumas Co., California, 1 August 1965, E. & I. Munroe (CNC). (p. 108).

56. *Euxoa satiens* (Sm.), ♂. Oliver 1,000′, British Columbia, 9 September 1953, D. F. Hardwick (CNC). (p. 108).

57. *Euxoa satiens* (Sm.), ♂. Summerland 1,400′, British Columbia, 10 September 1960, D. F. Hardwick (CNC). (p. 108).

58. *Euxoa satiens* (Sm.), ♂. 13 mi S. Watford City 2,000′, South Dakota, 1 September 1968, D. F. Hardwick (CNC). (p. 108).

59. *Euxoa satiens* (Sm.), ♀. Barton Flats 6,600′, San Bernardino Mts., San Bernardino Co., California, 29 August 1959, C. Henne (LACM). (p. 108).

60. *Euxoa satiens* (Sm.), ♂. 21 mi N Chiloquin 4,500′, Oregon, 2 September 1967, D. F. Hardwick (CNC). (p. 108).

1
2
3
4
5
6
7
8
9
10
11
12
13
14
15
16
17
18
19
20
21
22
23
24
25
26
27
28
29
30
31
32
33
34
35
36
37
38
39
40
41
42
43
44
45
46
47
48
49
50
51
52
53
54
55
56
57
58
59
60

PLATE 6

Noctuoidea

NOCTUIDAE

figs. 1–60

NATURAL SIZE 1:1

1. *Euxoa violaris* (Grt. & Rob.), ♂. E Wareham, Massachusetts, 20 September 1971, W. E. Tomlinson (CNC). (p. 109).
2. *Euxoa violaris* (Grt. & Rob.), ♀. E Wareham, Massachusetts, 27 September 1972, W. E. Tomlinson (CNC). (p. 109).
3. *Euxoa violaris* (Grt. & Rob.), ♀. Ocean City, Florida, 30 October 1962, H. O. Hilton (USNM). (p. 109).
4. *Euxoa cursoria wirima* Hdwk., ♂. Fort Smith, Northwest Territories, 10 August 1950, W. R. Mason (CNC). (p. 111).
5. *Euxoa cursoria wirima* Hdwk., ♂. Fort Smith, Northwest Territories, 5 August 1950, J. B. Wallis (CNC). (p. 111).
6. *Euxoa cursoria wirima* Hdwk., ♂. Fort Smith, Northwest Territories, 1 August 1950, J. B. Wallis (CNC). (p. 111).
7. *Euxoa cursoria wirima* Hdwk., ♂. Fort Smith, Northwest Territories, 10 August 1950, J. B. Wallis (CNC). (p. 111).
8. *Euxoa o. ochrogaster* (Gn.), ♀. 2 mi N Scout Lake 2,600′, Saskatchewan, 27 August 1968, D. F. Hardwick (CNC). (p. 113).
9. *Euxoa o. ochrogaster* (Gn.), ♂. 2 mi SW Caron 1,900′, Saskatchewan, 21 August 1968, D. F. Hardwick (CNC). (p. 113).
10. *Euxoa o. ochrogaster* (Gn.), ♀. Edmonton, Alberta, emerged 26 July 1906 (CNC). (p. 113).

11. *Euxoa o. ochrogaster* (Gn.), ♀. First Creek, Chelan Co., Washington, 2 July 1949, E. C. Johnston (CNC). (p. 113).
12. *Euxoa nostra* (Sm.), ♂. Cedar Pass 6,000′, Warner Mts., Modoc Co., California, 8 August 1965, E. & I. Munroe (CNC). (p. 113).
13. *Euxoa nostra* (Sm.), ♀. 7 mi WSW Lee Vining 9,600′, California, 12 August 1967, D. F. Hardwick (CNC). (p. 113).
14. *Euxoa siccata* (Sm.), ♂. Lethbridge, Alberta, 14 September 1945, G. A. Hobbs (CNC). (p. 114).
15. *Euxoa siccata* (Sm.), ♂. 12 mi N Scott City 2,800′, Kansas, 2 October 1968, D. F. Hardwick (CNC). (p. 114).
16. *Euxoa siccata* (Sm.), ♂. Lethbridge, Alberta, 18 September 1975, G. E. Swailes (CNC). (p. 114).
17. *Euxoa choris* (Harv.), ♂. 11 mi N Keremeos 2,600′, British Columbia, 3 August 1960, D. F. Hardwick (CNC). (p. 114).
18. *Euxoa choris* (Harv.), ♀. 5 mi W Lee Vining 8,900′, California, 7 August 1967, D. F. Hardwick (CNC). (p. 114).
19. *Euxoa obeliscoides* (Gn.), ♂. 12 km SW Ottawa, Ontario, 1 July 1986, J. D. Lafontaine (CNC). (p. 115).
20. *Euxoa obeliscoides* (Gn.), ♂. 12 km SW Ottawa, Ontario, 1 July 1986, J. D. Lafontaine (CNC). (p. 115).
21. *Euxoa obeliscoides* (Gn.), ♂. 4 mi N Lyle 1,500′, Washington, 21 August 1960, D. F. Hardwick (CNC). (p. 115).
22. *Euxoa obeliscoides* (Gn.), ♂. Lethbridge, Alberta, 31 August 1921, E. H. Strickland (CNC). (p. 115).
23. *Euxoa obeliscoides* (Gn.), ♀. 4 mi N Prescott, Yavapai Co., Arizona, 25 August 1972, Lloyd M. Martin (CNC). (p. 115).
24. *Euxoa oberfoelli* Hdwk., ♂. Sand Springs 3,100′, Montana, 25 August 1970, D. F. Hardwick (CNC). (p. 116).
25. *Euxoa oberfoelli* Hdwk., ♀. Stoughton, Saskatchewan, emerged 22 August 1973, J. R. Byers (CNC). (p. 116).
26. *Euxoa lillooet* McD., ♂. 6 mi E Mayfield 6,500′, Utah, 25 August 1965, D. F. Hardwick (CNC). (p. 116).
27. *Euxoa lillooet* McD., ♀. 17 mi W Pagosa Springs 6,600′, Colorado, 18 August 1971, D. F. Hardwick (CNC). (p. 116).
28. *Euxoa basalis* (Grt.), ♀. 9 mi N Coaldale 2,700′, Alberta, 4 August 1961, D. F. Hardwick (CNC). (p. 117).
29. *Euxoa basalis* (Grt.), ♀. 12 mi WNW Estes Park 11,600′, Colorado, 30 July 1967, D. F. Hardwick (CNC). (p. 117).
30. *Euxoa basalis* (Grt.), ♀. Currie 5,800′, Nevada, 9 September 1965, D. F. Hardwick (CNC). (p. 117).
31. *Euxoa costata* (Grt.), ♂. 2 mi E Golden 3,000′, British Columbia, 21 July 1960, D. F. Hardwick (CNC). (p. 123).
32. *Euxoa costata* (Grt.), ♀. 34 mi N Creston 1,900′, British Columbia, 30 July 1960, D. F. Hardwick (CNC). (p. 123).
33. *Euxoa castanea* Lafontaine, ♀. Paratype. Horseshoe Canyon 2,750′, Drumheller, Alberta, 17 July 1960, D. F. Hardwick (CNC). (p. 123).
34. *Euxoa castanea* Lafontaine, ♂. 2 mi SE Greens Peak 9,500′, White Mts., Apache Co., Arizona, 5 August 1962, E. & I. Munroe (CNC). (p. 123).
35. *Euxoa foeminalis* (Sm.), ♀. 5 mi N Prescott 5,450′, Yavapai Co., Arizona, 21 June 1975, Lloyd M. Martin (CNC). (p. 123).
36. *Euxoa foeminalis* (Sm.), ♂. 5 mi N Prescott 5,450′, Yavapai Co., Arizona, 29 May 1972, Lloyd M. Martin (CNC). (p. 123).
37. *Euxoa idahoensis* (Grt.), ♀. 5 mi NE Exshaw 4,225′, Alberta, 18 July 1961, D. F. Hardwick (CNC). (p. 124).
38. *Euxoa idahoensis* (Grt.), ♂. Writing-on-stone Provincial Park, Alberta, 8–10 July 1982, J. D. Lafontaine (CNC). (p. 124).
39. *Euxoa idahoensis* (Grt.), ♂. Big Buffalo Creek, N of Cedar Pass, Jackson Co., South Dakota, 8 July 1964, Douglas C. Ferguson (USNM). (p. 124).
40. *Euxoa idahoensis* (Grt.), ♂. Slate Creek 6,000′, 9 mi NW of Hill City, Black Hills, South Dakota, 15 July 1964, Douglas C. Ferguson (USNM). (p. 124).
41. *Euxoa idahoensis* (Grt.), ♀. 3 mi NE Sargents 9,700′, Colorado, 14 August 1971, D. F. Hardwick (CNC). (p. 124).
42. *Euxoa idahoensis* (Grt.), ♂. 6 mi N Woodland 8,200′, Colorado, 13 August 1971, D. F. Hardwick (CNC). (p. 124).
43. *Euxoa idahoensis* (Grt.), ♂. Eureka, Utah, 16 July 1911, Tom Spalding (CNC). (p. 124).
44. *Euxoa clausa* McD., ♂. Lethbridge, Alberta, 28 July 1929, H. L. Seamans (CNC). (p. 124).
45. *Euxoa clausa* McD., ♀. Allotype. Lethbridge, Alberta, 28 July 1929, E. H. Strickland (CNC). (p. 124).
46. *Euxoa brevipennis* (Sm.), ♂. 10 mi E Tonopah 6,000′, Nevada, 24 September 1969, D. F. Hardwick (CNC). (p. 125).
47. *Euxoa brevipennis* (Sm.), ♂. Lethbridge, Alberta, 18 August 1935, H. L. Seamans (CNC). (p. 125).
48. *Euxoa brevipennis* (Sm.), ♀. 2.5 mi SSW Valyermo 4,800′, San Gabriel Mts., Los Angeles Co., California, 27 October 1963, C. Henne (LACM). (p. 125).
49. *Euxoa servita* (Sm.), ♂. 1 mi S Beverley 2,400′, Saskatchewan, 22 August 1968, D. F. Hardwick (CNC). (p. 125).
50. *Euxoa servita* (Sm.), ♂. 11 mi NE Fernie 3,600′, British Columbia, 27 July 1960, D. F. Hardwick (CNC). (p. 125).
51. *Euxoa servita* (Sm.), ♀. 10 mi E Cranbrook 2,700′, British Columbia, 28 July 1960, D. F. Hardwick (CNC). (p. 125).
52. *Euxoa servita* (Sm.), ♀. Coldbrook, Kings Co., Nova Scotia, 23 July 1949, D. C. Ferguson (CNC). (p. 125).
53. *Euxoa servita* (Sm.), ♂. Greer 8,500′, White Mts., Apache Co., Arizona, 2 August 1962, E. & I. Munroe (CNC). (p. 125).
54. *Euxoa redimicula* (Morr.), ♂. 12 km SW Ottawa, Ontario, 8 August 1986, J. D. Lafontaine (CNC). (p. 126).
55. *Euxoa redimicula* (Morr.), ♂. Ithaca, New York, 16 August 1937, J. G. Franclemont (JGF). (p. 126).
56. *Euxoa redimicula* (Morr.), ♀. Ithaca, New York, 22 August 1937, J. G. Franclemont (JGF). (p. 126).
57. *Euxoa a. auripennis* Lafontaine, ♀. Holotype. 10 mi E Cranbrook 2,700′, British Columbia, 28 July 1960, D. F. Hardwick (CNC). (p. 127).
58. *Euxoa a. auripennis* Lafontaine, ♂. Paratype. 10 mi NW Choteau 4,050′, Montana, 12 August 1961, D. F. Hardwick (CNC). (p. 127).
59. *Euxoa a. auripennis* Lafontaine, ♀. 9 mi NW Leavenworth 2,200′, Washington, 24 August 1974, J. D. Lafontaine (CNC). (p. 127).
60. *Euxoa auripennis arizonensis* Lafontaine, ♀. Paratype. Greer 8,500′, White Mts., Apache Co., Arizona, 9 August 1962, E & I. Munroe (CNC). (p. 127).

1
2
3
4
5
6
7
8
9
10
11
12
13
14
15
16
17
18
19
20
21
22
23
24
25
26
27
28
29
30
31
32
33
34
35
36
37
38
39
40
41
42
43
44
45
46
47
48
49
50
51
52
53
54
55
56
57
58
59
60

PLATE 7

Noctuoidea

NOCTUIDAE

figs. 1–64

NATURAL SIZE 1:1

1. *Euxoa olivalis* (Grt.), ♂. Doolittle Ranch 9,800′, Mt. Evans, Colorado, 13 August 1961, E. W. Rockburne (CNC). (p. 127).
2. *Euxoa olivalis* (Grt.), ♀. 5 mi W Lee Vining 8,900′, California, 7 August 1967, D. F. Hardwick (CNC). (p. 127).
3. *Euxoa agema* (Stkr.), ♂. 10 mi W Bishop 10,000′, California, 11 August 1967, D. F. Hardwick (CNC). (p. 128).
4. *Euxoa agema* (Stkr.), ♂. 43 mi S Gallatin Gateway 6,700′, Montana, 16 August 1964, J. D. Lafontaine (CNC). (p. 128).
5. *Euxoa oblongistigma* (Sm.), ♂. 8 mi SE Laramie 8,400′, Wyoming, 28 August 1964, D. F. Hardwick (CNC). (p. 128).
6. *Euxoa oblongistigma* (Sm.), ♂. 13 mi S Watford City 2,000′, South Dakota, 1 September 1968, D. F. Hardwick (CNC). (p. 128).
7. *Euxoa citricolor* (Grt.), ♂. 11 mi W Fallon 4,100′, Nevada, 15 September 1969, D. F. Hardwick (CNC). (p. 129).
8. *Euxoa citricolor* (Grt.), ♀. 2 mi W Walsenberg 6,500′, Colorado, 11 September 1968, D. F. Hardwick (CNC). (p. 129).
9. *Euxoa tronella* (Sm.), ♀. 24 mi E Lund 6,000′, Nevada, 4 September 1965, D. F. Hardwick (CNC). (p. 129).
10. *Euxoa tronella* (Sm.), ♂. Coyote Summit 5,600′, 6 mi SSW Tempiute, Nevada, 26 September 1969, D. F. Hardwick (CNC). (p. 129).
11. *Euxoa tronella* (Sm.), ♀. 12 mi W Green River 6,200′, Wyoming, 26 August 1964, D. F. Hardwick (CNC). (p. 129).
12. *Euxoa tronella* (Sm.), ♂. Badlands National Monument 2,300′, 6 mi NW Interior, South Dakota, 6 September 1968, D. F. Hardwick (CNC). (p. 129).
13. *Euxoa teleboa* (Sm.), ♂. Sand Springs 3,100′, Montana, 25 August 1970, D. F. Hardwick (CNC). (p. 130).

14. *Euxoa teleboa* (Sm.), ♀. 6 mi SW Folsom 7,300′, New Mexico, 13 September 1968, D. F. Hardwick (CNC). (p. 130).
15. *Euxoa teleboa* (Sm.), ♂. 13 mi SW Quemado 7,400′, New Mexico, 10 September 1971, D. F. Hardwick (CNC). (p. 130).
16. *Euxoa difformis* (Sm.), ♂. Lehi, Utah, 24 September 1936, H. F. Thornley (CNC). (p. 130).
17. *Euxoa difformis* (Sm.), ♂. 5 mi SE Dayton 2,100′, Washington, 4 September 1970, D. F. Hardwick (CNC). (p. 130).
18. *Euxoa difformis* (Sm.), ♂. Sand Springs 3,100′, Montana, 25 August 1970, D. F. Hardwick (CNC). (p. 130).
19. *Euxoa difformis* (Sm.), ♀. 2 mi SE Austin 7,600′, Nevada, 13 September 1969, D. F. Hardwick (CNC). (p. 130).
20. *Euxoa difformis* (Sm.), ♀. Avalon, Santa Catalina Island, California, 27 September 1932, Don Meadows (CNC). (p. 130).
21. *Euxoa difformis* (Sm.), ♀. 7 mi WNW Orland 700′, California, 30 September 1970, D. F. Hardwick (CNC). (p. 130).
22. *Euxoa difformis* (Sm.), ♀. 5 mi SE Follyfarm 4,250′, Oregon, 15 September 1961, D. F. Hardwick (CNC). (p. 130).
23. *Euxoa difformis* (Sm.), ♂. Spring Mtn., Sonoma Co., California, 13 October 1939, E. C. Johnston (CNC). (p. 130).
24. *Euxoa difformis* (Sm.), ♂. Petaluma, Sonoma Co., California, 22 September 1938 (CNC). (p. 130).
25. *Euxoa difformis* (Sm.), ♂. Gorman, Ridge Route, California, 18 October (CNC). (p. 130).
26. *Euxoa moerens* (Grt.), ♂. 2 mi SE Austin 7,600′, Nevada, 13 September 1969, D. F. Hardwick (CNC). (p. 131).
27. *Euxoa moerens* (Grt.), ♂. 5 mi SW Ely 7,400′, Nevada, 6 September 1969, D. F. Hardwick (CNC). (p. 131).
28. *Euxoa moerens* (Grt.), ♂. 13 mi SE Lukachukai 7,300′, Arizona, 3 September 1971, D. F. Hardwick (CNC). (p. 131).
29. *Euxoa latro* (B. & Benj.), ♂. Blue Ridge 7,386′, San Gabriel Mts., Los Angeles Co., California, 8 September 1963, C. Henne (LACM). (p. 131).
30. *Euxoa latro* (B. & Benj.), ♀. 21 mi N Chiloquin 4,500′, Oregon, 2 September 1967, D. F. Hardwick (CNC). (p. 131).
31. *Euxoa murdocki* (Sm.), ♀. 21 mi N Chiloquin 4,750′, Oregon, 2 September 1974, J. D. Lafontaine (CNC). (p. 132).
32. *Euxoa dodi* McD., ♂. 2 mi N Scout Lake 2,600′, Saskatchewan, 27 August 1968, D. F. Hardwick (CNC). (p. 132).
33. *Euxoa infracta* (Morr.), ♀. 12 mi E Bozeman 5,700′, Montana, 14 August 1964, D. F. Hardwick (CNC). (p. 132).
34. *Euxoa laetificans* (Sm.), ♂. 13 mi N Wolf Creek 3,800′, Montana, 13 August 1961, D. F. Hardwick (CNC). (p. 133).
35. *Euxoa laetificans* (Sm.), ♀. 13 mi N Wolf Creek 3,800′, Montana, 13 August 1961, D. F. Hardwick (CNC). (p. 133).
36. *Euxoa q. quadridentata* (Grt. & Rob.), ♂. 11 mi NW Eureka 5,800′, Utah, 15 September 1965, D. F. Hardwick (CNC). (p. 134).
37. *Euxoa quadridentata flutea* (Sm.), ♀. 5 mi SE Follyfarm 4,250′, Oregon, 15 September 1961, D. F. Hardwick (CNC). (p. 134).
38. *Euxoa quadridentata flutea* (Sm.), ♂. 21 mi N Chiloquin 4,750′, Oregon, 2 September 1974, J. D. Lafontaine (CNC). (p. 134).
39. *Euxoa inscripta* Lafontaine, ♂. Holotype. 22 mi N Craig 7,100′, Colorado, 15 August 1965, D. F. Hardwick (CNC). (p. 134).
40. *Euxoa inscripta* Lafontaine, ♀. Paratype. 8 mi SW Kemmerer 6,800′, Wyoming, 23 August 1964, D. F. Hardwick (CNC). (p. 134).
41. *Euxoa inscripta* Lafontaine, ♀. 6 mi WNW Sierraville 6,700′, California, 21 August 1967, D. F. Hardwick (CNC). (p. 134).
42. *Euxoa unica* McD., ♂. Holotype. Saskatoon, Saskatchewan, 23 August 1937, Kenneth M. King (CNC). (p. 135).
43. *Euxoa niveilinea* (Grt.), ♂. Capulin Natl. Monument, 6 mi W Folsom 7,300′, New Mexico, 13 September 1968, D. F. Hardwick (CNC). (p. 135).
44. *Euxoa niveilinea* (Grt.), ♀. 4 mi NW Corona 6,400′, New Mexico, 18 September 1968, D. F. Hardwick (CNC). (p. 135).
45. *Euxoa dargo* (Stkr.), ♂. Swift Current, Saskatchewan, 5 September 1939, A. R. Brooks (CNC). (p. 135).
46. *Euxoa melura* McD., ♀. Satus Creek, Yakima Co., Washington, 30 May 1949, E. C. Johnston (CNC). (p. 136).
47. *Euxoa d. detersa* (Wlk.), ♂. White Point Beach, Queens Co., Nova Scotia, 29 August 1937, J. McDunnough (CNC). (p. 137).
48. *Euxoa detersa personata* (Morr.), ♂. Simcoe, Ontario, 3 September 1959, Freeman & Lewis (CNC). (p. 137).
49. *Euxoa detersa personata* (Morr.), ♀. 12 km SW Ottawa, Ontario, 7 September 1985, J. D. Lafontaine (CNC). (p. 137).
50. *Euxoa detersa personata* (Morr.), ♂. Hessville, Indiana, 7 September 1912, A. Kwiat (CNC). (p. 137).
51. *Euxoa cicatricosa* (Grt. & Rob.), ♂. 17 mi E Austin 6,600′, Nevada, 12 September 1969, D. F. Hardwick (CNC). (p. 137).
52. *Euxoa cicatricosa* (Grt. & Rob.), ♂. 8 mi S Tonasket 1,100′, Washington, 11 September 1960, D. F. Hardwick (CNC). (p. 137).
53. *Euxoa cicatricosa* (Grt. & Rob.), ♂. 3 mi NE Dixon 5,700′, New Mexico, 16 September 1968, D. F. Hardwick (CNC). (p. 137).
54. *Euxoa cicatricosa* (Grt. & Rob.), ♂. Swift Current, Saskatchewan, 27 August 1937 (CNC). (p. 137).
55. *Euxoa recula* (Harv.), ♀. Juniper Hills 3,500′, Mojave Desert, Los Angeles Co., California, 24 October 1972, C. Henne (LACM). (p. 138).
56. *Euxoa recula* (Harv.), ♂. 3 mi N Bend 3,350′, Oregon, 8 September 1974, J. D. Lafontaine (CNC). (p. 138).
57. *Euxoa recula* (Harv.), ♂. 12 mi NNE Trona 2,700′, California, 13 October 1966, D. F. Hardwick (CNC). (p. 138).
58. *Euxoa a. aequalis* (Harv.), ♂. San Francisco, California, 8 October 1920, E. P. Van Duzee (CAS). (p. 140).
59. *Euxoa a. aequalis* (Harv.), ♀. San Francisco, California, 6 September 1909, F. X. Williams (USNM). (p. 140).
60. *Euxoa aequalis acornis* (Sm.), ♂. Swift Current, Saskatchewan, 5 September 1933, A. R. Brooks (CNC). (p. 140).
61. *Euxoa aequalis alko* (Stkr.), ♀. Mt. Pinos 7,600′, Los Padres Natl. Forest, Kern Co., California, 21 September 1962, C. Henne (LACM). (p. 141).
62. *Euxoa aequalis alko* (Stkr.), ♀. 4 mi NW Cedarville 6,200′, California, 30 August 1967, D. F. Hardwick (CNC). (p. 141).
63. *Euxoa aequalis yukonensis* Lafontaine, ♂. Holotype. 2 km N Carcross 650 m, Yukon, 7 August 1980, Wood & Lafontaine (CNC). (p. 141).
64. *Euxoa aequalis yukonensis* Lafontaine, ♀. Paratype. 2 km N Carcross 650 m, Yukon, 7 August 1980, Wood & Lafontaine (CNC). (p. 141).

1 2 3 4 5
6 7 8 9 10
11 12 13 14 15
16 17 18 19 20 21
22 23 24 25 26 27
28 29 30 31 32 33
34 35 36 37 38
39 40 41 42 43
44 45 46 47 48
49 50 51 52 53 54
55 56 57 58 59
60 61 62 63 64

PLATE 8

Noctuoidea

NOCTUIDAE

figs. 1–60

NATURAL SIZE 1:1

1. *Euxoa conjuncta* (Sm.), ♂. Capulin Natl. Monument, 6 mi W Folsom 7,300′, New Mexico, 13 September 1968, D. F. Hardwick (CNC). (p. 141).
2. *Euxoa conjuncta* (Sm.), ♀. 6 mi E Canadian 2,000′, Texas, 28 September 1968, D. F. Hardwick (CNC). (p. 141).
3. *Euxoa cona* (Stkr.), ♂. Coyote Summit 5,600′, 6 mi SSW Tempiute, Nevada, 26 September 1969, D. F. Hardwick (CNC). (p. 142).
4. *Euxoa cona* (Stkr.), ♀. 15 mi ESE Warm Springs 5,300′, Nevada, 25 September 1969, D. F. Hardwick (CNC). (p. 142).
5. *Euxoa munis* (Grt.), ♂. 5 mi W Lee Vining 8,900′, California, 7 August 1967, D. F. Hardwick (CNC). (p. 142).
6. *Euxoa munis* (Grt.), ♂. Manyberries, Alberta, 17 August 1952, A. R. Brooks (CNC). (p. 142).
7. *Euxoa munis* (Grt.), ♂. 13 mi SE Salina 7,400′, Utah, 16 August 1965, D. F. Hardwick (CNC). (p. 142).
8. *Euxoa munis* (Grt.), ♀. 15 mi NW West Yellowstone 6,700′, Montana, 17 August 1964, D. F. Hardwick (CNC). (p. 142).

9. *Euxoa munis* (Grt.), ♂. Manyberries, Alberta, 1 August 1951, D. F. Hardwick (CNC). (p. 142).
10. *Euxoa misturata* (Sm.), ♂. 2 mi W Walsenberg 6,500′, Colorado, 11 September 1968, D. F. Hardwick (CNC). (p. 143).
11. *Euxoa misturata* (Sm.), ♂. 3 mi NE Dixon 5,700′, New Mexico, 16 September 1968, D. F. Hardwick (CNC). (p. 143).
12. *Euxoa misturata* (Sm.), ♂. 9 mi S Silver Springs 4,200′, Nevada, 20 September 1961, D. F. Hardwick (CNC). (p. 143).
13. *Euxoa misturata* (Sm.), ♂. 1 mi SW Portal 4,900′, Arizona, 8 October 1969, D. F. Hardwick (CNC). (p. 143).
14. *Euxoa misturata* (Sm.), ♂. 6 mi SE Maybell 6,200′, Colorado, 17 August 1965, D. F. Hardwick (CNC). (p. 143).
15. *Euxoa misturata* (Sm.), ♂. 6 mi E Canadian 2,000′, Texas, 28 September 1968, D. F. Hardwick (CNC). (p. 143).
16. *Euxoa misturata* (Sm.), ♀. 4 mi NW Corona 6,400′, New Mexico, 18 September 1968, D. F. Hardwick (CNC). (p. 143).
17. *Euxoa misturata* (Sm.), ♂. Juniper Hills 3,500′, Mojave Desert, Los Angeles Co., California, 14 October 1964, C. Henne (LACM). (p. 143).
18. *Euxoa melana* Lafontaine, ♂. Paratype. 6 mi E Canadian 2,000′, Texas, 27 September 1968, D. F. Hardwick (CNC). (p. 144).
19. *Euxoa atristrigata* (Sm.), ♂. Oliver 1,000′, British Columbia, 26 August 1953, D. F. Hardwick (CNC). (p. 145).
20. *Euxoa atristrigata* (Sm.), ♀. 15 mi N Helena 3,800′, Montana, 15 August 1961, D. F. Hardwick (CNC). (p. 145).
21. *Euxoa nevada* (Sm.), ♂. 9 mi S Silver Springs 4,200′, Nevada, 20 September 1961, D. F. Hardwick (CNC). (p. 145).
22. *Euxoa nevada* (Sm.), ♂. 14 mi NW Baker 7,300′, Nevada, 3 September 1969, D. F. Hardwick (CNC). (p. 145).
23. *Euxoa nevada* (Sm.), ♀. 2 mi SE Austin 7,600′, Nevada, 12 September 1969, D. F. Hardwick (CNC). (p. 145).
24. *Euxoa cinereopallida* (Sm.), ♂. 6 mi E Canadian 2,000′, Texas, 28 September 1968, D. F. Hardwick (CNC). (p. 145).
25. *Euxoa cinereopallida* (Sm.), ♂. 5 mi SE Follyfarm 4,250′, Oregon, 15 September 1961, D. F. Hardwick (CNC). (p. 145).
26. *Euxoa cinereopallida* (Sm.), ♂. 13 mi SE Lukachukai 7,300′, Arizona, 7 September 1971, D. F. Hardwick (CNC). (p. 145).
27. *Euxoa mitis* (Sm.), ♀. 4 mi NW Corona 6,400′, New Mexico, 18 September 1968, D. F. Hardwick (CNC). (p. 146).
28. *Euxoa mitis* (Sm.), ♂. 6 mi W McDermitt 4,600′, Nevada, 16 September 1961, D. F. Hardwick (CNC). (p. 146).
29. *Euxoa mitis* (Sm.), ♂. 2 mi WSW Eastend 3,200′, Saskatchewan, 26 August 1968, D. F. Hardwick (CNC). (p. 146).
30. *Euxoa luctuosa* Lafontaine, ♂. Paratype. Mount Loche 6,700′, Davis Mts., Texas, 27 August 1970, A. & M. E. Blanchard (CNC). (p. 147).
31. *Euxoa luctuosa* Lafontaine, ♀. Paratype. 16 mi W Kayenta 7,300′, Arizona, 5 September 1971, D. F. Hardwick (CNC). (p. 147).
32. *Euxoa luctuosa* Lafontaine, ♀. Paratype. 17 mi W Pagosa Springs 6,600′, Colorado, 18 August 1971, D. F. Hardwick (CNC). (p. 147).
33. *Euxoa nomas incognita* (Sm.), ♂. Firth River, British Mts., Yukon, 24 July 1956, R. E. Leech (CNC). (p. 149).
34. *Euxoa nomas incognita* (Sm.), ♂. km 416 Dempster Highway 750 m, Yukon, 22–28 June 1980, Wood & Lafontaine (CNC). (p. 149).
35. *Euxoa aberrans* McD., ♀. 2 mi W Savona 1,800′, British Columbia, 3 September 1960, D. F. Hardwick (CNC). (p. 150).
36. *Euxoa aberrans* McD., ♀. 15 mi N Helena 3,800′, Montana, 15 August 1961, D. F. Hardwick (CNC). (p. 150).
37. *Euxoa aberrans* McD., ♂. Manyberries, Alberta, 22 July 1951, D. F. Hardwick (CNC). (p. 150).
38. *Euxoa macrodentata* Hdwk., ♂. Base of Mt. Wallace 1,050 m, Kluane National Park, Yukon, 6 August 1980, Wood & Lafontaine (CNC). (p. 150).
39. *Euxoa manitobana* McD., ♂. 9 mi N Whiterocks 7,300′, Utah, 19 August 1965, D. F. Hardwick (CNC). (p. 150).
40. *Euxoa manitobana* McD., ♀. T15N R16W Sec. 31, Oceana Co., Michigan, 4 July 1967, E. H. Metzler (CNC). (p. 150).
41. *Euxoa montana* (Morr.), ♂. Pennsylvania Mtn. 3,775 m, Park Co., Colorado, 31 July 1980, D. Ford (CNC). (p. 151).
42. *Euxoa montana* (Morr.), ♂. Pennsylvania Mtn. 3,775 m, Park Co., Colorado, 31 July 1980, D. Ford (CNC). (p. 151).
43. *Euxoa perolivalis* (Sm.), ♀. 2 mi WSW Eastend 3,200′, Saskatchewan, 26 August 1968, D. F. Hardwick (CNC). (p. 151).
44. *Euxoa perolivalis* (Sm.), ♀. 13 mi W Milk River 3,400′, Alberta, 5 August 1961, D. F. Hardwick (CNC). (p. 151).
45. *Euxoa perpolita* (Morr.), ♂. West Dover, Halifax Co., Nova Scotia, 8 September 1953, D. C. Ferguson (CNC). (p. 151).
46. *Euxoa perpolita* (Morr.), ♂. 17 mi W Dupuyer 4,900′, Montana, 11 August 1961, D. F. Hardwick (CNC). (p. 151).
47. *Euxoa taura* Sm., ♂. Jefferson Co., Montana 1 September 1927, W. C. Cook (CNC). (p. 152).
48. *Euxoa taura* Sm., ♂. Jefferson Co., Montana 1 September 1927, W. C. Cook (CNC). (p. 152).
49. *Euxoa taura* Sm., ♂. 8 mi SW Kemmerer 6,800′, Wyoming, 23 August 1964, D. F. Hardwick (CNC). (p. 152).
50. *Euxoa flavicollis* (Sm.), ♂. Sunnydale, Lloydminster, Alberta, 26 July 1943, P. F. Bruggemann (CNC). (p. 153).
51. *Euxoa flavicollis* (Sm.), ♀. Lethbridge, Alberta, 2 August 1927, H. L. Seamans (CNC). (p. 153).
52. *Euxoa maimes* (Sm.), ♀. Swift Current, Saskatchewan, 20 August 1939, A. R. Brooks (CNC). (p. 153).
53. *Euxoa maimes* (Sm.), ♀. Saskatoon, Saskatchewan, 14 August 1940, A. R. Brooks (CNC). (p. 153).
54. *Euxoa ridingsiana* (Grt.), ♂. Black Foot Coulee, Alberta, 8 August 1940, A. R. Brooks (CNC). (p. 154).
55. *Euxoa ridingsiana* (Grt.), ♂. 5 mi W Lee Vining, 8,900′, California, 7 August 1967, D. F. Hardwick (CNC). (p. 154).
56. *Euxoa wilsoni* (Grt.), ♀. San Francisco, California, 10 July 1937, E. C. Johnston (CNC). (p. 154).
57. *Euxoa wilsoni* (Grt.), ♂. McClure Beach, Point Reyes Pen., Marin Co., California, 17 June 1951, W. R. Bauer (CNC). (p. 154).
58. *Euxoa riversii* (Dyar), ♂. Zuma Beach, Los Angeles Co., California, 19 May 1971, F. P. Sala (USNM). (p. 155).
59. *Euxoa riversii* (Dyar), ♂. Zuma Beach, Los Angeles Co., California, 5 May 1970, F. P. Sala (USNM). (p. 155).
60. *Euxoa riversii* (Dyar), ♂. Zuma Beach, Los Angeles Co., California, 6 May 1970, F. P. Sala (USNM). (p. 155).

1
2
3
4
5
6
7
8
9
10
11
12
13
14
15
16
17
18
19
20
21
22
23
24
25
26
27
28
29
30
31
32
33
34
35
36
37
38
39
40
41
42
43
44
45
46
47
48
49
50
51
52
53
54
55
56
57
58
59
60

NOTES

1. ABBREVIATIONS FOR COLLECTORS AND COLLECTIONS

ABK	Alexander B. Klots
AEB	A. E. Brower
AMNH	American Museum of Natural History, New York
ANSP	Academy of Natural Sciences, Philadelphia
BM	Bryant Mather
BMNH	British Museum (Natural History), London
CAS	California Academy of Sciences, San Francisco
CM	Carnegie Museum, Pittsburgh
CNC	Canadian National Collection, Ottawa
CU	Cornell University, Ithaca
FMNH	Field Museum of Natural History, Chicago
FSCA	Florida State Collection of Arthropods, Gainesville
GS	Gayle Strickland
HUMB	Museum Alexander Humboldt, Berlin
INHS	Illinois Natural History Survey, Champaign
JBH	John B. Heppner
JGF	John G. Franclemont
JN	John Newman
JRH	J. Richard Heitzman
KWP	K. W. Philip
LACM	Los Angeles County Museum of Natural History
LEM	Lyman Entomological Museum, Montreal
MCZ	Museum of Comparative Zoology, Cambridge
MNSA	Museu Argentino de Ciencias Naturales "Bernardino Rivadavia," Buenos Aires
MSU	Michigan State University, East Lansing
MSUS	Mississippi State University, Starkville
NHMV	Naturhistorisches Museum, Vienna
NCSU	North Carolina State University, Raleigh
NSM	Nova Scotia Museum, Halifax
PMBC	Provincial Museum of British Columbia, Victoria
RHL	Ronald H. Leuschner
ROK	Roy O. Kendall
ROM	Royal Ontario Museum, Toronto
SDNH	San Diego Natural History Museum
SIUC	Southern Illinois University, Carbondale
UA	University of Alberta, Edmonton
UAF	University of Arkansas, Fayetteville
UBC	University of British Columbia, Vancouver
UCB	University of California, Berkeley
UCD	University of California, Davis
ULK	University of Louisville, Kentucky
UM	University of Michigan, Ann Arbor
UMC	University of Missouri, Columbia
UMO	University Museum, Oxford
UWM	University of Wisconsin, Madison
USNM	National Museum of Natural History, Washington
VAB	Vernon A. Brou
WEM	William E. Miller
WPC	Wedge Plantation Collection, McClellanville
YPM	Yale Peabody Museum, New Haven
ZSBS	Zoologische Sammlung des Bayerischen Staates, Munich

NOTES

2. COMMON NAMES

The use of an asterisk "*" in the text denotes a name listed in *Common Names of Insects & Related Organisms 1982* published by the Entomological Society of America.

French-language common names have been taken from Auclair, J. L. et al., 1964, *French Names of Insects of Canada,* 3rd edition, published for the Quebec Society for the Protection of Plants by Department of Agriculture and Colonization, Quebec. The abbreviation "m." after a name indicates that it is masculine, "f." that it is feminine.

3. CITATIONS OF AUTHORITIES

Authors' names without parentheses indicate that the specific name is associated with the genus in which it was described.

Authors' names in parentheses indicate that the specific name has been transferred from the genus in which it was described to another genus.

4. WING LENGTH

Wing length is the measurement in millimeters from the base to the apex of the forewing.

5. LOCATION OF TYPE SPECIMEN

The current location of the type specimen is given by the appropriate abbreviation in square brackets immediately following the type locality. The words "type lost" indicate that it no longer exists. If no information is given, a type may exist; but its present location was not determined.

6. NOMENCLATURE FOR LARVAL SETAE

Hinton's 1946 terminology is used to refer to larval setae.

INDEX TO ANIMAL NAMES

Principal entries are given in bold face
Plate references are given as (1:5)

Generic names cited only in combination with specific names, whether in synonymy or text, are not given in the index. Look for such entries under the specific name. For example, *Euxoa tessellata* will be found under *tessellata,* but not under *Euxoa.*

INDEX

INDEX TO PLANT NAMES